C0-AWT-589

PERGAMON INTERNATIONAL LIBRARY
of Science, Technology, Engineering and Social Studies
The 1000-volume original paperback library in aid of education, industrial training and the enjoyment of leisure
Publisher: Robert Maxwell, M.C.

Raw Materials Purchasing

An Operational Research Approach

This book is the fourth volume in a new international book series

FRONTIERS OF OPERATIONAL RESEARCH
AND APPLIED SYSTEMS ANALYSIS

General Editor: Rolfe Tomlinson

The series is aimed at helping practitioners to broaden their subject base and to advise managers and specialists in related areas of important new developments in the field.

The scope of the books will be broad and their authorship international. The subject is interpreted widely to include any scientific, applied, interdisciplinary and systems-oriented approach to the study of real-life decision problems.

Rolfe Tomlinson is Professor of Systems and Operational Research at the School of Industrial and Business Studies at the University of Warwick, and is President of the European Association of OR Societies. He was for many years the Director of Operational Research at the National Coal Board and was then Area Chairman for Management and Technology at the International Institute for Applied Systems Analysis. The series reflects his deep belief that scientific systems-oriented research can be of direct practical use over a much wider range of topics than at present, but only if the work is problem-oriented and seen as a collaborative effort with the problem owner.

Volume 1
EDEN, C., JONES, S. AND SIMS, D.
Messing about in Problems: An Informal Structured Approach to their Identification and Management

Volume 2
TOMLINSON, R. and KISS, I.
Rethinking the Process of Operational Research and Applied Systems Analysis

Volume 3
STAHL, I.
Operational Gaming: An International Approach

A Related Journal

JOURNAL OF THE OPERATIONAL RESEARCH SOCIETY*
Published on behalf of the Operational Research Society Limited, London

Editor: Peter Amiry, Department of Engineering Production, University of Birmingham, P.O. Box 363, Birmingham B15 2TT, England

This journal publishes contributions on any matter relevant to the theory, practice, history, or methodology of Operational Research or the affairs of the Society. It is aimed at applications in any field and encourages the submission of accounts of good, practical case studies illustrating OR in action; of reviews of the state of development of fields of knowledge relevant to OR; and of controversial articles on methodology, technique or professional policy.

* Free specimen copy sent on request.

Raw Materials Purchasing

An Operational Research Approach

by

BRIAN G. KINGSMAN

University of Lancaster, U.K.

PERGAMON PRESS

OXFORD · NEW YORK · TORONTO · SYDNEY · PARIS · FRANKFURT

U.K.	Pergamon Press Ltd., Headington Hill Hall, Oxford OX3 0BW, England
U.S.A.	Pergamon Press Inc., Maxwell House, Fairview Park, Elmsford, New York 10523, U.S.A.
CANADA	Pergamon Press Canada Ltd., Suite 104, 150 Consumers Road, Willowdale, Ontario M2J 1P9, Canada
AUSTRALIA	Pergamon Press (Aust.) Pty. Ltd., P.O. Box 544, Potts Point, N.S.W. 2011, Australia
FRANCE	Pergamon Press SARL, 24 rue des Ecoles, 75240 Paris, Cedex 05, France
FEDERAL REPUBLIC OF GERMANY	Pergamon Press GmbH, Hammerweg 6, D-6242 Kronberg-Taunus, Federal Republic of Germany

First edition 1985

Library of Congress Cataloging in Publication Data
Kingsman, Brian G.
Raw materials purchasing.
(Frontiers of operational research and applied systems analysis; v. 4)
Bibliography: p.
1. Raw materials—Purchasing—Decision making.
2. Industrial procurement—Decision making.
3. Commodity exchanges. I. Title. II. Series.
HF1051.K56 1984 658.7′2 84–6479

British Library Cataloguing in Publication Data
Kingsman, Brian G.
Raw materials purchasing.—(Frontiers of operational research and applied systems analysis; 4)
1. Raw materials—Purchasing
2. Operations research
I. Title II. Series
658.7′2′072 HF1051
ISBN 0–08–029976–8 (Hardcover)
ISBN 0–08–029975–X (Flexicover)

Printed in Great Britain by A. Wheaton & Co. Ltd., Exeter

To Margaret, Karl and Claire,
whose value, unlike prices,
was never uncertain

Editor's Preface

This fourth volume in the Series "Frontiers of Operational Research and Applied Systems Analysis" covers a topic of major commercial importance. Indeed, in the face of continuing uncertainty in world markets, there can be few topics which are more critical than the purchasing of raw materials. Because of the uncertainties involved, the field is very much one in which 'expert' dealers, operating by experience and hunch, hold court. At first sight the opportunities for 'rational' analysis would seem minimal; yet here, as in other unexpected areas of commercial management, analysis has proved not only to be useful but highly profitable. No-one has done more in this field than Dr Kingsman with his colleagues at the University of Lancaster.

Much of the merit of this book lies in the fact that it is based on a series of practical studies. It is replete with illustration and sound advice for the practitioner, at the same time as providing technical information for the respective analyst. Thus, not only has the author broken through to new frontiers in the field, he has provided both a map and a manual to those coming after him who would wish to profit from this newly discovered territory.

ROLFE TOMLINSON
General Editor

Preface

The importance of the general purchasing function to a company's well-being has often not been sufficiently appreciated in the past. Surveys have shown that on average over half of the final product cost is made up of bought-in supplies and materials. A good or bad purchasing policy will thus have a significant effect on a company's performance. The goods bought by the purchasing function divide naturally into two classes, manufactured and semi-manufactured items, and secondly raw materials. The major characteristic of the first class is that usually the prices paid are the result of negotiations between the buyer and his suppliers and that prices are kept constant for an appreciable time into the future. The major characteristic of most raw materials purchased is that the prices are not preset by the individual buyer and seller and that the prices are continuously changing, from day-to-day and even from minute-to-minute in some cases. The prices are determined by a market of all buyers and sellers worldwide. Individual buyers and sellers generally have little direct influence by themselves on the prices quoted by the market at particular times. The problem of purchasing raw materials whose prices are continually changing is one faced by many companies in countries all over the world.

The term 'commodities' is often used to describe such materials. However, this term is often only interpreted as referring to those materials whose prices are determined through international commodity exchanges in public trading sessions. Many other materials have fluctuating prices, but are bought on the basis of telephone conversations between buyers and sellers since no open central commodity exchange exists. Furthermore, many economists use the term to describe anything that is bought and sold. Because of all these ambiguities the term raw materials will generally be used in this book. Because the limited published work in this area has considered only those materials traded on international commodity exchanges the problem is usually called the commodity purchasing decision problem. Although this label will be used frequently it should always be remembered that it refers to the purchasing of any material whose price is continually fluctuating over time whether bought through international commodity exchanges or more informal arrangements.

It is probable that more than half of the items bought on a worldwide basis by the purchasing function of companies are raw materials as that term is used here. There are large variations of course between one company and another. The extreme cases where the purchasing costs make up 75% or more of the final product cost tend to be where most of the purchasing is for raw materials, for example animal feeds and the nonferrous metal industries. Indeed some manufactured products are 'raw materials' in the sense that the term is used here, for example copper cable, copper tubing, electric light plugs, etc., since their prices to customers will vary as the price of copper varies.

Much work has been carried out and published on various aspects of the problems of purchasing manufactured and semi-manufactured items. However it is surprising that so little published work has considered the problem of purchasing raw materials in any real depth, in view of the everpresent possibility of large losses or great gains arising from the uncertain large fluctuations in future prices. This book is an attempt to remedy that deficiency. It is based on many man years of research and consultancy by the author and his colleagues. The book contains an analysis of the commodity purchasing problem from an operational research viewpoint. The underlying structure of the problem is identified and discussed. Models for the component parts are then developed to provide decision making policies. These are illustrated via numerous case studies to show both the problem situations that occur and the potential improvements that might be made. In general the case studies described are ones carried out by the author or where he has had some personal involvement.

The book is mainly written for operational research practitioners who are involved in studies in the purchasing area or are contemplating such studies in co-operation with the buyers and for purchasing managers who are responsible for buying raw materials. The book is aimed at this dual audience since success in improving purchasing performance will only come from the purchasing management and the management scientist working hand-in-hand on a joint study. Because of this approach the mathematical and statistical aspects have, wherever possible, been put into appendices to each chapter. A nontechnical explanation is given where this is not appropriate. It is recommended that those readers without a good statistical knowledge should jump over the technical appendices, at least on a first reading. Several case studies which illustrate the application of the methods discussed to particular aspects of the problem have been included. Since the book is mainly aimed at those directly involved with the raw material purchasing problem, the case studies are described in some detail so as to give better guidance on what such studies entail and how to cover the various problems and difficulties that arise in practice.

It is hoped that the book will also be of interest generally to operational research practitioners and theorists, particularly to students on OR courses, as an example of operational research methodology applied to a currently neglected but very significant new problem area. Many parts of the book should be of interest and use to economists, particularly those concerned with the trading of materials between countries and with international commodity agreements. Any researcher studying speculative time series should also find much of relevance to his work. Many parts of the book are of interest to students taking management courses, particularly those leading to a purchasing management qualification, since it highlights an important area of management activities currently somewhat neglected.

I would like to record my appreciation of the help of a number of my colleagues at Lancaster, both in the preparation of this book and in the research leading up to it. Professor Alan Mercer first aroused and then encouraged my interest in this area. Colin Jex worked jointly with me on the copper buying study described in Chapter 12. Dr Steve Taylor, firstly as a research student and then as a colleague, developed the new general model for daily commodity price movements presented in the appendix to Chapter 2. He also produced for me several of the figures and tables of Chapter 2. Brian Parker provided many useful comments and suggestions on the first draft, as did Professor Rolfe Tomlinson, the editor of this series. Dr Rui Guimaraes, of Oporto University, whilst a research student at Lancaster, co-operated in some of the work on maize price forecasting. I also owe a debt of gratitude to many Masters students on the Lancaster courses, who worked with me over the years in applying the ideas presented in this book on their practical projects, only a subset of which have been included. Finally, but not least, my thanks to Margaret Threlfal and my wife, Margaret, who between them typed and retyped the manuscript. Because of the decision to preserve anonymity it is not possible to mention the many purchasing personnel who co-operated in the research in the companies of the case studies described. I record my thanks to them for the help they gave to the research.

Contents

Summary

The opening chapter introduces the problems that companies and purchasing managers have in coping with the commodity purchasing problem. To provide an appropriate background to the purchasing situation, a brief description is given of the different types of market in which raw materials are bought and sold and in which their continually changing prices are generated. The type of price movements that occur and the sheer size of the price fluctuations that can exist are illustrated for several materials in Chapter 2. The 'conventional wisdom' among many researchers is that raw material prices follow a random walk. This means that the price movements are entirely random with no systematic components or causes. If this were so then an immediate consequence is that there is little point in trying to determine improved purchasing policies since the attempt is bound to fail. Chapter 2 concludes with a brief description of some recent research which establishes conclusively that the random walk model is an inadequate approximation to the underlying process generating daily commodity prices. Systematic trends can and do exist and persist for moderate periods of time.

The various components and stages in the purchasing decision can best be seen by examining practical examples in some detail. This will also show to what extent a common structure exists in the purchasing decision problem. Four examples of actual buying situations are described and discussed in detail in Chapter 3. The materials and the practical detail in each example are taken from real life situations. However, for reasons of commercial confidentiality, the names of the companies used in the examples and the later case studies are fictitious. Each problem is described together with the constraints placed on the buyer's freedom of action by company policy. The general way in which buyers and their companies have organised themselves to cope with their purchasing problems is also described.

These buying examples all show that the commodity purchasing decision contains five major stages. The first stage is to determine what are the future consumption usages for each material that the buyer needs to purchase. If there are alternative grades of each material or alternative materials which are partially substitutable for each other then this stage contains very complex decision problems indeed. The second stage is to batch consumption requirements into delivery order quantities to be placed with the supplier, specifying size and delivery date. One important conclusion determined for this stage is that classical Statistical Inventory Control methods are inappropriate for purchasing raw materials, because of the way pricing, supply and delivery of the materials is organised. A Materials Planning Requirements approach must be used. A company needs to forecast the future demands for its final products and produce a schedule of the demands it will meet in each future period. Parts explosion or other methods can then be used to give the raw material requirements for each future time period. The third stage is the determination of the absolute limits on the freedom of action allowed to the buyers. Discussion of this problem leads to the concept of the 'Strategic Buying Period', the period of time over which any future period's requirements may be purchased, and maximum and minimum values for the 'Purchase and Stock Cover', the total amount of material held by the company at any time either in stock or as forward contract orders. Fixed values are determined for these factors which depend on the company's general ability to make good forecasts of prices over differing future time periods, the company's overall strategic policies on capital investment in raw material stocks and its attitude to the risks of occasional large losses relative to expected overall net gains compared to a back-to-back buying policy. The fourth stage is the determination of the 'Active Buying Period' or 'Purchase and Stock Cover'. This is the value the buyer uses in practice and is buying performance related rather than concerned with company strategy. It depends upon whether the buyer feels able to make confident forecasts of future prices in all the varying situations that can occur in the market. If not then a single constant value for the active buying period should be used for all time. If successful price forecasts can be made then it can be an adaptive value, sometimes leading to buying requirements for several months ahead over the next 3 or 4 weeks whilst at other times buying only to cover current requirements. An adaptive buying period requires the development of a medium-term econometric/behavioural price forecasting model which 'explains' why particular price movements have occurred and enables reversals of an existing trend to be anticipated in advance. The final fifth stage is the tactical day-to-day buying problem, to decide on each particular day or at which particular price offer, within the limits imposed by the early stages, a purchase or purchases for each delivery order should be made. The tactical buying problem manifests itself as one of two basic forms; the continuous supply and continuous usage situation with material being required for usage at every buying opportunity, mainly relevant to the metals, or that of buying a known amount of material before a known or specified deadline, mainly applicable to the agricultural materials. Both problems are essentially that of trying to measure how high or how low a particular price offer is relative to all the other alternative buying opportunities at which that purchase could have been made.

All of these stages are discussed in detail in Chapter 4. This chapter is essentially a discussion of the methodology of the commodity purchasing decision problem. Much work in the past has wrongly treated the tactical day-to-day buying problem as the complete commodity purchasing problem. This has often also been treated as that of trying to predict accurately the price at some specific single point in future time. This is of course an insoluble problem in that form. This book shows the naivety of these simplifications of the real problem and demonstrates the complexity of the total commodity purchasing problem and all the various aspects that must be covered in any meaningful and useful purchasing policy. The ways in which different types of forecasting systems for predicting future prices constrain the decision making policies is investigated. In addition the most appropriate manner in which to combine the longer term forecasts from an econometric/behavioural type model and short-term statistical price forecasts within the purchasing policy is explored and determined.

One difficulty in this problem area has been the lack of procedures to assess the past purchasing performance and hence set targets or standards for future use. Given the methodology discussed in Chapter 4 some answers to this problem can be derived. Measures of the purchasing performance are proposed and discussed in Chapter 5 for the tactical buying problem. Their use is illustrated via a case study.

The remainder of the book describes the approaches that can be used to develop models and policies for each of the various aspects of the purchasing decision identified in Chapter 4. The models and policies are illustrated by case studies.

In Chapter 6 work on the problem of purchasing partially substitutable raw materials, the blending problem as it is commonly called, is described and reviewed. Alternative approaches that have been used to partition this problem into feasible submodels to give practical policies are discussed for three widely differing industries.

A dynamic programming model is developed for scheduling linseed oil deliveries in Chapter 7. This is an example of the Materials Requirements Planning approach to ordering and stock control. It is shown that the extra complexity and sophistication of this model compared to standard simpler heuristic procedures is well worth while. The case study shows how the model and the derived ordering and delivery policy should be modified as circumstances change and the forecasts of future usages are revised.

As discussed in Chapter 4, a key component in determining the active buying period is the ability to make reliable medium-term price forecasts. This requires an understanding of the underlying processes that explain past price behaviour. In Chapters 8, 9 and 10 models are described and discussed to explain commodity market price behaviour in a suitable form for forecasting future prices. The general concepts and methodology are described in Chapter 8. Chapter 9 illustrates their use for price determination in the world linseed market, with particular reference to the impact of floating currency exchange values. In Chapter 10 a study to develop such a model for explaining and predicting world maize prices on the Chicago market is described in detail. This shows the various stages that must be covered in such studies. The value of the models developed is demonstrated by the price forecasts made for the following 4 years after the end of the study period. The forecasting of future commodity prices must be a continuous process since market circumstances change over time. This is illustrated by showing how the models were actually used in practice to make price forecasts month by month over the 1977/78 crop year.

The approach taken by the author is that the price fluctuations that occur in the market are caused by the uncertain knowledge that the 'market' has of the current and near future state of supply and demand. As a collectivity the people in the market make rational decisions based on the current 'expectations' of supply/demand factors. There exists at any time a majority expectation which can be measured in the form of a 'market barometer'. There is then a (1–1) relationship between the market price and the 'market barometer'. If the same value of the market barometer occurs now as occurred several years ago then so will the same price. The price factor, for the post 1971 period, may need to be adjusted over time to take account of changing relative currency values, inflation etc. This adjustment reflects other changes in international economies, not changes in the state of supply and demand of that raw material. It is asserted and then demonstrated that the market expectations of supply and demand factors can be measured in quantitative terms, either based on estimates circulating in the market from respected sources or via simple formal models using past usages. It is what the market thinks is the situation that matters in determining the prices over the short- to medium-term future, not the actual situation which is only known in hindsight at a much later period.

A decision-making model for the tactical daily buying problem is presented in Chapter 11, after other simpler approaches have been reviewed. This is based on a dynamic programming formulation of the problem. A sequential decision policy is derived in a general form requiring price forecasts to be input in the form of probability distributions. No constraints need be imposed on the method by which these forecasts are generated. They can be subjectively made by the buyers themselves, be the output of a statistical forecasting system or the output of an econometric/behavioural model of the sort described in the preceding three chapters. To generate a set of practically solvable recurrence relations requires that the future price distributions are serially independent. In practice this condition will not be completely satisfied. The decision price break policy derived from the dynamic programming model then has to be used as an heuristic or approximate policy. The practical problem is then to determine appropriate price forecasting methods and to see whether the assumption of independence is adequate for most practical situations.

Chapters 12 and 13 describe two case studies in detail. Price forecasting systems are derived and tactical purchasing policies developed for the cases of buying wheat, maize, and copper. In both cases the forecasts used depend only on past prices as input. It is shown that some form of the adaptive exponential smoothing method of Trigg and Leach is probably the best statistical forecasting method for volatile time series such as raw material prices. Improvements on the previous purchasing performance can be achieved which give appreciable savings on a back-to-back policy of buying regular amounts of material each day.

1

The Commodity Purchasing Context

1.1 THE DIFFICULTIES IN COMMODITY PURCHASING MANAGEMENT

The major part of the raw materials, both metals and agricultural products, consumed by industry have to be bought in markets whose prime characteristic is that the prices are continually fluctuating up and down in an uncertain manner. The same situation applies to the purchase of foods for human consumption. The only major exceptions are fuel products and iron ore, although some oil is sold on a fluctuating free price market in Rotterdam, with consequent effects on the prices set by the oil producers. In general the markets are international in character influenced by the behaviour of buyers and sellers worldwide. Thus the problem of purchasing materials in uncertain fluctuating price markets is common to buyers and consumers in most countries of the world.

One of the largest international confectionery manufacturers founded its fortunes on its ability to purchase cocoa. Indeed it has been alleged that the company made more money from buying cocoa than from selling chocolate. At the other extreme a major U.K. confectionery manufacturer reported enormous losses from its operations in the cocoa market only a few years ago. An international tyre manufacturer was almost ruined because of large and disastrous rubber purchases. This illustrates that a company can be an extremely good or a very bad buyer in commodity markets. Numerous further examples can be collected from the annual reports of many companies over the years.

With such raw materials for some industries making up 70% of the final product cost the effects of a good or bad purchasing policy are obvious. Variations in purchasing performance from one company to another and from one year to the next are hardly surprising, because the year-to-year and day-to-day variability in price make commodity purchasing decisions one of the most complex problems that a company must face. Because of the possibility of such large losses the top management of some companies have attempted to impose buying policies that avoid the problem. The possibility that equally large gains can be achieved for the company by good purchasing policies is ignored.

Some of the difficulties that exist in commodity purchasing management are inherent in the market situations in which buyers must operate, whilst others are due to the way buyers have organised themselves to operate in the commodity purchasing situation. The reasons for these difficulties fall into several general categories.

1. Purchasing decisions must be made under great uncertainty as to the future state of the market, in particular the future level of prices. The industrial purchaser, whatever his size, cannot control the prices, neither currently nor in the future. He essentially is a spectator who can only watch prices go by and can only say yes or no to a particular price at a particular time. The international commodity markets in which the prices of many materials are generated are usually organised so as to prevent any one person being able to dictate price levels. Although he may buy his material direct from a producer the price will usually be determined by the commodity market. The commodity market is run by brokers who act on behalf of the producers, consumers, gamblers, speculators or themselves. The purchaser cannot himself take part directly in the formation of prices. Obviously the decisions he makes will eventually have some effect on prices, but only as part of the total group of consumers. The commodity market situation is ever changing. Although buyers are aware of the possibility of large unexpected changes, in many instances they cannot be predicted or anticipated. An outbreak of maize blight in mid August can reduce the U.S. maize crop by 20 million tons over 3 or 4 weeks; very heavy frost can destroy half the Brazilian coffee crops overnight; the destruction of the bridges on the Benguela railway in the Angolan civil war halted the flow of copper from Zaire for several weeks. Event such as these will all affect prices immediately. Yet such events cannot be precisely predicted. Nobody ever knows precisely, until after the event, how much has been produced and how much of it will be consumed this year even in more mundane circumstances. People's views and estimates of these factors differ and change over time.
2. Large price fluctuations can occur both in the long-term and the short-term. The uncertainty in future events discussed above is made worse because the events manifest themselves in very large changes in price. The commodity markets act through the price mechanism as a means of rationing the supply of a material amongst the potential consumers. There are many time lags in the system and several different uses for the material with different alternative materials for each use. Prices depend more heavily on the gap between the demand and the supply than on the absolute levels of either factor.

   The result is that very large price changes are required to reduce consumption and ration out a shortage in supply, or alternatively, increase consumption to use up a bumper harvest. Prices become very sensitive at times to marginal changes in supply and demand factors. Since people can form differing views on the size of these factors at any one time, even over a few days prices can fluctuate substantially as market interest sways between the alternative views.

3. The performance of the commodity purchasing department in most companies has not been systematically monitored and measured because appropriate ways of doing so have not been developed and evaluated. It is thus difficult to know whether a buyer is doing a good or bad job. It also makes the buyer vulnerable to uninformed outside criticism, particularly from hindsight. In buying, 'chickens come home to roost' rather quickly! Whether a particular purchase was at a good or bad price is known within several days or weeks. One recent poor purchase can be easily overexaggerated out of context, compared to several earlier good to moderate purchases, without some objective measure. It is not unknown for a director to criticise a buyer on the basis of a current newspaper article, written by some pundit stating, with the benefit of hindsight, that it was obvious that prices would come down as they did because of this and that reason. At the earlier time of the purchasing decision, the situation is rarely so clearcut and obvious. At times the buyer can make exactly the right decision in the prevailing circumstances but unfortunately achieve the wrong outcome. This is because of unexpected market changes, as mentioned earlier, which could not have been foreseen beforehand. Because of all the complexities and the uncertain price fluctuations it is necessary to accept occasional bad purchases or losses in order to make savings on average over the medium- and long-term. Furthermore, without some measure the conscientious buyer himself does not really know how good his performance has been and whether there is much room for improvement. Learning from past good or bad decisions is also impossible without systematic monitoring and measurement. In the past, few buyers have even cursorily examined their previous purchases to see if they can develop any guidelines for future decisions. A decision once made, like the past itself, is all too often forgotten.
4. A further reason for many of the difficulties found in this area in many companies, is that the various stages in the process that results in the final detailed purchasing decision have not been formally analysed. The first two categories of problems were an essential part of the commodity purchasing environment. This and the previous category reflect inadequacies and mistakes on the part of management. Little attention has been given to these problems by management scientists. Traditional economists have also contributed little to assist purchasing managers understand and cope with these problems. Formal training for the

commodity-purchasing role is virtually nonexistent. It is a prime example of learning by doing, or by watching the way an existing buyer performs the role. Buyers are expected to develop a 'feel for the market' an intuition of the ways prices will move in the future. Specific guidance on how to do this is usually lacking.

In most instances price forecasting and taking buying decisions are regarded as synonymous. Hence price forecasting is wrongly considered as presenting the one 'big' problem to be solved. Efforts to improve purchasing efficiency are virtually entirely concentrated on trying to forecast future prices accurately. All too often this is implicitly based on the misconception that prices at some particular point in future time can be forecast accurately, if only the secret were known. Any such forecast will contain substantial errors to a greater or lesser extent. Purchasing decision-making systems must explicitly take into account the inevitable inaccuracies that must occur, when trying to predict the future outputs (prices) of an economic system (the commodity market), whose inputs (the supply and demand factors) are continually changing. What is the best form of forecast for making purchasing decisions has rarely been considered in such forecasting studies.

Buyers and operators in commodity markets develop 'saws' and 'maxims' and *ad hoc* rules of thumb to help them forecast and buy. These are rarely tested on a long price series to evaluate properly their effectiveness, but rather have been found to work on one or two occasions. Most of them are in fact untestable. They are qualitative in nature, and usually imprecisely defined. Thus some judgement is required, so that two people using the same 'maxim' in the same situation will often make different assessments of the market movements, and hence different purchasing decisions. Setting up precise operating purchasing rules to evaluate past price series is thus impossible. The vast majority of these rules of thumb are designed for the use of speculators and gamblers in the markets, who are only interested in whether prices will rise above or fall below the present price at *some* time in the future. They have no need to be in the market unless they are fairly confident they will make a profit. A buyer has a continuing raw material requirement, so he cannot avoid the market whenever he feels unsure of its state and is unable to make any real meaningful forecast. He also wants to know how long prices will stay higher than the current price, rather than merely that they will be higher at some time in the future.

An example of market folklore, met in one case study on copper buying, was that the lowest price during a week always occurred on the Monday. For once this could be tested. The average cash prices for each day of the week over a 2-year period (25 April 1966–18 April 1968) for copper were calculated and are shown in Table 1.1 below. The different

TABLE 1.1 Average daily copper price

Day	Number of prices	Mean price in £5 per ton
Monday	94	518.7
Tuesday	102	517.7
Wednesday	102	517.9
Thursday	102	516.7
Friday	99	515.1
Av. Daily price	—	517.2

number of prices for each day of the week arise from bank holidays when the market is closed.

If anything, contrary to the popular belief, Monday has the highest price of the week. The maximum difference between the highest and lowest daily mean values (518.7–515.1) is approximately two-thirds of the standard error of the means. The differences between the daily mean values are thus not significant. Hence it cannot be asserted that the prices of copper are consistently lower on any one particular day of the week.

Each raw material commodity differs significantly from others because of its physical properties, its means of production, its places of origin, its uses, its distribution channels etc. Such differences produce specialisation in the trading of the materials, with separate markets developing for each commodity. Even though a commodity exchange may cover several materials (e.g. the London Metal Exchange, for copper, lead, zinc and silver), the different materials are traded separately on physically distinct markets in the same building or at different times of the day. The brokers and dealers in these markets must become specialists. They become a 'sugar' man, or a 'copper' man or a 'maize' man etc. In the U.K., with mystique and respect surrounding the financial institutions in the City of London, the necessary specialism of the commodity markets and the brokers with whom the buyers deal produces the popular misconception amongst many industrial buyers that knowing all the details concerning the production, distribution, etc., of a material *is* commodity buying. Much of this detailed knowledge is necessary but, by itself, it is not sufficient to make a good buyer. This attitude tends to foster strongly the idea that each commodity purchasing problem is unique. Thus no guidelines or approaches developed from one commodity purchasing situation assist in another. In extreme cases, it also leads to the belief that each new situation for the same commodity is unique and that no lessons can be drawn from past behaviour.

5. Partly as a consequence of the difficulties mentioned in (3) and (4) above, the purchasing of raw materials function is an area where few if any management control systems have been imposed. There has been little setting of targets for purchasing management to achieve nor the devising of budgetary or standard cost controls that must be adhered to. Even such rudimentary controls as limiting the amount of money a buyer or a purchasing manager can spend before requiring the permission of a higher director are rarely imposed in practice. Most deals to buy materials are agreed over the telephone by word of mouth, with the buyer only having a very short period to make up his mind. It is thus obviously necessary to keep an up-to-date record of the purchases made relative to the consumption needs and of the total expenditure incurred. The amount of company money at risk if prices fall should be continuously monitored. The lack of such simple controls and up-to-date knowledge has undoubtedly contributed to some extent to the large losses in their raw material buying that companies have incurred. One temptation that buyers are sometimes prey to, is to believe that they are right in thinking prices will go up, or down, and the market is wrong. In such cases if the buyers are allowed to resell material on the futures market then they should be encouraged to cut their losses by reselling and await a better opportunity to buy later.

   In certain companies actual primary raw material is the biggest part of expenditure, more than labour, fuel etc. If this is not controlled then the business is not being controlled.

1.2 COMMODITY MARKETS

The markets, where the prices are determined at which raw materials will be bought and sold, can take many forms, ranging from the large international commodity exchanges in London, New York and Chicago to the weekly cattle auction of the local farmers in a small market town. In a sense we all, as consumers, are involved in a commodity market, since the vegetables for example, that we buy in the grocer's shop or the local market are changing in price from one week to the next. Some markets are highly organised and centralised with commodity exchanges, where prices are determined via an open-call system based on the interplay of world-wide supply and demand. Others are informal, with a few buyers and sellers communicating only by telephone to negotiate prices between themselves for a particular amount of a particular grade of some raw material.

In all markets it is possible for a buyer to negotiate a price for the delivery of some specified grade of a raw material at some future time at some specified location. Such markets are generally called cash or actuals markets. The time ahead for which a price can be fixed depends on the particular materials and market, and the general market conditions. In some instances this may only be up to three months ahead, whilst in others it can be up to 18 months. Usually buyers can either fix a contract including delivery to their works, or fix delivery to some specified port in the producing country and then arrange shipping and land transport in their own country themselves. Most private companies will choose the first option.

The organised international commodity exchanges also offer a further option of trading in 'futures' contacts. A futures market is essentially one that trades in paper contracts rather than the actual physical raw material. A futures contract specifies that a standardised quantity of material of a standardised quality will be delivered at some specified time to some standardised destination. The particular details differ from commodity to commodity. Typically the delivery time period is within some particular month of the year. The seller of the futures contract may deliver the material at any time during the month specified. In some market futures contracts exist for all months but more usually only a subset of months are available; for example, five months on Chicago for wheat and maize and seven for soyabeans. In all cases one contract for delivery in March 1979, for example for maize, is exactly the same as any other contract for maize for delivery in March 1979. On the London Metal Exchange, trading is more restricted, to a 3 months contract only, where delivery is specified for the particular day three months after the day of the purchase of the contract. Usually the exchange itself acts as a guarantor to both the buyer and seller of the futures contract to ensure that all the contractual obligations are met and also acts as an arbitrator in any contract disputes that may arise.

In most instances the buyer of a futures contract can take physical delivery of the amount specified in the contract when it matures, if he so wishes. However in order to ensure a smooth running of the exchange, the seller of the contract can deliver material of a different quality to that specified in the futures contract itself, with a predetermined premium or discount payable for the particular grade. The Chicago market for example allows four deliverable grades of soyabeans and twenty-five deliverable grades of wheat. In the same way, alternative delivery destinations are also allowed. Thus the buyer can end up receiving a different grade of material to that which he desired originally. The exchange predetermined premiums or discounts may not be the same as the actual savings or extra costs to the buyer of using the different grades in his production process. For this reason, only a small proportion of futures contracts are settled by physical delivery of the actual commodity; typically 1% or 2%. Normally the buyer resells his futures contract back into the market and at the same time makes a purchase in the parallel cash or actuals market for the actual quality and quantity of raw material that he requires. Generally as a futures contract approaches its period of delivery, its price moves closer to the price in the cash market. There may be a small difference between them, but the buyer has been able to secure the supply of his future raw material requirements at a fixed-price well in advance, instead of waiting to the very last moment when prices may have risen much higher than at the earlier time.

Several cash markets may exist at different geographical locations within the same country. Their prices and the prices on the commodity 'futures' market naturally are very closely correlated over time. If this were not so, then profits could be made by selling in one market, buying the same amount in another and then physically transporting the material to the first market to meet the sale. The differences between markets in the same country tend to reflect internal transportation and storage cost differences. A potential buyer within the country would thus tend to buy his maize, for example, at whichever market gave the lowest transportation cost to his place of usage. Obviously small temporary local market variations in price occur at the different markets, but for this maize the general price levels are determined by the major world market for maize, the commodity exchange at Chicago.

For international buyers of maize outside the U.S., the situation is more complicated. There exists a cash market for maize on a Gulf Ports pricing basis, which gives the price charged for U.S. maize as loaded on board a ship in one of the Gulf Ports. This is an informal market without a physical central exchange, so prices are not determined in public. It is essentially a telephone or telex market where sellers or brokers quote a price directly to a potential buyer. A buyer can thus negotiate a contract for maize of a desired quantity and quality to be put on board a ship at some future date, up to at least 12 months ahead, at a price fixed now. The Gulf Ports prices over time follow the price movements on the Chicago futures market. The difference between them is the cost of transportation within the U.S. from Chicago to the Gulf Ports, plus some storage costs and loading costs at the port. This margin varies over time because of fluctuations in internal transport costs and the relative availability of rail or barge transport to carry maize to the ports. Often in winter, parts of the Mississippi are frozen over so that barges cannot move the maize. In such circumstances a temporary shortage of physical maize actually in the Gulf Ports for export can arise, with a consequent rise in price.

In general, sellers of maize will quote a Gulf Ports price as the Chicago futures price on that day, for the particular month of shipment required, plus a 'margin' reflecting the current extra 'transport costs'. The overseas buyer can accept that complete Gulf Ports price if he wishes. Alternatively he can agree a contract to guarantee his requirements for future usage by accepting the 'transport costs' margin for the difference between the Gulf Ports and Chicago prices. He is then allowed to wait until a later day to fix the Chicago price for the relevant future month. At some later date he informs the seller that he wishes to buy at that day's Chicago price. The total contract price is then the Chicago price on that day plus the Gulf Ports margin fixed at the earlier time. The Gulf Ports 'margin' fluctuates in value over time, generally independent of the fluctuations in the Chicago futures prices over time. Thus the lowest value for this 'margin' may not occur at the same time as the lowest price on Chicago. Hence fixing the two components of the total price on different days can give a lower total purchase price. Although a lower total price may be obtained by delaying the choice of the day for the Chicago price, the buyer has fixed the actual supply for his requirements at the earlier date and hence does not have to worry unduly about possible future shortages. The overseas purchaser must arrange shipment of his maize from the Gulf Ports across the Atlantic himself and also the inland delivery within his own country.

European purchasers can also buy U.S. maize via markets in Europe, at London and Rotterdam for example. This is for maize already shipped across the Atlantic and on a basis for collection from a silo in the particular port. The prices in London and Rotterdam follow closely the U.S. prices. The monthly prices in those ports, together with the U.S. Gulf Ports price, are shown in Fig. 1.1 for the early seventies. This shows how closely these

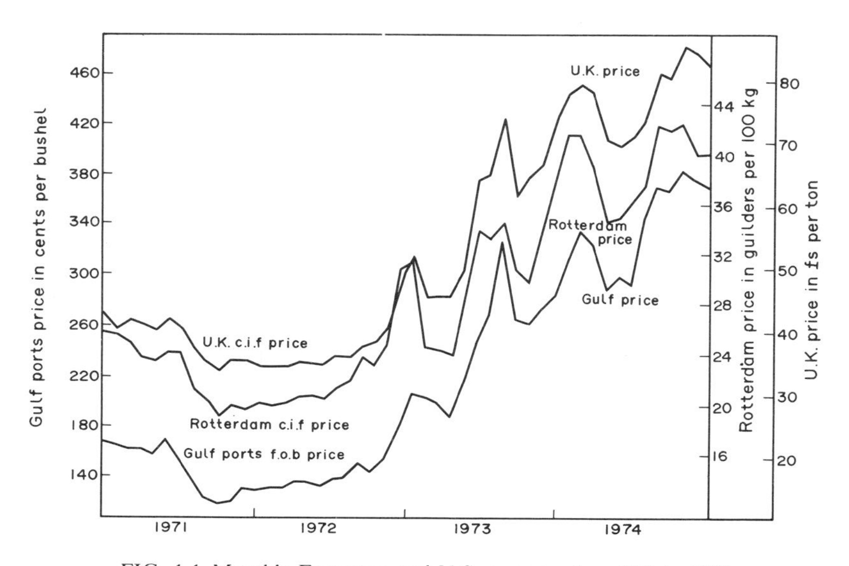

FIG. 1.1 Monthly European and U.S. export prices 1971 to 1974

prices move together. The differences between the series reflect the relative currency values between the pound sterling, Dutch guilder and the U.S. dollar, and the freight rates across the Atlantic. Note freight rates in 1974 were four times the levels of 1972. Occasionally European prices are slightly higher than the Gulf Ports plus a freight rate reflecting relative shortages of maize in Western Europe. Since the ending of the era of fixed currency exchange rates in 1972 buyers outside the U.S. have also had to take account of future changes in value of their currency relative to the U.S. dollar. These alternative methods for fixing the total purchase price generally exist for overseas buyers of all U.S. agricultural products, since usually a futures market for that product exists somewhere within the U.S.

1.3 HEDGING AND SPECULATION

Futures markets developed to facilitate trading in raw materials and to try to smooth out the time gap between the production and consumption of the materials. For agricultural materials, production occurs on a regular seasonal basis at some well-defined time of the year yet consumption generally occurs throughout the year. Transportation times from producer to consumer are often quite lengthy. Considerable uncertainty can exist concerning future levels of production and consumption, which can lead to considerable variations in prices, as we have seen earlier. Someone has to carry the physical stocks of materials over these time gaps. They face possible losses because of price changes between the time they acquire the stocks and the time they dispose of, or use them. The future markets are often justified as a mechanism for minimising this price risk by means of a method known as 'hedging'.

The classical view of hedging, as put forward in the literature and brochures published by the exchange authorities or the brokers operating on the exchanges, is that a processor of the physical commodities can protect himself against price risks by undertaking an offsetting transaction on the futures market whenever he buys or sells the physical material. For example, consider a miller who buys wheat which he processes and resells in the form of flour. Let us assume he buys 10,000 bushels of wheat on 18 August at $2.00 per bushel* because he thinks prices will rise. At that time the wheat is for usage to produce flour in December but he is unable to sell the flour. Possibly the consumers of flour think that prices will fall or alternatively have a policy of not buying their requirements so far ahead. The miller's flour pricing policy is based on the cost of the wheat in the cash market on a day plus some manufacturing margin. Assume that the price of wheat in the event falls and that the flour miller finally sells all of the flour

*A bushel is a measure used in the international grain trade. One imperial ton contains 40 bushels whilst one metric ton is equivalent to 39.37 bushels.

from the August purchase on 5 November, on the basis of the prevailing wheat price of £1.84 per bushel. The miller has thus incurred a loss of £0.15 per bushel on each of the 10,000 bushels that he bought on 18 August. If the miller has used hedging, then he would have sold 10,000 bushels of wheat in March futures, say, on the same day that he made his purchase in the cash market. The March futures price would have been higher than the cash price he paid, say $2.15 per bushel. Assuming a perfect correlation between the movements of the futures price and the cash prices, the March futures price on 5 November will be $2.00 per bushel. The miller buys back his March futures on that day. His total transactions will be

Cash or physical transactions	Operations in the futures market
18 August	
Miller buys 10,000 bushels of wheat in cash market at $2.00 per bushel.	Miller sells 10,000 bushels in March wheat futures and receives $2.15 per bushel
5 November	
Miller sells his flour and receives $1.85 per bushel equivalent for the wheat used to make the flour	Miller buys 10,000 bushels of March wheat futures and pays $2.00 per bushel.
Loses $0.15 per bushel	Gains $0.15 per bushel

It is seen that the profit that the miller achieves in his hedging operations in the futures market compensates for the loss incurred on the physical operations due to the fall in price. His profits will then be achieved from his manufacturing ability. If, on the other hand, prices had risen instead of falling then he would have made a profit on his physical transactions but a loss on his hedging operations in the futures market. He is thus theoretically in the same situation whether prices rise or fall. Hence hedging in this way is recommended as a policy for companies who do not wish to try to understand the market and forecast prices but regard themselves solely as manufacturers. This is an example of a 100% hedge. The miller could decide only to hedge 50% of his wheat requirements or even less. There is no obligation to hedge either all or nothing. Any intermediate value can be hedged. Such hedging policies are very common in the U.S., particularly for cereals and soyabeans. At the present time they are little used by U.K. companies.

The real practical situation is however, much more complex than this simple example would suggest. Cash and futures prices do not maintain the same difference over time. It fluctuates in value with a general tendency to decrease as the future approaches maturity. Thus 100% protection against falls in price can never be achieved. For example, in practice the difference on 5 November will be nearer to $0.10 than $0.15 per bushel. Because of fluctuations or special circumstances, the difference could fall to zero at times. Thus dangers do exist. A large active market of many buyers and sellers helps to mitigate many of these dangers. The lack of such active futures markets for many commodities in the U.K. compared to the U.S. is a partial reason for the small use of hedging by British companies.

'Hedging' is used to describe many different types of policies more complex than this simple example. Debate occurs at intervals in the academic literature as to what precise forms of hedging exist and the underlying purposes and motivations. It becomes difficult to distinguish clearly between some hedging policies and open trading in the futures contracts to make a profit unconnected with any physical processing operations. Such trading is generally called speculation. Speculation is a word that has acquired rather bad connotations because it conjures up visions of rather shady characters, wheeler-dealers, 'spivs', 'parasites' on the decent members of society such as 'good honest' manufacturers and producers. Yet manufacturers are also really traders who buy basic materials and sell their skill in transforming the material into some finished product. The connotations of the word speculation arouse a distaste and to some extent fear of commodity markets amongst many manufacturers in the U.K. This attitude has unfortunate consequences. It can mean that policies for purchasing raw materials based on similar methodological approaches and statistical methods as, for example, those used in planning marketing strategies, are dismissed or not considered because they are conceived to be speculation.

One real option a futures market gives to a buyer is the possibility of reselling an earlier purchase if prices unexpectedly fall. If he thinks prices will fall even further he could sell an amount equal to his original purchase in the futures market and then buy it back again later, when prices have fallen. He will thus have paid a lower overall price for his materials. However he also risks forecasting prices wrongly a second time and thus incurring a second cash loss.

Another point to be borne in mind is that some speculation is necessary for a futures market to work. Someone must bear the price risk that the flour miller of our simple example is trying to avoid. The traditional classical view would be that speculators act to stabilise the markets, buying heavily when prices are low and selling heavily when prices are high, so that they can resell or buy it back later when markets are stabilised. They thus stabilise prices and improve the intertemporal spread of the supply of material. However because of the ease of trading in futures markets, the speculators (the word is used here for anyone who both buys and sells the commodity, not in a pejorative sense) can buy and sell the paper contracts without having any physical stocks whatsoever or any intention of accumulating physical stocks. Thus they have little interest in absolute prices but only in the differences in price levels over time. It has been suggested that speculators generally practice 'movement trading'. In this they always go with the market. Whenever a price movement starts, the speculators follow the movement and reinforce or exaggerate it. Thus speculators act to destabilise prices rather than stabilise them. The view of the 'man in the street' is certainly that speculation is a destabilising influence on prices. Although numerous attempts have been made, it is impossible by any formal statistical analysis to prove which of the hypotheses are true. Obvious problems exist in identifying speculators and speculation since they may also be producers or consumers, who can add any losses on to their actual physical purchasing costs and pass them on eventually to the consumers of their final products. Asserting that speculators exaggerate upward or downward movements of prices more than would otherwise have been the case, does not imply asserting that speculators consistently make profits. Thus proving that 'speculators' as a group make losses does not mean disproving the former assertion at the same time. Sometimes speculators make very large losses, as the fairly recent example of the Hunts in the silver market illustrates. The exaggerated upward and downward price movements may merely be a result of sellers and buyers generally panicking in the face of extreme shortages or surpluses. Some markets have, on rare occasions, existed for certain times both with and without futures markets. The general statistical findings of the very few studies of such markets is that futures trading does not increase price variability but if anything reduces it. (See Naik (1970) and Working (1963)). A very good detailed discussion of the various aspects of commodity futures markets covered in this section is given in the book by Goss and Yamey (1976).

1.4 INFORMAL MARKETS AND FUZZY PRICES

From the point of view of the buyer, the international commodity exchanges provide a mechanism for determining a daily price which is acceptable to both buyers and sellers on that day. Most importantly, the price is determined in an open manner and is available to any buyer or seller throughout the world. Thus a buyer contemplating a purchase knows what other buyers think is a reasonable price at that moment. In an unorganised informal market a buyer only knows what price is offered to him and has little knowledge of what price other buyers are paying. Examples of such markets are the rare metals, titanium and cadmium, and the spices amongst agricultural materials. It should be kept in mind that some of the informal markets within a particular country are based on the prices quoted on an international exchange in a far away different country, taking account of shipping costs, port handling costs, inland transport costs and currency exchange rates.

On an international commodity exchange, a buyer must pay whatever is the price currently being quoted in the market plus the broker's commission charge. If he doesn't like the price he must wait until later in the day or another day. In an informal market, even if based on the prices in an

TABLE 1.2 Reductions achieved on Brokers' initial price quotes

Broker's initial quote	Price paid by company	'Discount'	Broker*
109.2	104.5	4.7	A
106.7	104.5	2.2	A
103.5	103.5	0.0	B
104.4	103.5	1.1	B
103.9	102.0	1.9	B
101.4	101.0	0.4	A
101.1	101.0	0.1	B
100.3	100.0	0.3	A
98.8	97.0	1.8	A
100.3	97.0	3.3	B
93.5	93.5	0.0	A
96.5	93.5	3.0	A
106.2	103.1	3.1	B
107.7	107.7	0.0	A
154.9	154.4	0.5	B

*A and B are alternative brokers used by the company concerned. The prices are in £s per ton.

	Average 'Discount'	Standard deviation of 'Discount'
Broker A	1.55	1.59
Broker B	1.43	1.27

international market in an overseas country, part of the price offered by the broker to the buyer is under the control of the broker rather than the market; transport costs if the broker uses his own ships for example. In addition the broker may believe prices will be lower in the future and therefore may delay purchasing the amount he sells to the buyer. The price he quotes will also include some profit margin for himself. The prices may thus be described as 'fuzzy'. Instead of waiting for another day, the buyer can attempt to negotiate with the broker to reduce his original price offer.

This is particularly so if the buyer deals with more than one broker. In practice, 'discounts' on the original price offer by the broker can often be negotiated. They depend on both the market's and the broker's individual circumstances but are sometimes quite large. Some examples for a particular commodity are shown in Table 1.2. The average 'discount' was £1.4 per ton; approximately 1.5% of the quoted price. The values shown in the table are for those occasions when the company concerned actually made a purchase. The purchases were equally split between two different brokers, as indicated. A cursory glance shows that there is little difference between the discounts offered by the two brokers. Thus in an informal market with fuzzy prices a buyer should always seek to reduce the broker's first offer price, even if he has decided to make a purchase on that day.

2

Commodity Price Movements

2.1 LONG-TERM PRICE MOVEMENTS

A major problem as mentioned in Chapter 1, is the sheer size of the price changes, upwards and downwards, that can occur for raw materials from one year to the next. This year-to-year variability is illustrated in Figs. 2.1 and 2.2, which show the annual average prices for cocoa, copper and sugar on the London commodity exchanges. Cocoa and copper prices are measured on the left-hand vertical scale in both figures, whilst sugar prices are measured on the right-hand scale. Figure 2.1 shows the prices from 1952 to 1970, now regarded by many as a 'golden age' of low stable commodity prices. Figure 2.2 shows prices from 1967 to 1978. (Note that 1 cm on the vertical price scale in Fig. 2.2 covers three times the price range of Fig. 2.1. This was necessary to give a clear indication of the relative movements of cocoa and sugar prices over the last 25 years.) This trivial point illustrates strongly the difference in price levels, at least for cocoa and sugar, before and after 1973, which was a traumatic and radical turning point year for commodities and the world economic system. The prices shown are the real prices. No adjustment has been made for the 14%

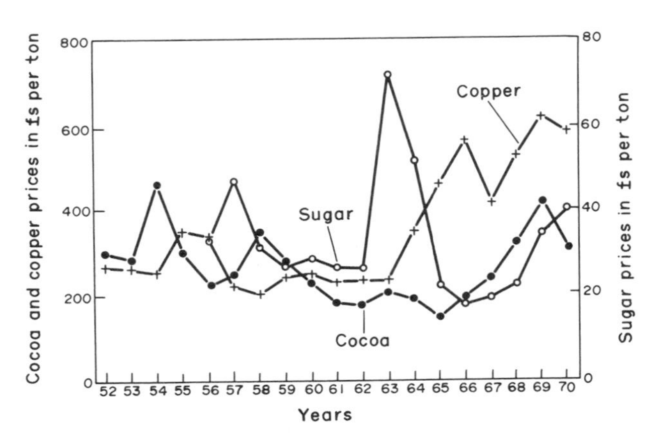

FIG. 2.1 Annual average prices pre 1970

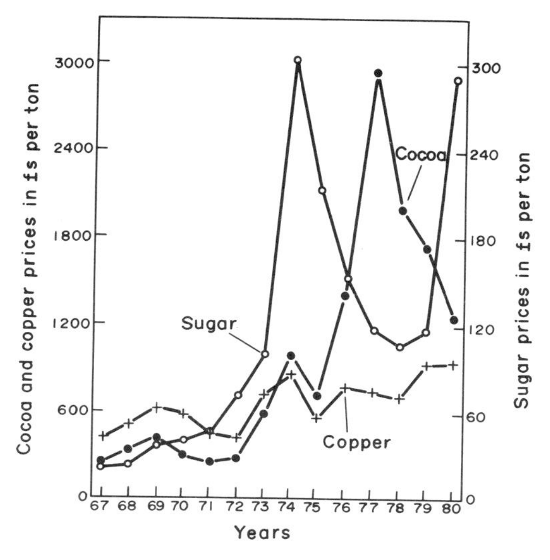

FIG. 2.2 Annual average prices post 1967

sterling devaluation in 1968 nor the floating of the pound sterling after 1971.

However, although the absolute price levels are higher after 1973, the annual changes in relative terms pre and post 1973 are similar. Figure 2.2 shows that the annual average sugar price fell from a peak of £305 in 1974 to only 38% of that value 3 years later. The peak year for sugar prices in Fig. 2.1 was 1963 at £72. Three years later, sugar prices had fallen to only 25% of the peak value. From January 1973 to February 1974 copper prices rose from £450 to £1400 per tonne and then fell back to £600 by the end of 1974. Similar relative-sized fluctuations occurred in the sixties for copper. On 25 April 1967 the price of copper was £347 per ton. Less than a year later at the beginning of March 1968 it had risen to £818. Over the next 2 months, it fell back to £441 per ton. Hence relative changes from year-to-year in the fifties and sixties are as bad as, if not worse than the changes in the seventies. People have been mesmerised by the high absolute price levels in the seventies to conclude that the earlier years were ones of price stability. The problems imposed on buyers by the large price changes over the transition from 1972 to 1974 can easily be imagined.

To comprehend the nature of the price movements that can occur and to understand better why there were such large changes in price from year-to-year over the 27 years covered in Figs. 2.1 and 2.2, it is useful to consider separately each of the three commodities illustrated in the figures.

2.2 COCOA PRICES

Cocoa is not a food essential for survival in any country in the world. Its main use is in confectionery and cakes in the form of chocolate. It is thus a 'luxury food' bought out of disposable income after the basic needs have been met. As can be seen in the figures, cocoa prices rose from £290 per ton in 1952/53 to peak at £450 in 1954, but then fell back to below the earlier levels over the next 2 to 3 years. Prices rose to a further small peak in 1958 followed by a long period of falling prices to reach a minimum in 1965. Prices rallied over the next 4 years to a peak in 1969, which was lower than the price achieved in 1954. Subsequently prices fell to a minimum in 1971/72. A long period of rising prices began, culminating in the highest ever recorded price in 1977. The average over the year was £2944 per tonne.

The major cause of these price changes from year-to-year was the varying surplus of the supply of cocoa over the amount consumed. These values year by year are shown in Fig. 2.3. The supply of cocoa in a year is made up of that year's production (crops from all growing countries) plus the stocks carried over from the previous year. World grindings (which measures cocoa consumption) rose strongly from 1946 to 1950 due to large crop increases. It remained steady or fell back over the next 4 years. Thereafter it grew at a roughly constant rate up to 1973.

A large supply of cocoa in a year can arise from a very good crop or high carryover stocks from the previous year or a combination of these two factors. A poor crop has an immediate impact on the relative supply/demand situation and hence on prices, although the effect on price can be mitigated

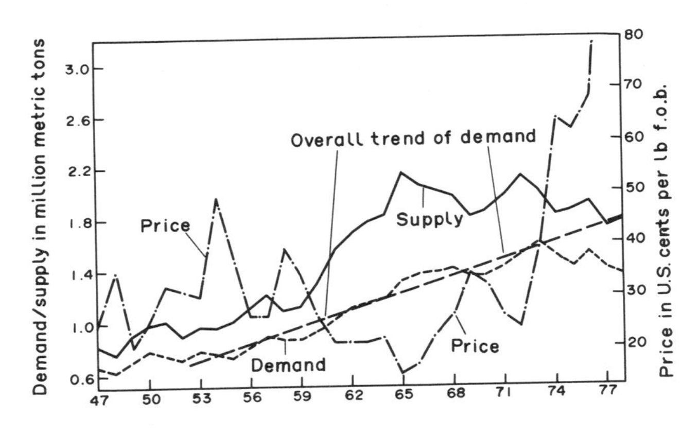

FIG. 2.3 Postwar supply, demand and prices for Cocoa. (Supply is opening stocks plus season's production, demand is season's grindings.) (Reproduced from 'Cocoa Statistics, December 1977' published by Gill and Duffus Ltd)

if the carryover stocks are high. Thus often 2 years, not too far apart, of poor cocoa crops are required to bring about high prices. Over the fifties cocoa production generally increased at a similar rate to consumption. There were however poor crops in 1951/52, 1953/54 and 1957/58 which resulted in local price peaks in 1954 and 1958. Cocoa consumption fell in both these years. Cocoa production then grew rapidly at a faster rate than consumption up until 1965/66. The bumper crop of 1964/65 gave large stocks of cocoa for carryover into following years. Production fell back and remained below consumption until 1969/70. Prices rose continuously over these 4 years. Bumper crops were harvested in 1969/70, 1970/71 and 1971/72. Production always exceeded consumption and prices continuously fell back. Production fell drastically in 1972/73 but rallied over the next 3 years. However production had apparently reached a maximum output level, below the record crops of 1971/72. Consumption achieved its maximum level in 1972/73, which was above the record production of 1971/72. Due to the shortages resulting from the low crops, consumption had to fall. A drastic crop failure occurred in Ghana and Nigeria in 1976/77. Consumption had to be further reduced.

The difference between the cocoa supply, the crops plus the carryover stocks, and the grindings shows the actual cocoa surplus each year. Grindings (the actual cocoa consumed) have fallen almost continuously from their peak value in 1973 to the present. 'Demand' for cocoa presumably remained in existence but it could not be satisfied because there was not sufficient cocoa available. Prices increased and acted as a rationing device to spread the available supply amongst the various demands. Because of the steady growth of consumption from 1953 to 1973, a better idea of the scarcity situation in the late seventies is obtained by extrapolating the 1953/73 pattern of grindings as shown in the Fig. 2.3. The gap between the cocoa supply and this extrapolated 'demand' is lower in relative terms than in early years and actually becomes negative in 1977 and 1978. Thus 1977 and 1978 are much worse in terms of relative shortages of supply than at any time in the past. This factor plus the decline in the value of the pound sterling relative to other currencies over the past 30 years 'explains' why these 2 years had the highest cocoa prices ever, almost four times the values of the earlier peak prices in 1954.

2.3 COPPER PRICES

Copper by contrast over the period up to 1970 had its highest prices in 1966 and its lowest prices in 1958. For most of the period up to 1963, copper prices were around £250 per ton. They then rose strongly to fluctuate generally between £500 and £600 per ton from 1965 to 1970. At the end of the fifties copper mines throughout the world, in Chile, Zaire,

Zambia as well as the U.S.A. and Canada (the major producing countries at that time), were controlled by international mining companies. The companies feared that the relatively large price fluctuations over the previous 10 years would lead to the loss of traditional usages to substitutes. They decided in 1960 to attempt to stabilise prices by agreeing to sell all of the copper that they mined at a fixed price of around £236 per ton. As can be seen in Fig. 2.1, this policy appeared to work for almost 5 years. Copper was sold direct to consumers at this fixed price, bypassing the international copper market which was mainly centred in London, and to a lesser extent in New York. Unfortunately for this price stabilisation exercise, reclaimed scrap copper also provides a major part of the supply of copper for industrial use. This copper was sold or at least priced through the international markets, e.g. the London Metal Exchange for U.K. copper scrap. In 1962 and 1963, demand for copper increased due to a growth in industrial output in the developed economies. From 1962 to 1964 world consumption of refined copper rose by nearly 1 million tonnes (20%). Mine production of copper cannot be increased significantly except over a period of years so that a copper shortage developed. Indeed mine production of copper between 1962 and 1964 could only be increased by 500,000 tonnes, only half the increase in consumption. The free market price, for mainly scrap copper, increased to reflect this shortage and rose more than 50% above the fixed price of the mining companies. The governments of Chile, Zaire and Zambia, who derived much of their revenue from taxes on their copper exports, a major part of which was proportional to the price of copper, saw themselves foregoing potential extra revenue and so put pressure on the mining companies to increase their fixed prices. Newspapers reported that some unscrupulous consumers were buying copper from the mining companies at their fixed price and then immediately reselling it as scrap on the free market for a vast profit. Economists argue that the very high prices on the free market were because it dealt only in marginal supplies of copper, i.e. the reclaimed scrap. If the mining companies abandoned their fixed-price strategy and sold all their copper through the free market then world copper prices would become stabilised again, at only moderately higher prices than the companies' existing fixed price but substantially lower than the prevailing free market price. Under all of this pressure the international mining companies abandoned their strategy gradually and by mid 1966 all copper was sold through the free market, priced daily according to the daily free market prices. Figure 2.1 shows the result of this change in policy. Unfortunately copper prices continued to increase and fluctuated even more over the remainder of the sixties than they had done in the early fifties. These fluctuations were as high as anything seen in recent years.

In general the peaks in copper prices have occurred in periods of rapid growth in consumption and/or interruptions to the flow of new copper from the mines because of strikes or other causes. For example, 1966 saw a long strike of Chilean copper miners and the start of the Rhodesian UDI which temporarily stopped Zambian copper being exported. In 1969 there was a strike of copper miners in the U.S. which lasted for 9 months. The highest copper prices in late 1973 and early 1974 were essentially due to a rapid growth in demand for copper in all the industrialised countries and an inability of the mines to respond quickly. For the first time post war the economies of Western Europe, North America and Japan entered a 'boom phase' of the economic cycle at the same time. Industrial production soared everywhere. The growth in demand for copper caused shortages and panic in the markets so prices increased to record levels. As the economies went into recession demand for copper fell as did copper prices. The copper prices over 1975 (allowing for the continuing decline in the value of the pound sterling) were at very similar levels, or even lower, to those in the late sixties. Prices subsequently rose to around £750 per ton over 1967–78.

If one allows for the decline in the value of the pound sterling against other currencies of over one third since 1971, during the era of floating exchange rates, plus the 14% devaluation in November 1967, the average copper price over 1978 is equivalent to about £400 per ton in pre 1967 terms. Looking at the prices over the early sixties this is not exceptionally high. Thus, unlike recent cocoa prices, copper price movements over the later seventies are very comparable in size to those of the fifties and sixties.

2.4 SUGAR PRICES

Unlike cocoa, sugar is for many countries an essential food as a major source of energy. It has extensive uses in food manufacturing as well as direct consumption. Whenever they can afford it, all countries seek to increase their consumption per head of population, presumably eventually to the saturation levels of the richer industrialised countries. Sugar prices have also had similar relative-size fluctuations in value from one year to the next over the postwar period. Prices fell from around £48 per ton in 1957 to the £30 per ton level in the following year and remained there for almost 5 years. The high prices of 1957 appear to be a direct result of the 'Suez crisis' of November 1956. Curiously sugar appears to be remarkably sensitive to wars and crises in the Middle East, much more so than any other commodity. From 1958 to 1962, price fluctuations were small and prices did not fall much below £30 per ton. This was a result of an International Sugar Agreement to maintain prices above a minimum level,

£27.5 per ton, by imposing agreed export quotas on the producing countries. Successive good crops led to the accumulation of very large stocks of sugar. The agreement collapsed in the middle of 1962 as countries tried to get rid of these accumulated stocks at very low prices, below the agreed minimum of the Sugar Agreement. At the end of the year and in the early months of 1963, news of crop failures in Cuba and the Soviet Union, two of the largest sugar producers, transformed the situation into one of a serious sugar shortage. Prices then rapidly increased and stayed at these levels over 1963 and the first half of 1964. Bumper crops in Europe and more normal crops in Cuba brought about a return to a situation of large sugar surpluses. Prices fell and remained low, near to £20 per ton over the remainder of the sixties. A further Russian crop failure then brought about a moderate shortage so that prices rose to around £40 per ton in 1970. Note that the highest and lowest prices for sugar in Figure 2.1 occurred in 1963 and 1966. Neither of these were the same years as for cocoa and copper.

From 1970 onwards production remained below consumption. So the large stocks accumulated at the end of the sixties were reduced year-by-year. The rate of growth of consumption slowed down. The lack of sugar and the increase in prices together caused a rationing of use. The beet crop in Europe fell by 3 million tonnes in 1974, mainly due to a Russian crop failure. Although the world sugar cane crops were up, overall world production was 1 million tonnes down. Furthermore, the largest shortfalls were in the richer countries who could afford to pay high prices to import sugar. The average annual growth of consumption over the early seventies would have reduced stocks to the absolute minimum pipe line inventory levels. Very high prices thus occurred as the very limited sugar available was rationed out between the competing demands.

2.5 SHORT-TERM PRICE MOVEMENTS

The annual average prices shown in Figs. 2.1 and 2.2 conceal significant fluctuations over the short term, which can exceed the year-to-year changes. The annual average price of copper over 1968 was £525 per ton, £126 higher than that of 1967 but £98 lower than that of 1969. The highest price quoted on the market during 1968 at £825 per ton was almost twice the lowest price offered at £420 per ton. The difference between the highest and lowest prices was thus almost four times the annual price change for the years before and after 1968. This range of price variation within the year was, in fact, larger than the changes from one year to the next in the annual average prices over the whole 27-year period covered in Figs. 2.1 and 2.2. The size of the fluctuations from day-to-day in prices can be seen from Figs. 2.4 and 2.5. Figure 2.4 shows daily cash copper prices on

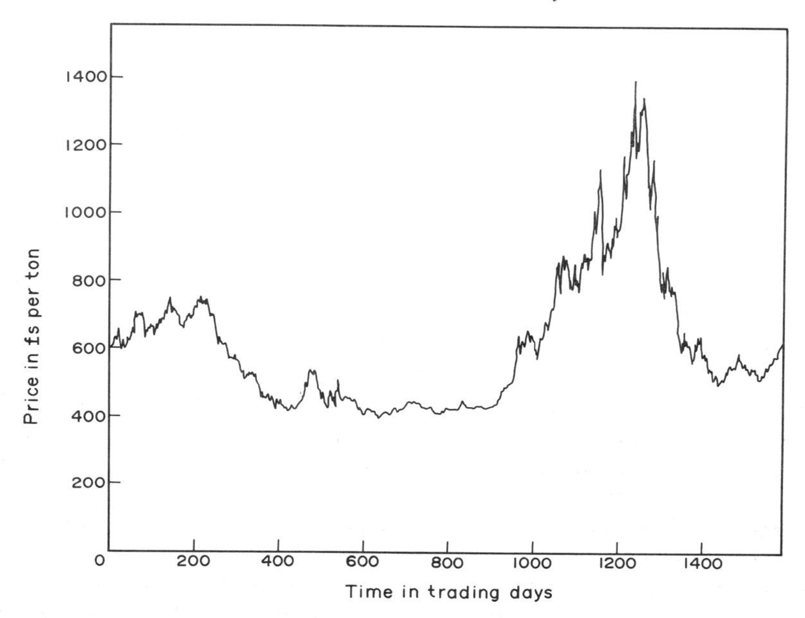

FIG. 2.4 Daily copper prices on the London Metal Exchange in £s per ton from 28 May 1969 to 27 August 1975

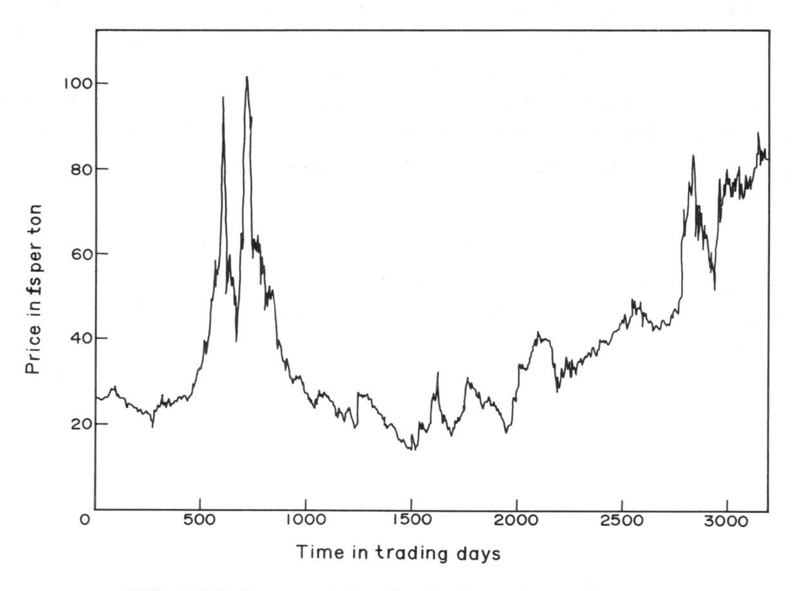

FIG. 2.5 Daily sugar prices for the December future on the London Market 1961 to 1973

the London Metal Exchange from May 1969 to August 1975. Figure 2.5 shows daily sugar prices for the December future price, on the London market, over a longer period, 1961 to 1973. Note that there are approximately 250 market trading days per year. There are of course large differences in the size and type of price fluctuations from one year to the next. At times, as for example for most of 1972 for copper, prices can fluctuate very little from one day to the next about some constant price level. An even more stark contrast between price behaviour from one year to the next is shown in Fig. 2.6. This gives the daily prices for sugar over 1963 and 1966. The former year has very violent price variations whilst the latter is very stable with almost no fluctuation about the small negative trend that persisted over the whole year. A similar stark contrast is illustrated in the behaviour of tin prices from June 1970 to May 1974; see Fig. 2.7. Prices fluctuated narrowly about £1500 per ton for the 3 years up until May 1973. Over the next year they rose rapidly to £4000 per ton, with substantial temporary reversals of price. The price increase is much steeper than that of copper. This figure shows clearly the abrupt and unusual nature of the start of the commodity price explosion and world economic crises of 1973/74. As a matter of interest silver, lead and zinc prices over 1970 to 74 on the London market had a pattern of price movement of exactly the same form as for tin.

Thus short-term price movements over several months fluctuate even more than the variations in the annual average prices would suggest. The commodity purchasing problem is trying to choose the lowest possible

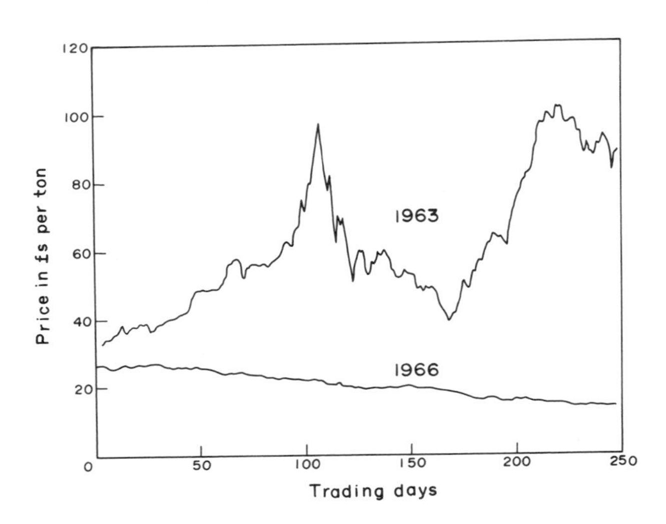

FIG. 2.6 Contrasting daily price movements for two December futures sugar contracts

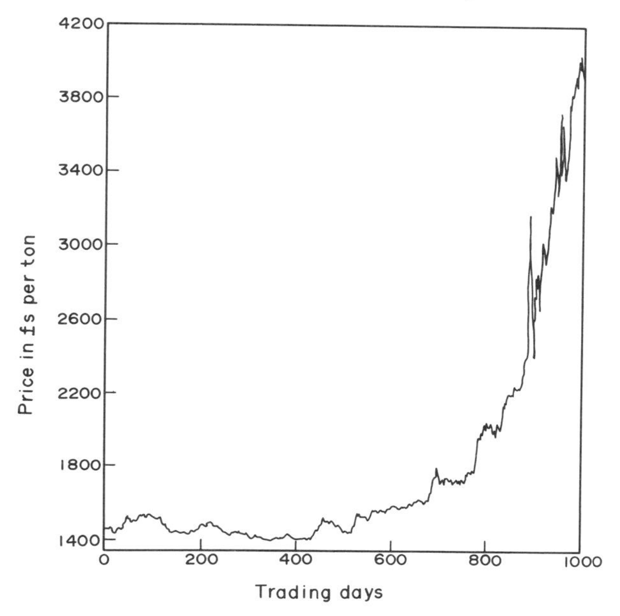

FIG. 2.7 Tin prices at London from mid 1970 to mid 1974

prices over particular periods of such price movements, ranging from days to months. The problem facing a buyer in picking out good prices can be imagined from the various examples presented. Even over the very short-term, large price variations exist. Table 2.1 gives the percentage price change that occurred on the London commodity markets for a sample of commodities over the month of January 1970. Note that this is in the earlier period of Fig. 2.1, when prices were believed by many to be more stable than at present. This month was originally a random choice and not picked as a month of unusually high or low-price movements. The table shows that rises or falls in price over the month of around 7% are quite common. A confectionery manufacturer would have found that the price for his cocoa fell over the month but that his sugar price increased by a similar amount. Coconut oil and groundnut oil are partially substitutable materials used in the manufacture of margarine, yet the price of one increased whilst the other fell. The profit contribution for a company typically amounts to 7% of the final product cost, whilst raw materials often make up 70% of the final product cost. Thus making wrong purchasing decisions over January 1970, with the size of the price changes that occurred, could have almost wiped out the profits. On the other hand good purchasing decisions could have increased profits by almost 75%.

TABLE 2.1 Examples of price changes over a month (Taken from *Financial Times* Commodity Reports—January 1970)

Commodity	Price change	Commodity	Price change
Copper	−7%	Coconut oil	−11%
Lead	−5%	Groundnut oil	+6%
Zinc	−5%	Groundnut seeds	+10%
Silver	−6%	Rubber	+4%
Maize	+8%	Sugar	+10%
Cocoa	−12%	Peppers	−3%

The open-call system generally used in the international commodity markets often appears organised in such a way as to facilitate large day-to-day price movements and exaggerate the effect of rumours. On the London Metal Exchange for example, trading in copper occurs in two 5-minute dealing sessions, one in the morning and one in the afternoon. The twenty or so dealers, sit in seats arranged around a circle facing each other. When the 5-minute trading session opens they call out to each other the amounts they want to buy and sell and the prices they are prepared to take. Dealing is initially slow and then becomes hectic in the last minute of trading. To an outsider it looks like disorganised bedlam and the dealer with the loudest voice seems to have an advantage. The final price reached at the end of the 5 minutes then becomes the basis generally for all copper bought and sold worldwide on that day. There is no time available for rational thinking in this type of market during the trading sessions. Rumours circulating immediately prior to the start of the session cannot be verified nor can their effects on future production, consumption and price be properly assessed. When later in the day the rumour is found to be untrue, prices in the afternoon session are rarely reduced or increased by exactly the same change as occurred in the morning because of the false rumour.

The twenty largest daily price changes, in percentage terms, for copper over 1966–1976 and sugar over 1961–1973 are shown in Tables 2.2 and 2.3. The causes of these changes as conjectured at the time by market commentators are also shown in the tables. Some of these causes are rational, others are best classified as rumours, and occasionally the 'experts' could find no explanation. Virtually half of the changes are ascribed by the trade press itself to rumour or unknown cause. Many of the changes classified as rational are due to political events or their imagined effects rather than direct changes to the production, consumption or trade in the commodity. Note also the irrationality of the rumours at times. Firstly, item 4 of the copper list, Table 2.2, is an unfounded rumour that

TABLE 2.2 The largest copper price changes

Rank	Magnitude (%) and Sign	Date	Cause	Classified as: Rational	rumour	unknown
1	+16.94	19. 5.67	Single dealer corners market	x		
2	−12.77	25. 7.66	Zambia breaks Rhodesia blockade	x		
3	−12.24	14.12.73	Three-day working week announced	x		
4	+10.98	1. 8.73	Rumour of Allende killed; 'midsummer madness'		x	
5	−10.92	23. 2.68	American dockers black copper shipments	x		
6	+10.74	20.11.67	Sterling devalued	x		
7	+10.33	26.10.67	Misjudged speech by LME chairman		x	
8	+ 9.99	18. 4.74	Problem in finding sellers		x	
9	+9.50	5. 4.68	—			x
10	−9.48	7. 6.67	Israel proposes cease-fire	x		
11	−8.85	27. 4.66	Zambia returns to free market pricing	x		
12	+8.47	26. 2.68	Cf. 5: ban lifted; 'a farce'	x		
13	+8.10	29. 4.66	Cf. 11: Reaction	x		
14	+8.09	20. 6.68	'Supply squeeze'		x	
15	+8.04	12. 7.73	Purchases by China and Japan	x		
16	−7.75	13. 7.71	Peace move in U.S. strike	x		
17	−7.59	22. 5.67	Cf. 1: End of squeeze	x		
18	+7.49	8. 6.71	'Chartist buying'		x	
19	−7.27	6. 9.66	Verwoerd assassinated	x		
20	−7.26	17. 7.74	Uncertainty over U.S. miners' wage agreement		x	
			Totals	13	6	1

the president of Chile had been assassinated whilst item 19 is a true report of the assassination of the president of South Africa. Chile is a major copper producer whilst South Africa is not. Secondly, the start of the Arab–Israeli 'Six-Day' war of 1967 appears in the sugar list (Table 2.3) as causing a large price change whilst it is the end of the war that occurs in the copper list. The only other item common to both lists is the 14%

TABLE 2.3 The largest sugar price changes

Rank	Magnitude (%) and Sign	Date	Cause	Classified as: Rational	rumour	unknown
1	+13.62	6. 6.63	—			x
2	−12.92	4. 6.63	Market 'shake out'			x
3	+11.39	4.12.72	Hopes of Russian purchases	x		
4	+11.30	5. 6.67	Commencement of Arab–Israeli war	x		
5	+11.26	18. 6.63	Investigation of speculators launched	x		
6	+11.03	31. 1.62	Cuba withdraws from the world market	x		
7	+10.36	23.10.68	World sugar agreement is approved	x		
8	−9.98	30. 1.64	Rumours of Soviet purchases		x	
9	+9.68	29. 8.63	—			x
10	−9.22	23. 3.72	Panic selling by speculators		x	
11	+8.61	23.10.62	Challenge to Cuba blockade anticipated	x		
12	−8.55	29. 3.63	Nervous 'technical reaction'		x	
13	+8.52	21.11.67	Sterling devaluation	x		
14	+8.51	3. 1.72	Rumours of Soviet purchases		x	
15	−8.43	7. 4.72	Extra-market sales by Brazil	x		
16	+8.38	11.12.72	—			x
17	−8.28	6. 6.67	Cf. 4: War rumours	x		
18	+8.13	15.12.71	'No rational cause'			x
19	−8.12	14. 6.63	—			x
20	−8.07	19. 6.67	Extra-market sales by Argentina	x		
			Totals	10	4	6

devaluation of the pound sterling in November 1967; this caused a price increase of 10.74% for copper and 8.52% for sugar in one trading day.

Some further examples of the daily price movements are shown in Figs. 2.8 and 2.9. These figures illustrate an important facet of commodity price behaviour. Figure 2.8 shows the daily price movements for maize to be delivered in July over the period April to July and the prices for delivery in

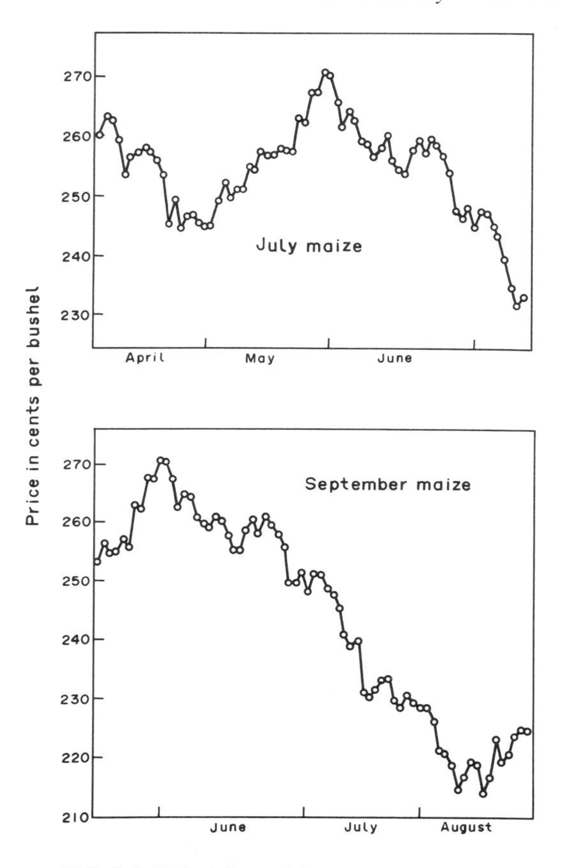

FIG. 2.8 U.S. daily maize prices on Chicago

September over the June to September period. These prices are for 1978 as quoted on the Chicago Board of Trade, the major market for world maize, and are in U.S. cents per bushel. Figure 2.9 shows maize prices for the same two delivery months over the same periods. However the prices in this case are for U.S. maize delivered to Liverpool in 1967, that is 11 years earlier than the prices shown in Fig. 2.8. Disregarding the absolute levels of maize prices in the 2 years, the pattern of price movements over the same periods of the year is incredibly similar for both delivery positions for 1967 and 1978. Even more interesting is that the latter year is after the price explosion and traumatic events of 1973/74, whilst the former year is the earlier era of lower absolute price levels. Figures 2.8 and 2.9 illustrate the point that history does repeat itself. Buyers all too often regard each situation that faces them as completely new and unexpected, different in

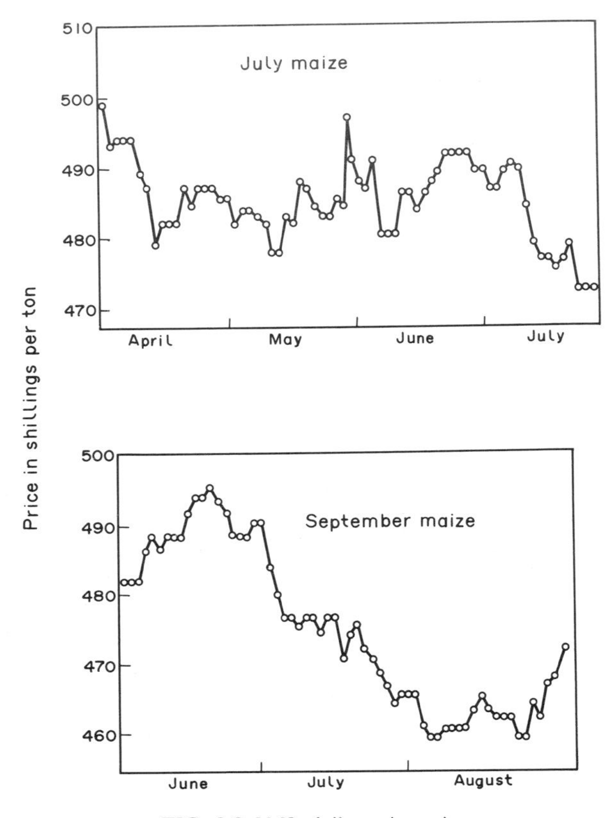

FIG. 2.9 U.K. daily maize prices

kind as well as detail, to the past. Their attitude to some extent gives the impression that buying implies living from one crisis to the next, with it being impossible to learn anything useful from the past.

2.6 TRENDS OR RANDOM WALKS IN COMMODITY PRICES

Over the last 20 years a considerable effort has been devoted to statistical analysis of all kinds of daily speculative price series. Much of this work has been carried out on stock price movements, but in recent years this has been extended to cover daily commodity prices. The earliest of these analyses are collected in Cootner (1964), whilst later surveys are given in Granger (1973) and Labys and Granger (1970). The conclusion generally accepted amongst academic researchers for some time is that daily prices follow a process known as a random walk. This states that tomorrow's price will be today's price plus or minus some random amount.

This random amount will be completely independent of the change that took place from yesterday to today, or indeed of the changes between any earlier successive pairs of days. This means that systematic trends do not occur and persist in the data over time. Whatever has been happening to prices in the recent past, tomorrow's price is equally likely to be either above or below today's price. Indeed the expected change in price from today to tomorrow is always zero. The best forecast of future prices will always be the current day's price. It is theoretically impossible, if the random walk assumption is true, to say that at the end of the next few days the price is much more likely to be above today's price than below it, for example.

An important consequence is that no trading rule, involving buying and selling the commodity at different times, can consistently make money, nor even consistently lose it. For purchasing the random walk assumption means that no policy can exist that will enable purchases over the long term to be bought at an average cost below the average market price. If the random walk is true for daily commodity price movements, then there is absolutely no point in carrying out research to try to determine improved buying policies. The random walk implies that such research will always fail. It is thus necessary to show that systematic trends do exist or at least cast considerable doubt on the validity of the random walk model before trying to derive daily purchasing policies. If a buying policy can make purchases in the long term at lower than the average market price in practice, then this will disprove the random walk model. However the proponents of the random walk model would say that such a result for one commodity over one time period could occur by chance quite naturally. They would doubt that such results would occur over a much longer time period or for other commodities very frequently. Showing conclusively that the savings achieved in a particular example are statistically significant is a time consuming and complex task. It requires the collection of daily prices over many past years, usually at least 10 years, which is often extremely difficult to achieve for many commodities. It is preferable to try to demonstrate that the random walk model is not a proper description of the process that generates daily prices.

Contrary to the views of the theoretical researchers, most purchasing managers and operators in the commodity markets would reply that you only need to examine a graph of daily prices by eye to see that trends do exist and that price movements are not random. Figures 2.10, 2.11 and 2.12 are plots of daily prices over a long period, starting and ending with the same price. Each figure appears to have significant price trends persisting for long periods of time. However only one of these figures contains real commodity prices. The other two are artificially generated from a random walk model adjusted to take account of the changing variance of the price

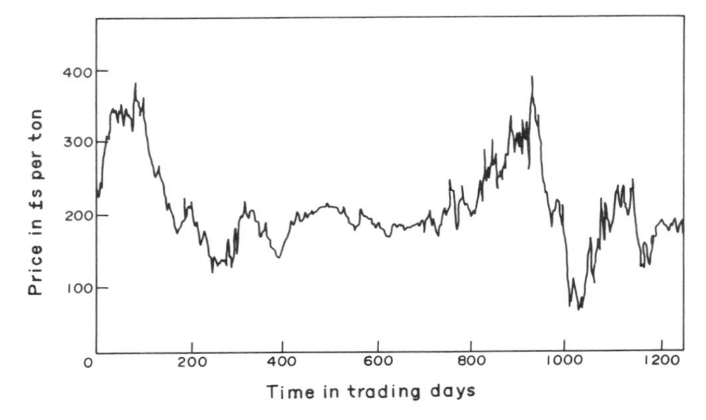

FIG. 2.10 Five year's commodity price history

changes in the real series. The reader is invited to say which is the real price series; the answer is given at the end of this chapter after the appendix. The visual impression of trends is caused by the aggregation of random variables, which just happen randomly to be predominantly positive or predominantly negative for long periods of time. The figures show how deceptive random walks are to the eye, especially when nonnormality and a fluctuating variance are present in the stochastic process. Thus, unfortunately, visual evidence cannot be relied on to disprove the random walk. The figures are taken from research by Taylor.

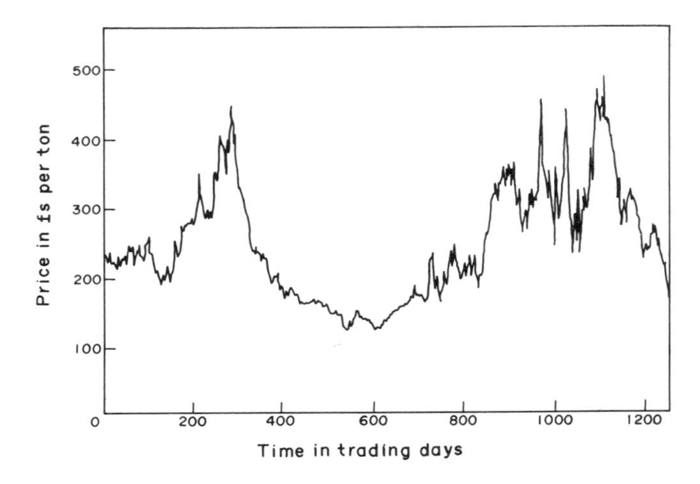

FIG. 2.11 Five year's commodity price history

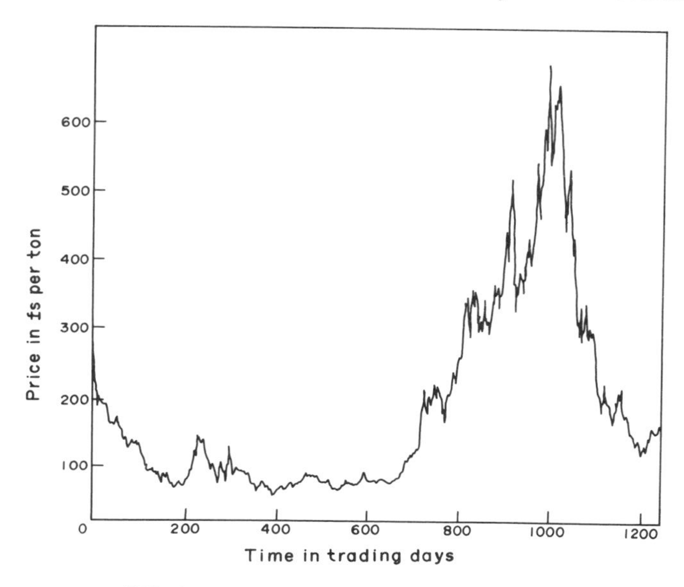

FIG. 2.12 Five year's commodity price history

Many attempts have been made using statistical tests or trading rules to disprove the random walk model. Unfortunately many of the studies that have claimed success in this aim have suffered from statistical or methodological errors. However Leuthold (1972), and Cargill and Rausser (1975) have shown that daily price changes of some commodity contracts are not independent. Until now, no models for the observed weak dependence have been proposed. This somewhat lessens their impact on the validity of the random walk model.

Econometric studies of medium-term price movements, discussed later in Chapters 8, 9 and 10, show that supply and demand expectations do influence prices, leading to trends in prices over time. Systematic changes in price over the medium-term exist which can be ascribed to changes in the general market supply/demand situation. There is thus a contradiction between the results of research on prices as examined from the medium-term view as opposed to the very short-term. The results of the medium-term models cast considerable doubts on the random walk model. However it might be argued that the systematic changes perceived in the medium-term models are not trends in the same sense as one thinks of trends in daily prices but some other form of adjustment in price over time.

Some recent research, see Taylor and Kingsman (1978), has demonstrated the existence of trends in daily price movements and shown that the random walk model is an inadequate approximation to the process generating daily commodity prices. An alternative model has been proposed for daily price movements which explicitly includes a trend term. This model is shown to fit the actual data better than the random walk model. The problems in identifying the real process generating daily prices have arisen mainly because the daily trends (price changes) are very small compared to the prices themselves and to the size of the day-to-day fluctuations about any underlying systematic movement that may exist They are also due to the limitations of the classical statistical models developed for time series analysis. The statistical analysis leading to this new model is necessarily complex. The main points are summarised in an appendix to this chapter.

This section is included for the benefit of those actually carrying out research in this area. Any reader who does not have a good grounding in statistical methods is recommended to miss out this section. It is an interesting diversion which is not essential for following the methods described in subsequent pages. However we can now proceed in our search for ways to develop policies for improved purchasing of raw materials, confident that, although the task will be difficult, it is not theoretically doomed to end in failure before we start.

Appendix

A New Model for Daily Commodity Price Movements

As mentioned earlier in the chapter the random walk model has generally been accepted amongst academic researchers for some time as the stochastic process generating speculative time series. However recent research now strongly suggests that the random walk model is an inadequate approximation to the process underlying commodity price movements. It has quite a long history behind it. As long ago as 1900 Bachelier submitted a thesis modelling French security prices. He deduced that the price of a security should follow a Brownian motion, a process now known as a random walk. For a sequence of regular observations over time, z_t, the random walk model is

$$z_t = z_{t-1} + u_t$$

or alternatively

$$\log z_t = \log z_{t-1} + u_t$$

The errors, u_t, are assumed to have zero expectation and form a sequence of independent random variables. Bachelier originally proposed that these errors were normally and identically distributed. The popularity of the logarithmic transformation is due to Fama (1968), who claimed that this transformation stabilised the variance of the u_ts over time. Bachelier's assumption of normally distributed errors has however been superseded. The distributions of the logarithmic price changes have 'fat tails, high peaked centres and are hollow in between. They have much more probability than the normal within $\pm\frac{1}{2}s$ (where s is the standard deviation), much less between $\pm s$ to $\pm 2s$ and much more at greater than $\pm 2\frac{1}{2}s$.' (See Praetz). Praetz (1975), Clark (1973), and Blattenberg and Gonedes (1974) have considered alternative finite variance distributions to the normal. A Student-t distribution appears to be a likely candidate. Another alternative put forward by Mandelbrot (1963) and (1967) is the Pareto-

Levy family of distributions, particularly those members with an infinite variance.

The results obtained from research into developing medium-term econometric price forecasting models, mentioned earlier in the chapter, suggest that a drift or varying trend term, μ, should be included in the above model to give

$$z_t = z_{t-1} + \mu + u_t.$$

It is assumed that μ results from the general state of the market and will change over time as market conditions change. Since such changes may occur at infrequent intervals relative to the daily changes in prices, μ may remain constant for longish periods of time.

Many studies have been performed using autocorrelation analysis and the Box-Jenkins method of time series analysis. As mentioned earlier, the distribution of the price changes is not normal, as is generally assumed in time series theory. The long-tailed property of the distributions can increase the sampling error of the autocorrelation coefficients. An apparently significant coefficient may be due to two coincidently separated large price changes. A further central assumption of the Box-Jenkins method is that the variance of the model residual errors is constant. Table 2.4 shows the standard deviation and variance of the daily price changes for sugar prices over each of the 13 years, 1961–1973. The ratio of the highest to the lowest variances is 100:1, demonstrating the instability of the variance of the price changes over time. This instability causes autocovar-

TABLE 2.4 Comparison of the price change variance by years

December contract	Standard deviation (£)	Variance (£^{2})	Variance of log price change
1961	0.23	0.05	0.08×10^{-3}
1962	0.37	0.14	0.23
1963	2.29	5.23	1.22
1964	1.11	1.23	0.50
1965	0.38	0.14	0.24
1966	0.26	0.07	0.18
1967	0.63	0.40	0.78
1968	0.44	0.20	0.32
1969	0.56	0.31	0.28
1970	0.38	0.14	0.10
1971	0.58	0.34	0.14
1972	1.89	3.59	0.71
1973	1.21	1.46	0.23

iance analysis to become 'inefficient'. The autocorrelation coefficients do not equally summarise the interrelationships of all price changes, but over emphasise the relationships for those periods when the variance is highest. The sugar price data had exceptionally high price change variance for 1963 and 1972. An autocorrelation coefficient computed over the whole 13 year period would in effect only describe behaviour over these two high variance years. It is all of these types of problem that explain many of the difficulties and 'curious results' that occur in time series analysis.

The Box-Jenkins model formally includes a general constant term. Since on a daily basis the trend will be very small it has always been neglected and the constant term set equal to zero in published research. However a persistent small daily trend compounded over several weeks can become very significant. This is demonstrated in Table 2.5, which again shows some statistics on sugar price movements over the 13 years, 1961–1973. Columns 2 and 3 record the first and last observations for each year, and column 4 the average daily percentage price change. These averages are of course small, but the trend effect over a year, given in column 5, is frequently very large. To test the significance of these trends, for each year the t-ratio of the average logarithmic transformed price change is calculated as

$$t_i = \frac{\text{average logarithmic transformed price change } \bar{x}_i}{\text{estimate of the standard error of } \bar{x}_i}$$

TABLE 2.5 Summary of trend statistics

December contract	First price	Last price	Day-to-day average % price change	Full period % price change	t-ratio
1961	26.00	21.35	−0.08	−18	−1.38
1962	22.55	35.10	+0.18	+56	+1.85
1963	32.50	86.50	+0.39	+166	+1.77
1964	61.25	26.45	−0.33	−57	−2.35
1965	27.45	20.50	−0.12	−25	−1.17
1966	26.55	13.80	−0.26	−48	−3.09
1967	16.50	25.40	+0.17	+54	+0.98
1968	30.15	30.50	0.00	+1	+0.04
1969	33.65	31.85	−0.02	−5	−0.21
1970	34.05	44.80	+0.11	+32	+1.74
1971	43.70	58.50	+0.12	+34	+1.54
1972	56.40	97.60	+0.22	+73	+1.29
1973	74.50	86.80	+0.10	+17	+0.82

The thirteen t-ratios are given as the final column of Table 2.5. The 1973 value is computed from 154 observations, the others from a minimum of 243 observations. If the process generating the prices is trendless then the t-ratios are all approximately normal random variables with a mean of 0 and a variance of 1. Inspection of the observed t-ratios shows that they are most unlikely to be a set of standardised normal variables.

A formal test of the trendless random walk hypotheses,

$$H_0 : \mu_1 = \mu_2 = \text{- - - -} = \mu_{13} = 0$$

against the alternative hypothesis that some of the μ_i (the individual year trends) are nonzero, is performed by considering the test statistic

$$T = \sum_{1}^{13} t_i^2.$$

Under the null hypothesis H_0, each t_i is essentially a standardised normal variable, so that T has distribution χ^2_{13}. The simplest justification for the choice of this test statistic is to observe that the likelihood of the t-ratios under H_0 is:

$$L(t_1, t_2, \text{- -}, t_{13}) = \left(\frac{1}{\sqrt{2\pi}}\right)^{13} \exp\left(-\frac{1}{2}\sum t_i^2\right) \sim \exp\left(-\frac{1}{2}T\right).$$

Thus H_0 should be rejected if $\exp(-\frac{1}{2}T)$ is small, that is when T is large. When the t-ratios are squared and summed the value $T = 33.62$ is obtained. The probability of a χ^2_{13} variable exceeding this observed value for T is less than ½%. Thus it is concluded that H_0 must be rejected. There is clear statistical evidence for trends in sugar prices. The argument above has been for the existence of trends over periods of as long as a year. In reality the trends will change more often than once per year. However if this is so, then the autocorrelation function for the sugar price series should not be zero at all lags.

The actual autocorrelation coefficients for lags 1–10 for sugar are shown in Table 2.6 for each year. The values calculated are all very small. For a random walk with identically distributed normal errors, the 95% confidence interval for each coefficient is approximately ± 0.13 for the years 1961–1972 and ± 0.16 for the final year 1973, i.e. $\pm 2/\sqrt{n}$. If the variance changes, or the errors are nonnormally distributed with high kurtosis, both of which apply to the sugar price data, then the approximation underestimates the confidence interval. Of the listed 130 coefficients, five exceed $2/\sqrt{n}$ and four are less than $-2/\sqrt{n}$. The total of nine outside the 95% confidence interval is only negligibly greater than the expected number of

TABLE 2.6 Autocorrelation of log price changes

Contract	Lag 1	2	3	4	5	6	7	8	9	10	Average over lags
1961	0.08	−0.08	−0.17	0.03	−0.01	0.09	0.14	−0.06	−0.02	−0.02	−0.002
1962	−0.05	0.06	0.01	0.00	0.06	−0.20	0.03	−0.00	0.04	−0.11	−0.016
1963	0.19	0.01	0.04	0.05	−0.03	0.06	−0.00	0.15	0.01	0.04	0.052
1964	−0.06	0.01	−0.21	0.07	0.01	0.02	−0.14	−0.01	0.00	0.03	−0.028
1965	0.07	−0.09	0.03	0.04	−0.05	−0.06	0.03	−0.00	0.13	0.01	0.011
1966	−0.07	−0.13	−0.05	0.04	0.02	−0.10	−0.03	−0.01	0.03	0.10	−0.020
1967	−0.03	0.05	0.13	−0.03	0.06	−0.02	−0.09	0.13	0.07	0.03	0.030
1968	0.01	−0.02	−0.00	0.07	0.18	0.03	0.05	0.01	0.10	0.01	0.044
1969	0.07	−0.06	−0.10	0.13	0.09	−0.06	−0.12	0.00	0.11	−0.09	−0.003
1970	−0.00	−0.10	−0.07	0.14	−0.09	−0.10	−0.06	0.01	−0.02	0.09	−0.020
1971	0.05	0.08	0.08	0.04	0.00	−0.08	−0.04	0.02	0.07	0.10	0.032
1972	0.10	−0.07	0.00	0.05	0.15	−0.02	−0.11	−0.08	0.05	−0.00	0.007
1973	−0.14	−0.07	−0.03	0.01	0.11	−0.16	−0.06	−0.02	−0.00	−0.03	−0.039
Average over contracts	0.017	−0.032	−0.026	0.049	0.038	−0.046	−0.031	0.011	0.044	0.012	

$130 \times 0.05 = 6.5$. Bearing in mind the effects of nonnormality, the hypothesis of zero values for the autocorrelation coefficients cannot be rejected. As the values are all insignificantly different from zero the only Box-Jenkins model that fits is the random walk

$$\log z_t - \log z_{t-1} = c + u_t$$

where the constant term, c, has to be estimated as simply the mean log price change over the whole period. Note that the Box-Jenkins model above does not allow for the constant, c, to change at infrequent intervals of time relative to the period between observations of the series.

We thus have a conflict between two types of statistical analyses, one confirming the existence of trends, the other denying it. However the denial of trends is really a denial that the general Box-Jenkins class of models adequately describe commodity price movements. What we need is a new type of model, which permits the existence of trends yet is consistent with very small autocorrelation coefficients.

Taylor has proposed an alternative model to describe the stochastic process that generates commodity prices which does precisely this. Suppose that on each trading day there is a small probability, equal to $1-p$, that new information circulates around the market causing a change in the price trend. Suppose further that each trend is independent of all the past values and is drawn from the same probability distribution, which has mean zero and variance v^2. The price trend now merits a subscript and the model becomes:

$$\log z_t = \log z_{t-1} + \mu_t + u_t$$

where

$$\begin{aligned}\mu_t &= \mu_{t-1} \text{ with probability } p \\ &= \text{some new value with probability } 1-p.\end{aligned}$$

The logarithmic transformation is retained because constant variance is required in the derivation of the theoretical autocorrelation function. Denote the variance of the residuals u_t by σ^2 and assume that they have zero mean. Then it can be shown that the theoretical autocorrelation function for this stochastic process is:

$$\rho_i = \frac{p^i v^2}{v^2 + \sigma^2} \ .$$

Providing that the ratio $v{:}\sigma$ is small the theoretical autocorrelation function is negligible and will not be identified by inspecting the sample coefficients individually. Consequently, for a small ratio, there is no contradiction between this model and the observed coefficients listed in Table 2.6. Studies of an autocorrelation function formed by pooling the separate yearly functions have given parameter estimates of $1 - p = 0.0294$ and $v^2 = 0.0369$, for which the implied mean trend duration is $1/(1-p) = 34$ days. The theoretical autocorrelation function is estimated as $\rho_i = (0.0369)(0.9706)^i$, so that for example, $\rho_1 = 0.036$.

The conclusion of this detailed statistical analysis is that trends do exist in commodity prices and change over time. The random walk model is only an inadequate approximation to the stochastic process generating prices. This means that the search for improved purchasing policies for buying at prices less than the average market price over the long term is a worthwhile and potentially rewarding activity. One methodological advantage of the new model is that it eliminates the contradiction between the results of earlier studies of short-term price movements and the longer term econometric studies.

THE REAL COMMODITY PRICE SERIES

The real commodity price series is Fig. 2.12. The discerning reader may have noted that it is a repeat of Fig. 2.4, which showed the daily copper prices from 1969 to 1975.

3

Some Practical Buying Examples

3.1 A COPPER BUYING SITUATION

Bowland Metal Products is a company buying refined copper and transforming it into manufactured products, such as specialist electrical cables, which are sold to other industries. Sales are made on contracts either to large industrial consumers or to large wholesalers. On occasions when contract orders are low, the manufacturing output is maintained by producing to stock, which is stored to meet future contract orders. The main aim of the company production planning is to stabilise production levels. Thus the demand for the refined copper can be specified for the foreseeable future. Daily consumption is small at around 20 tons per day.

The refined copper is purchased mainly in the form of copper wire from a secondary producer, who purchases copper initially in the form of rod or bar copper and draws it out into the wire form. BMP can buy copper daily from the secondary producer on the basis of the London Metal Exchange morning settlement price for copper wirebars. Each day the supplier rings BMP at 2.00 p.m. telling them the prevailing copper settlement price. BMP then have until 5.00 p.m. to ring back to place a definite order at that price. They can buy as much as they wish each day at that price. The supplier will then deliver the copper as BMP wishes, spreading the delivery over a number of days if required at 20 tons per day. Delivery can be made the very next day if requested, so that BMP can use it in their manufacturing process on the following day. There is no effective limitation on the supply, so that the company can buy as much or as little as it wishes. Payment is not made until physical delivery takes place so that BMP do not incur storage charges or interest charges on the capital invested in copper bought in advance for delivery at some later date.

In this purchasing situation, a buying opportunity occurs each day at the single price offered on that day. A regular consumption of the raw material takes place at each buying opportunity (the 20 tons consumed each day). A delivery of the material to cover one day's consumption can be made at each buying opportunity. The future consumption requirements that the buyer has to meet are thus the 20 tons consumed each day. At each buying opportunity the buyer must decide how much copper to buy to cover his future consumption or alternatively how many future buying opportunities he should cover with his purchase at today's price.

One other aspect of this particular problem is that the buyer could decide to purchase at a regular rate of 20 tons per day. He can forget all about trying to forecast prices or pick out good buying points. It is the same as just having a clerk who regularly sends an order for 20 tons each day to the supplier. This is known as a 'back-to-back' policy and will result in the buyer paying the average market daily price for his material. Because of the difficulties in commodity purchasing many companies purchasing nonferrous metals or metal scraps have compelled their purchasing departments to operate such back-to-back policies. In cases where companies have a more varied production over time than BMP, they price the individual customer's orders for the final product individually, according to the morning closing price for copper and zinc on the London Metal Exchange on the day the customer's order was received. The buyers are then compelled to buy exactly that same amount of material within 24 hours either in the form of virgin metal or scrap as is appropriate to their circumstances.

The actual purchases made by BMP's purchasing director, who himself bought the copper for the company, over a 9 month period of relatively small copper price fluctuations are shown in Fig. 3.1. The period of downward prices over February to April illustrate the difficulties facing the buyer. He made three purchases all of which were because he thought prices had reached a bottom point and that the trend would reverse to

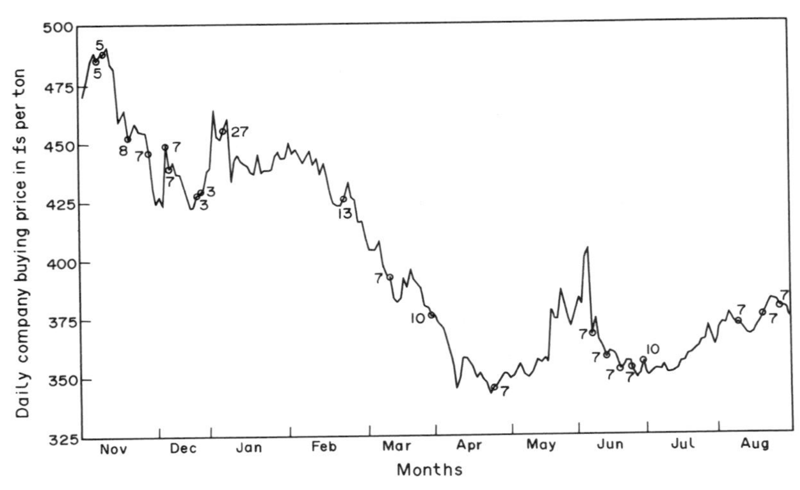

FIG. 3.1 The purchases (in days requirements) actually made by the Company Buyer

bring further price increases. He finally found the bottom of the market with his fourth purchase in late April. Although, with hindsight these are bad purchases, if you note that prices had been falling continuously from their high of £550 per ton in November and cover up all prices to the right of the day of purchase his decisions are understandable. These first three purchases cost the company approximately £10,000 (about 4%) more than a hand-to-mouth policy of buying a regular 20 tons per day. The purchase in April however was very good, at the lowest but one price over the whole period. The later purchases of Fig. 3.1 over June, August and September are all reasonable, saving an average of about £3 per ton on a hand-to-mouth policy. In total this amounts to about £5,000, which is however less than was lost over the earlier 2 months.

Because of the problems in predicting and coping with such price movements, particularly when prices are falling, the buyer limits the maximum amounts that he does purchase. Although actual payment for the copper does not take place until after delivery, a large purchase immediately commits the company to a large capital expenditure. For these reasons the company had specified a general policy that the maximum inventory of actual stocks of copper plus undelivered orders should not generally exceed 15 days copper usage i.e. 300 tons. This referred only to stocks of copper that had not entered into production. In practice the purchasing director seemed to operate with a maximum of 20 days usage, and on one or two occasions he even exceeded this level, unfortunately doing so at the wrong time. Generally his purchases were in the range 7–13 days usage, whenever the stocks fell to 1 or 2 days usage. Interestingly he rarely bought small amounts on a day-to-day basis.

3.2 A LINSEED OIL BUYING SITUATION

Cartmel Floor Coverings Ltd. manufacture linoleum, using linseed oil as the basic raw material. Although vinyl floor covering has captured most of the market since the mid sixties, there is still a small steady market for linoleum for public offices etc., where a hard-wearing floor covering is required. CFC use about 3400 tons of linseed oil per year but there are large variations in usage from one month to another, in the range 200 to 450 tons. The linseed oil can be purchased from two international brokers, who also arrange delivery of the oil direct to CFC's factory. The oil is shipped across the North Sea from Rotterdam. The bulk of the linseed oil imported into Western Europe from U.S.A. and Argentina, the main producers, is sent to Rotterdam in large vessels. There are large storage tanks there and dealers can enjoy the benefits of a free port, low-carrying charges and quick service. Rotterdam is the major market and commodity exchange for Western Europe. Linseed oil is then transhipped in smaller vessels to the actual consumers. This requires the use of specialist oil carrying vessels for transport across the North Sea to the U.K. One major complexity for CFC is that only ships of either 400 or 600 tons are available to do this. This implies that CFC receive between 6 and 8 shiploads per year and hence, that a delivery does not take place in every month of the year.

The linseed oil market at Rotterdam is an actuals market so there are no facilities for 'futures' trading. Prices can usually be fixed for the delivery of linseed oil for up to 6 to 9 months ahead, depending on market conditions. The two brokers supplying CFC take the prices quoted on the Rotterdam market by oil sellers or other brokers, in U.S. dollars per ton, and convert them into sterling using the appropriate forward dollar–sterling exchange rate for the month of delivery. The brokers then add on any import duties that are payable and insurance and freight charges for delivery to different destinations. An example of a broker's daily price sheet is shown in Table 3.1. The top two rows are different ex-tank Rotterdam prices quoted in the market, the third row is the dollar–sterling exchange rate and the final four rows include transportation and other costs to give a price in £s per ton delivered to different ports on the East Coast of the U.K. ranging from London up to Perth in Scotland. Although the brokers prepare a price quotation every day using the above procedure, they usually ring up CFCs buyer only twice per week. This keeps the buyer in touch with market trends. The buyer tends to contact the brokers himself only if he definitely wishes to make a purchase in a particular week, although he could telephone them every day if he wished. The broker's price is normally quoted to the buyer in the early afternoon when they have seen how the Rotterdam market has moved during the morning. The broker is prepared to make a sale at that price up to five o'clock even though the Rotterdam

TABLE 3.1 Example of broker's daily price sheet

November	December	January	February	March	April	May
215	216	219	221	223	225	227
216	217	219	221	223	225	227
2.35	2.34½	2.33¾	2.33¼	2.32¾	2.32¼	2.31¾
95.25	95.75	97.50	98.50	99.50	100.50	101.75
109.75	110.50	112.50	113.50	114.75	116.00	117.25
111.00	111.75	113.75	114.75	116.0	117.25	118.50
111.25	112.00	114.00	115.00	116.25	117.50	118.75

prices may change during the afternoon. If the buyer has not decided to make a purchase by 5.00 p.m., he must wait until the next day. Since this is an informal actuals market, once he decides to buy on a particular day he attempts to negotiate with the broker to reduce the initial price offered by the broker.

A major constraint is that the buyer must buy a 600 ton (or 400 ton) lot in a single purchase at the price agreed with the broker on that day. He cannot place an order for 100 tons of linseed oil to be delivered at the beginning of June, say, on the understanding that he will place orders for the remaining 500 tons over the next 2 months to make up a full boat load. There is no way that he can spread his purchases out over time, as for the copper buying of the first example, to try to buy at the average price over some period. If he decides that he wants a lot of 600 tons of oil to be delivered at the beginning of June, he must choose one price on one particular day to make a purchase of 600 tons. He must try to pick out that day with the lowest price offer amongst all the alternative days on which he could have made the purchase.

Each quarter, the company makes a forecast of the current year's quarterly demands for linoleum and also updates its predictions of the annual sales for the following 2 years. These sales forecasts are converted into annual raw materials' requirements and then into an average daily rate, taking account of holidays, likely overtime working in production etc. The linseed oil requirements for any month forward in time are obtained by multiplying the forecast daily usage rate by the number of working days in that month's period for linoleum production. The linseed oil requirements that the buyer has to purchase are defined on a monthly usage basis. Because of variations in the number of working days in a month, seasonal variations in product sales and hence production levels, the linseed oil usages vary quite considerably from month-to-month as mentioned earlier.

Furthermore there is some slight seasonality in sales over the year, for example, sales over the months of November, December and January are roughly 20% less than the rest of the year. Just under 30% of sales are made on long-term forward contracts. These two factors have to be taken into account in modifying the basic forecasting procedure just described. They further add to the variations in the month-to-month linseed oil usages. Production is planned to follow these month-to-month changes and there is no policy of smoothing production levels over time by making extra linoleum for stock when sales for that month's production are low.

Linseed oil contains about 1% impurities, known as 'foots'. These interfere with the manufacturing process. The oil must be stored in tanks for about 3 weeks after its delivery from the brokers to allow the 'foots' to settle out. Thus the oil to be used in production in any month must have been delivered to the company by the first week of the previous month to allow sufficient settling time. Linseed oil kept in closed tanks can be stored almost indefinitely. The current storage capacity is about half the annual requirements.

The supplying brokers sell the linseed oil on the basis that delivery will occur within some agreed specified month. According to the terms of the contract this may be any business day of that month that the supplying broker wishes to choose. In practice, deliveries over the past few years have usually arrived at CFCs premises in the first half of the month. It is possible for the buyer to place an order for 400 tons to arrive on the 2nd day of the month and another on the 26th day of the month. Both orders must be placed for delivery *sometime* during the month. They will most likely both arrive during the same week at the beginning of the month. It is generally impossible to cancel an order once it has been agreed with the broker. However in practice delivery of the order can be postponed to a later month or brought forward on payment of a small premium provided sufficient notice, say 2 months, is given to the brokers. The conditions of the contract are that CFC must pay the broker by the end of the month in which the oil was delivered.

Because of the limited delivery lot sizes and because the supplier can deliver an order on any day of the month that suits him, the buyer can neither have each month's usage delivered to him in separate lots at the beginning of the previous month, nor can he have his annual requirements of 3400 tons delivered in lots of 400 tons at regular six and one quarter weekly intervals. He must choose a subset of months of the year in which to have 400 or 600 tons delivered. Thus his first major problem is to determine the scheduling of his delivery quantities over future time in the most economic way, taking account of prices and storage costs. Having chosen in which months he wants deliveries to occur, he then has to decide when and at what price to place an order for each month's delivery. Prices, as shown in Table 3.1, are quoted separately for each forward month of delivery. The margins between the prices for different delivery months can change over time as the absolute price levels are changing. Also, comparing a purchase now for delivery in May with a purchase for delivery in July, say, there is a longer period of time in which to buy July than May and in the extra period of time prices may fall. The decision to buy the May delivery on a particular day does therefore not necessarily imply that the July delivery should be purchased also. Each delivery month and its own series of prices must be considered separately. Each separate chosen delivery month must be purchased as a separate decision and so must be formally treated as such in any purchasing policy. Of course on occasions two delivery months may be purchased on the same day but this will be the result of two separate decisions, not the single decision that today is a good day to buy everything.

There are obviously limitations on how far ahead of its actual delivery an order can be purchased. The market itself imposes a limit since trading does not take place in the market for more than 9 months ahead of the present time. Generally brokers and sellers are very reluctant to quote firm prices for months beyond the scope of the market. The company policy of CFC allowed the buyer complete freedom of action to buy as far ahead as he thought appropriate to give as low raw material costs as possible, even up to 9 months. In other markets, months can be bought and sold for up to 18 months ahead, although very few companies would allow their buyers to purchase material so far in advance of usage. Indeed most companies would regard CFCs 9 months as a rather risky policy. They would doubt that their buyers could forecast prices so far into the future. Even though the actual cash outflow will not occur for several months, it means that the company is committed *now* to a specific, large and unalterable capital expenditure. If the market falls instead of rising then a purchase for delivery so far ahead could be very costly. Another fear arises from uncertainty that that amount of material will actually be required. In CFCs case, the manufacturing process from taking the linseed oil out of the tanks to producing the final product ready for sale is about 3 months. Allowing for the settling time of the oil in the tanks and for the turnover time of stocks of the finished product, means that the oil purchased for delivery in nine months time will be used to meet sales that will take place in 13 months time. Forecasting sales accurately so far ahead is difficult. Indeed for some industries changes to products may occur so frequently that the demand for the product may not exist 13 months ahead. It is this sort of worry which leads companies to limit the buyer's freedom of action.

In order to ensure he has sufficient oil to keep production going the buyer cannot leave the purchase of any scheduled month's delivery until the very last moment. An unexpected temporary shortage could occur, which would result in his being unable to buy any oil or alternatively having to pay very large price premiums. The policy followed by the CFC buyer is to ensure he has bought each month's delivery by the beginning of the month that is two months prior to the month when delivery will actually take place. For example a delivery scheduled for July must be bought by the first day of May.

Putting these two constraints together means that each scheduled delivery month lot has a 'potential buying period' of 7 months in which it may theoretically be bought, starting 9 months in advance of the month of delivery and ending one month before delivery. Equal-sized 'potential buying periods' exist for all those months when deliveries are scheduled to occur. This is often expressed in an alternative fashion in terms of the 'purchase and stock cover'. At any point in calendar time, the buyer can have placed purchase contracts which, together with his stock of oil in the factory, will cover cumulative consumption up to 9 months ahead, whilst on the other hand he must never allow his purchase cover to fall to under 2 months. If he reaches the lower level he must immediately place an order for linseed oil to be delivered in one month's time. If there were no constraints on the size of the delivery order from the supplier, then the 'purchase and stock cover' means that his oil already in stock at the factory plus the orders placed with the supplier must cover between 2 and 9 months future consumption. If usage were completely regular over the year, the purchase stock cover limitations would be between 575 and 2500 tons. Note that in addition there will be about 200 tons of oil 'settling' in the tanks.

Specifying a 'potential buying period' is the same as specifying maximum and minimum limits on the 'purchase and stock cover' and vice versa. Unless usage occurs at a constant regular rate the latter must be specified in terms of future months' usage, since with the irregular monthly usages its absolute value in tons will change over time.

Thus for each scheduled delivery month, the buyer will have some 'potential buying period' within which to make a purchase. He has to choose one day, out of the seven months in CFCs case, on which to make a purchase at what he thinks is the lowest price. As time progresses he must compare each day's price with the prices he expects over the *remaining* days of the buying period. He must not compare them with the prices that were offered in the past, during the part of the buying period that has elapsed, since he cannot go back and buy at one of the earlier prices. He must always look forward, although the earlier prices may assist him to make forecasts of price levels over the remainder of the buying period.

A potential buying period of 7 months means that he has approximately 150 days of different prices on which to make a purchase. In making a decision in practice a buyer does not and cannot predict a price for each of the 150 days. He forms some idea of the general trend of prices over the whole buying period, e.g. prices will generally move upwards or downwards etc. He then assesses which part of the whole buying period, the first or second half, the first, second, third or fourth quarter etc., is likely to have the lowest prices. In essence he breaks the whole buying period down into a set of sub-buying periods and chooses one of them, as the one in which he will *actively* be considering making a purchase. Of course as time progresses and the market situation changes, he may modify his view of which is the most appropriate sub-buying period in which to make a purchase. The sub-period may be 1 or 2 months in length but rarely longer. When he enters his active sub-buying period he may even break it down further, deciding that he will make a purchase this particular week rather than another week, taking account of his view of the way prices will move over the rest of the active sub-buying period. He will then try to pick the day during that week with the lowest price, taking account of the day-to-day price fluctuations.

This process leading to the actual buying decision depends entirely on the buyer's forecasts of how prices will move. Such forecasts are rarely formal or quantitative. They are generally based on hunch and experience. Buyers usually try to 'read' rather than analyse the market. The buying decision is a sequential decision process gradually refining downwards the time period in which a purchase will be made. Within the overall company's strategic buying period or purchase and stock cover policy, the first stage requires medium-term price forecasts to determine the 'active buying period' within which the buyer will definitely make his purchase. Then within this active buying period, taking account of the day-to-day and week-to-week price fluctuations, a policy is required to assist the buyer to pick out the lowest price offer.

3.3 MAIZE BUYING SITUATIONS

The maize used by U.K. consumers must all be imported since U.K. farmers do not grow maize for commercial use. The companies in the case studies both bought all of their maize from brokers who sold it on a basis of delivery to some U.K. port. They did not try to buy maize in the U.S. and arrange their own sea transportation to their factories in the U.K. The various maize brokers in the U.K. may act on their own behalf or as agents for the large international grain shippers. The brokers or shippers will charter ships for a single voyage, or for a whole year, or have their own fleets to carry grain from the U.S. to Europe, for example. They will buy maize in the U.S., load it on to their ships and then transport it. They may either buy a whole ship load of grain well in advance and then attempt to sell that grain to customers near to a particular port or alternatively they may first try to sell grain to customers up to the total carrying capacity of the vessel and then go and buy that amount of grain within the U.S. The economics of sea transport mean that any vessel used must be loaded to its fullest cargo carrying capacity. There are economies of scale in using relatively large vessels, although these must be balanced against the demands of the customers around particular European ports and the cost of storing large quantities of grain in port silos, grain warehouses. Maize from overseas will be shipped directly to the larger ports, where there are several competing large consumers, in very large ships of 20,000 tons or more. An example in the U.K. would be shipment to Liverpool, which until recently had its own active Corn Exchange where brokers and buyers could meet daily at certain times of the day to sell and buy maize. For smaller ports with only a few consumers shipment may be directly from the producing country in smaller ships of under 10,000 tons say, or by transhipment from larger European ports, such as Rotterdam, which has become the central market for the grain trade for Western Europe.

Maize is shipped into the larger ports such as Liverpool on the basis of delivery sometime at the shipper's choice in either the first half or second half of the month. In other smaller West Coast U.K. ports, delivery would be on a monthly rather than half monthly basis. Delivery cannot be guaranteed more precisely than this to allow for delays at the port of loading or at sea.

Maize can be bought for delivery at each such future time period up to several months ahead. Prices are quoted separately for each different delivery period. Thus at any point in time, several prices are quoted for the different delivery dates e.g. first half April, second half April, first half May, second half May, etc. The prices for each delivery date will be different, although at the times the difference may be small. Often the prices for delivery in the first or second half of a month will be identical. Usually in the larger ports, price quotations are available daily for all delivery positions up to some months ahead, although on some days particular positions may not be quoted at all. In the larger ports, the buyers may obtain such prices from the local 'corn exchange' or by being contacted by the brokers on behalf of sellers. The number of days over which delivery positions are quoted varies widely, typically over a range 30–100 days, although certain months such as November are always offered over a much longer period of time. In the smaller ports with few consumers and only a single broker, the buyer himself will usually have to contact the broker to obtain a price quotation on a particular day, although the broker will probably telephone him once a week.

Skein Conglomerates has two subsidiary companies that must buy substantial amounts of maize, MacLonsdale Distillers Ltd. (MDL), who manufacture whisky, and Lancastrian Farm Feeds (LFF) who produce several different types of animal feed for cattle, poultry and pigs. MDL's major raw material is malt which the company produces in its own malting kilns using barley purchased directly from U.K. farmers. In addition the company requires about 40,000 tons of maize per year. The company normally uses U.S. No. 3 Yellow maize, but French maize or South African white or yellow maize are possible alternatives. The different types of maize cannot be blended together for the whisky-making process so that each 'production lot' must contain only one type of maize. Each alternative to No. 3 U.S. Yellow maize incurs a known premium of discount on the manufacturing costs. Subtracting this from the price offered enables a direct comparison to be made between the prices for the alternative types of maize. It implies that a 'standard company' maize price time series exists. In the main, this had been the price for U.S. No. 3 Yellow maize, although on about 10% of occasions it has been the adjusted French or

South African maize prices. MDL import their grain through Glasgow. It is purchased through a local U.K. broker on an ex-Port of Glasgow price basis. MDL collect the maize from the port silo and transport it by road from the port to its distilleries.

Generally over the past, the brokers which MDL have used have imposed a minimum order size constraint of 6,000 tons on the amount they will sell and deliver at any one time. This represents the smallest 'economic' ship size. Often the brokers/shippers will charter a larger ship than this and try to sell the grain it can carry to say, two, three or four customers. Because of this, the brokers sometimes have allowed MDL to buy in small order sizes of between 4,000 or 5,000 tons. The 6,000 ton constraint is thus a usual constraint but not always an absolute one.

In general MDL must make one single purchase for each order of 6,000 tons, although on occasions they have been able to make two or three separate purchases that total up to 6,000 tons. It appears that two different situations can arise. If MDL decide they wish to make a purchase and ask the brokers for a price quotation it must be for a 6,000 ton minimum-sized order at that price on that day. An alternative, less frequent situation, is that the brokers themselves contact MDL with an offer for a certain quantity of grain. The brokers have sold most, but not all, of the grain on a ship scheduled to arrive at some particular time, and contact potential customers trying to sell the remaining grain. The size of the quantity then depends on the particular circumstances but will be lower than 6,000 tons. In addition, once MDL have placed an order for 6,000 tons, then they can increase it by a later purchase of 1,000 or 2,000 tons etc., at a different price.

Since this minimum order constraint imposed by the suppliers exceeds the average monthly usage, deliveries cannot be scheduled to occur at the beginning of each month. Typically over the past 5 years, five to eight shipments per year have actually been delivered. The delivery conditions in the purchase contract allow the supplier some latitude in the delivery date. Deliveries can occur any time in the first half of a month or the second half of a month. Thus there are twenty-four delivery periods during the year. MDL must therefore choose delivery periods so as to minimise the cost of capital tied up on stock and the differing premiums or discounts for later or earlier deliveries, taking account of the different monthly consumption requirements.

The present stated company strategy is to keep a 5-week buffer stock of maize and to buy 6 to 10 weeks ahead. Virtually all purchases within the previous 5 years were consistent with this policy. With only a few exceptions, a purchase order scheduled for shipment in a particular month had been bought by the beginning of the previous month and not more than 5 months in advance of the shipment month. This implies, for example, that a purchase order for shipment in November 1973 cannot be purchased before the beginning of July nor later than the end of September. Thus the November 1973 shipment has a 'buying period' of 3 months from July to September within which a purchase must be made.

LFF produce several different types of animal feeds for cattle, pigs and poultry. The company consumes several hundred thousand tons of imported cereals, oilseeds, and oilseed meals, imported maize being particularly important. The major company factory is situated in Liverpool, very close to the docks. The company was originally founded in Liverpool because it was the major port of the U.K. for the countries from which the raw materials for animal feeds were imported. At that time also, inland transport of large amounts of bulky cargo was difficult and expensive. Furthermore the western half of the U.K. is the most important for the rearing of animals. Liverpool has remained the centre of the company's activities, although it has taken over several small animal feed companies in other parts of England, mainly near to ports on the west coast. Maize is important for animal feeds because of the energy and protein it contributes to the feed mix. The company uses U.S. No. 3 Yellow maize or French maize in large quantities. These can be interchanged relatively easily. It is never worth paying any premium to use better quality types of maize such as South African White for example.

At the time of the study, the purchasing of raw materials in LFF was partially decentralised. There was a branch buyer at each factory who was executively responsible to the branch factory manager except at the largest factory in Liverpool, where the company head office was also situated. There was a company chief buyer who had all the branch buyers functionally responsible to him. The chief buyer was also executively responsible for the buyer at the Liverpool factory. In addition he purchased certain important raw materials centrally for all the company's factories. The chief buyer sent out recommendations to buy, including qualitative prognostications and a discussion on future price movements regularly to the branch buyers. In most cases unless there were important local market factors present, the branch buyers were expected to follow the chief buyer's advice.

Maize can be partially substituted by feed wheat, barley or sorghum or some combination of all three. For each brand of animal feed produced, the raw material requirements can easily be determined by the use of linear programming methods if the prices of all the materials used are fixed. For those unfamiliar with the term, linear programming (or LP for short) is a particular mathematical technique for determining the best course of action amongst many alternatives where there are numerous restrictions on the choices to be made. It will be described and discussed in Chapter 6 of the book. The amount of maize that should be used to make a certain quantity of the particular animal food will depend on the actual price of maize and its price relative to the prices of the competing materials. The price of maize and its competing materials are continually changing from one day to the next. The amount of maize that one should plan to buy depends on the price that is paid for the maize used, but the price (or the average price) is not known until after the required amount of maize has been bought. A classical 'egg and chicken' situation prevails!

The advice on how much of which raw material to buy was generated by the interchange of information between the chief buyer and the formulation manager. These two people were executively responsible to different senior directors of the company. The formulation manager took into account the nutritional, production and other constraints on each brand and set up a formal quantitative method to attempt to minimise the raw material cost of each brand separately, in the manner described in Chapter 6 on the blending problem. The average prices that would be paid for raw materials to be bought in the future obviously plays a vital role in determining the right blend or mix of materials to use. The prices used by CFC in their blending models were based on discussions between the chief buyer and the formulation manager. The current practice was for the formal advice sent to the branch buyers on future raw material requirements, to assume that the blending formulae in operation at the time the advice was given would continue in use over all the time each delivery position was offered.

The chief buyer thus had the problem of anticipating significant price movements, which will affect the formulation manager's formulae and therefore future raw material requirements (i.e. what amounts of which material to buy), and advising the branch buyers on the timing of their purchases of each raw material within the framework of the 'known' requirements. In practice, changes to the notional prices used in the formulae were only changed whenever the chief buyer predicted significant price changes. This did not necessarily occur each month. It should be mentioned that this study was performed in the late sixties, when cereals prices were rather more stable than today. The chief buyer also had day-to-day responsibility for those materials bought centrally. The branch buyers have the simpler problem of timing the purchases to meet the 'known' requirements, within the constraint of the day-to-day management of raw material stocks.

The same considerations that made LFF build their factory in Liverpool near the docks motivated competitors and many other industries processing large amounts of bulky raw materials, e.g. cane sugar refiners, to build their factories in the same place. This has resulted in limited storage space being available to each company. Renting storage facilities outside the company premises is expensive and therefore never used by LFF. Several other materials are used in large quantities as well as maize, so there are high storage requirements just for the current week's production programme. Because as mentioned, internal storage space is limited and outside storage is expensive, it is always uneconomic for LFF to buy raw materials for delivery in advance of usage and store it themselves until required. Physical stocks and deliveries are thus arranged to keep production going and to guard against only a few days delay in delivery.

Because of the way shipping of the maize into the U.K. has been arranged, LFF have adopted a half monthly (for Liverpool) and monthly (for the other factories) time unit as the basic planning period for future production planning. The nutritional manager produces a schedule of requirements for maize (and the other materials) for each time period up to 3 months ahead, usually for each brand. The maize requirements to be bought for each time period's use will be the sum of the calculated maize requirements for each of the separate brands of animal feed for that time period. They can easily amount to over 10,000 tons per half month of maize for LFF, with almost 70 different animal foods being made. The buyer's role is thus to purchase these requirements at the lowest possible cost.

Since, as mentioned, it is uneconomic to store maize any longer than absolutely necessary, maize for a particular period's usage is bought for delivery in the immediately preceding period, month or half month. A delivery will be scheduled for each period, month or half month, of the year. The purchase will be made on the basis of the prices quoted for delivery in that period. The buyers will start to consider purchasing a particular period's requirements about 3 months in advance of its usage. From that point onwards, they will actively collect and consider the daily prices. In practice, they rarely make a purchase that far in advance of usage. Since so much maize is shipped into Liverpool and since there are several large consumers, prices for maize to be delivered in a particular period are frequently quoted for maize on that shipment right up to the day that particular ship docks at Liverpool. LFF's buyers have never left their purchases to that late a date but have, on rare occasions, bought maize being carried on a vessel that has already set sail across the North Atlantic from the U.S.A.

The central buyer makes only one, two or three separate purchases for the requirements for each usage period. The third purchase usually, and the second purchase sometimes, were made because a change to the chief buyer's view on how future prices would move, led to a change in the brand formulae resulting in an increased maize requirement. Unlike Glasgow, the maize sellers have not in the past imposed maximum and minimum limitations on the amount of maize that LFF can buy at any one time for each particular delivery period. This is presumably because of the competition among the sellers and because LFFs purchases have always been of a reasonable size. There are several potential customers for the maize, so there is less risk to the brokers of being left with a part shipload of maize unsold in Liverpool than in Glasgow. The brokers also know roughly what LFF's consumption needs will be and that there are advantages to LFF in having all their maize for some period delivered in one lot on the same ship. Thus if they accept a moderate first purchase the brokers have some confidence LFF will come back for a second purchase to raise the total to an acceptable quantity. Although in theory no restrictions on the purchase size apparently exist, in practice it is likely that if LFF tried to buy a few tons of maize each day over their buying period to make up 5,000 tons in total, say, the sellers would refuse to accept their order and impose formal constraints as in the Glasgow situation.

4

The Components of the Commodity Purchasing Decision

4.1 INTRODUCTION

The discussion of the purchasing decisions described in Chapter 3, shows that there are several different aspects of the problem that the buyer must consider before making a particular decision to buy some particular amount of some material. The various stages of the decision process are listed below. Working out the operational procedures for each stage in totality gives the purchasing policy. Unless all of these stages are considered the specific particular buying decision is likely to be wrong.

The various component stages are as follows:

1. Determine the future consumption requirements for each material of interest that are not covered by the current physical stocks nor by any orders placed earlier in time which have not yet been delivered. These will then be the amounts for which the buyer must consider making purchases.
2. Convert the uncovered forward consumption requirements into a schedule of future purchase order delivery quantities specifying their size and timing.
3. From the financial and operational constraints imposed by the overall company strategy determine the strategic buying period to give the maximum and minimum time ahead of delivery that each scheduled purchase order must be bought. This also gives the maximum and minimum values on the combined total purchases and stocks at any time.
4. Within the fixed constraints of the strategic buying period determine the active buying period within which the buyer will actually operate. This may be fixed at the same value as the strategic period or it may be varied from time-to-time, within the strategic buying period, if reasonably accurate medium-term price forecasts can be made. In this case it distinguishes the periods when active purchasing should take place from those when the buyer should merely live off his current stocks.
5. Develop tactical buying policies for day-to-day operations to give the

most appropriate decision for each scheduled delivery purchase order at each successive price offer.

This chapter presents a general methodology for the purchasing decision, a basic framework of connected problems, rather than prescribing operational solutions for each stage. Perhaps the first rule for a successful purchasing policy is for the purchasing manager to recognise that all of the above component stages must be explicitly or implicitly considered. Each of the stages of the purchasing decision identified are discussed in detail in the same order as above. The discussion aims to clarify the particular problems that arise in each stage and then to suggest the general methods that should lead to solutions to the problems identified. The remainder of the book then considers specific ways of coping with the problems occurring at each stage to give detailed operating rules and illustrates them with appropriate case studies. Putting a formal structure on the commodity purchasing problem and identifying the important problem or problems at each stage should enable the purchasing manager and his buyers to see the problem in the 'round', to see in which area their own most critical problems lie and then to decide how to organise and allocate their limited resources, money, manpower and time, to give effective improvements in their purchasing performance. It is likely that even if the detailed decision rules and operating methods discussed and recommended later in the book are not adopted, the general formalisation described in this current chapter, should assist the purchasing manager to better manage his purchasing function.

The first stage of determining future consumption requirements is fairly straightforward if only a single material is being purchased. Problems arise when a blend of partially substitutable materials or of different grades of the same materials are needed to make the final product. The complexities that this imposes on the situation are considered.

The second stage implies the determination of the size of order to be placed on the supplier and the timing of their delivery. Initially the general ordering and stock control procedures used elsewhere in purchasing are outlined. The various alternative approaches are then considered for the raw material purchasing situation. It is shown that Statistical Inventory Control methods are completely inappropriate and that some form of the Materials Requirements Planning methods (MRP for short) must be adopted to determine the size and timing of future delivery orders. In addition safety stocks must be calculated to cover the demands that occur when deliveries are delayed after the scheduled date of arrival. It is recommended that if only a small number of different materials need to be purchased, a formal model for each individual situation should be derived. These models will be of a general form known as dynamic programming (discussed later in the book, see Chapter 7). Although the solution procedures for these models are somewhat complex and relatively lengthy they give reductions in purchasing costs over the approximation methods often used in MRP models elsewhere in purchasing and production planning.

The third stage considers setting the absolute limits on the freedom of action allowed to the buyers. As we have seen in Chapter 3 this implies setting maximum and minimum limits on how far ahead of its delivery a purchase can be made to cover an order scheduled as in stage 2. This will give the earliest time in the future at which purchases to cover an order can be made and the latest time by which all that order's requirements must have been bought. The difference between these two times is the *strategic buying period*.

The length of the buying period fixes the maximum period ahead at any time for which price forecasts will need to be made. Price forecasts further ahead than the end of the buying period are completely irrelevant to current purchasing decisions. For example if the strategic buying period is fixed as 6 months ahead of the delivery of an order, then forecasting that a sharp price drop will occur 8 months hence, to below the current price level, after 7 months of consistent price increases is of no assistance. A purchase must be made within the next 6 months of increasing prices. The buyer just cannot wait until the price falls in 8 months time to buy since his production processes will be stopped for two months because of a lack of raw materials. Indeed forecasting beyond the length of the strategic buying period may be positively harmful. It could foster a delay in the purchase until later, in the forlorn hope that prices will fall several months earlier than predicted. Hence it follows that the accuracy with which price forecasts can be made over different time periods into the future is an important component in determining the length of the buying period. In general the further ahead the time for which price predictions are made, the larger the price forecasting errors and hence the bigger the losses that will be incurred from wrong forecasts.

An alternative concept for fixing the freedom of action allowed the buyer is the 'Purchase and Stock Cover', also discussed in Chapter 3. The 'Purchase and Stock Cover' is the amount of future consumption covered by a purchase at the present price plus the current physical stocks plus orders placed earlier but which have not yet been delivered by the suppliers. There is obviously a direct correspondence between the cover and the buying period. The demand from the earliest time in advance of any order's delivery that a purchase can be made to the scheduled delivery time and the demand from the latest time by which purchases must be completed to the scheduled delivery time give the maximum and minimum values for the cover. The former concept is more appropriate for those situations where consumption and deliveries take place on a daily basis whilst the latter is more appropriate if consumption and deliveries occur on a monthly basis.

The purchase and stock cover concept focuses attention on the level of stocks and the financial aspects, as for example the money that needs to be invested in stocks. The larger its maximum value the greater the potential savings from good forecasts and good purchasing decisions. This follows since the larger the purchase and stock cover the larger the purchase that can be made at what is predicted to be a low price. This aspect has to be balanced against the effects mentioned above, that the longer the buying period, which follows from the larger purchase and stock cover, the greater become the price forecasting errors. Furthermore the larger its value the fewer the purchases that will be made in a year. Hence there will be large variations over time in the purchasing expenditure and capital tied up in stocks. Company overall strategies and policies tend to control departmental activities by financial constraints. Such constraints would impose limitations on the maximum investment in stocks at any time and monthly expenditures on purchases. They also tend to prefer a more regular expenditure from month-to-month rather than very large variations. Comparing the price paid with what could have been achieved with some type of averaging policy, the higher the probability or risk of large losses at times, then the even higher the expected annual net gains required. The attitude of the company to risks is thus an important constraint on the purchasing operations.

All of these different aspects to the determination of the strategic buying period or stock cover are discussed in some detail in the fourth section of this chapter in order to clarify the particular problems that arise.

The next section contains a discussion of the *active buying period*. This is the length of the buying period (or the equivalent purchase and stock cover) that the buyer himself will use in practice. This must basically take account of how well he can make purchases as his freedom of action increases. It is thus buying performance orientated rather than concerned with overall company strategies. It will generally be less than the strategic buying period. There is an obvious interaction between the active and strategic buying period. A good purchasing policy or a good buyer may be able to purchase well over longer periods than the current strategic buying period. After due consideration of the factors that determined the choice of the strategic buying period, such a situation may lead to its modification.

There are two alternative forms for the active buying period. The first is to use one single constant value for a long period of time, whilst the other is to make it adaptive as the price situation and the price forecasts change. The choice is shown to depend upon the nature of the type of price forecasting system used.

The final stage is to derive a tactical day-to-day buying policy for use on each successive day of the active buying period, to assist the purchasing manager and his buyers to decide whether or not to make a purchase on that day. All too often the purchasing policy used in practice only really covers this final stage of the purchasing decision. The solution to this problem is frequently considered to be the accurate forecasting of prices on some specific day in the future. This however is an impossible forecasting problem. The inevitable errors that must exist in future price forecasts, made by whatever method, together with the inevitable variations in price from day-to-day must be considered explicitly to decide the appropriate operating rules for making daily purchase decisions as well as the general overall trend of the price forecasts.

The day-to-day purchasing problem occurs as some variant of either of two alternative forms which basically depend on whether delivery of new supplies and consumption occurs on a regular daily basis or whether they are on a more intermittent monthly or longer basis. Very broadly the two situations arise for the cases of buying semiprocessed metals, such as copper, and agricultural materials. However further analysis shows that these two different forms of the problem are very closely connected to a single decision situation. The day-to-day purchasing decision problem is to find some way of defining a measure of how low or high a particular price is, compared to the unknown prices that can be expected to occur over the remaining buying opportunities of the active buying period, at which alternative purchases can be made for the requirements considered for a purchase at today's price offer. The important result is that the measure of how high or low a price offer is, must basically be in the form of the *number* of successive future price offers (or buying opportunities) that are expected to occur before a lower price offer will be made.

A final benefit of formulating the commodity purchasing decision in the manner discussed in this chapter is that quantitative measures of past purchasing performance can be established that enable the purchasing manager to assess how well he has done. These measures are described in detail in the next chapter. They measure his performance relative to the situation he actually faced rather than relative to some earlier forecast of what the situation would be. Thus different buyers' skills in purchasing different materials can be compared. Initially such measures should be used by the purchasing manager and his buyers to learn about their past performance and see how improvements may be made. Later they can become the basis of setting standards for purchasing performance.

4.2 DETERMINING FUTURE CONSUMPTION REQUIREMENTS

The initial stage of specifying what future requirements are needed is very simple for the case of a single material of a specified quality. It merely requires a comparison of the current physical stocks, plus any forward contract orders placed that have not yet been delivered, with the forecast future consumption requirements. The stocks and forward orders are allocated to future consumption sequentially over time to determine a schedule of uncovered consumption requirements, for which purchases either now or in the future will have to be made. In some situations there are alternative grades of a particular material that can be used, which are offered at premiums or discounts on the price of some standard material that is the basis of trading and price fixing in a commodity market. If the different grades are completely substitutable, it is necessary to estimate what reductions or increases in production cost to achieve the final product specification are incurred by using the better or poorer grades of material, taking account of the differing yields and processing work required. In this way the prices offered for each different grade can be converted into a 'standard grade' price as used in the particular company's factory. Comparing these alternative standard grade prices directly will then tell the purchasing manager which grade he should plan to buy.

The problem of determining the future maize consumption requirements for making animal feeds for LFF is much more complex. In animal feeds manufacturing there are many partially substitutable materials available that can replace maize to a greater or lesser extent. The amount of maize that will be used depends upon the prices paid for maize relative to the prices of all the other materials. Their prices also fluctuate over time like maize. In addition the amount of maize used depends on the nutritional constraints for the particular animal feed. Different animal feeds, for cattle or pigs for example, will use different quantities of maize at the same relative prices. It is a very complex interactive problem. Its solution, as mentioned in Chapter 3, requires the use of large scale linear programming models. (For those unfamiliar with the term linear programming, it is described and discussed in some detail in Chapter 6). Such models require an input from the fourth and fifth stages of the commodity purchasing problem, as to what prices per ton will be paid for the various materials that can be used. In practice this input will be some reasonable prediction of their values. Various simplifications, which imply the ignoring of many of the interactive links, must be made in order to devise manageable subproblems for practical operations. Such a procedure gives specific maize requirements for each month, which purchasing must buy for delivery during the preceding month. This animal feeds problem is one example of the general 'blending problem', which occurs in raw material purchasing in many different industries. The problems that arise and the solutions adopted for various 'blending problems' are discussed in Chapter 6.

If only one material is required but different grades exist which are only partially rather than completely substitutable, then the situation becomes a blending problem, fortunately in a more simplified form than the animal feeds situation. It will have to be solved in the same way as the animal feeds problem using the methods discussed in Chapter 6.

4.3 DETERMINING PURCHASE ORDERS AND DELIVERY SCHEDULES

4.3.1 Ordering policies in general for purchasing

The raw materials that are purchased by industry are processed and transformed into products for resale to outside customers. The demands for, and hence sales of, these manufactured products are changing over time, usually in a partly systematic and partly stochastic fashion. The demands for the final products must therefore in some way be forecast. All companies then organise themselves to meet this forecast demand, or some chosen part of it, by planning their future production. The demand for the materials that the purchasing function has to procure is thus generated internally by the requirements of the production plan over time.

Purchasing has to decide how to organise the ordering and delivery of the raw materials to meet these production demands in the most effective manner. This implies determining what size of orders to place on what supplier and when to place the orders. Because of the ever changing prices that have to be paid for materials, these, perhaps more mundane, aspects of the commodity purchasing problem are underestimated by buyers at times. However if, for example, a 6,000 ton order for maize for MDL, see Chapter 3, Section 3.3, were delivered and paid for one month earlier than necessary, then at current prices around £9,000 in extra bank interest charges could be incurred for that one month's unnecessary storage.

The methods commonly used in purchasing to decide the ordering and stockholding policy divide into two general categories. Both have been devised to cover the case of buying manufactured and semimanufactured items at fixed constant prices. The choice between them depends upon the type of final product produced, the nature of the demand for the final product, the way in which it is sold, and whether for several months ahead the actual production can be maintained close to the preplanned levels. The two alternative approaches are usually titled 'Statistical Inventory Control' and 'Materials Requirements Planning'. The contrast between these two approaches can be broadly stated as follows. The former approach tries to plan production and stockholding of materials or goods so as to try to meet the fluctuating demands as they occur, or the demands per unit of time up to some chosen level. The second approach forecasts the demand for the final products. It then schedules a set of demands that it will aim to meet in each particular period of time. Production and stockholding are then planned to meet all of this scheduled demand. Many books and articles have been published on these topics, for example Brown (1967), and Hadley and Whitin (1963) on the former approach and Orlicky (1975) on the latter.

Statistical Inventory Control in its classical form broadly assumes that the demand to be satisfied is roughly constant over the near to medium-term future, but that it fluctuates over time, possibly by large amounts, in a random stochastic manner about this constant level. A particular general form of policy for placing orders is chosen. The best trade-off between the expected (or average) stockholding costs, ordering costs and stock-out costs is calculated using mathematical methods to give the best control procedures for implementing the chosen policy. The price of the item stocked has usually been considered to be constant and independent of the size of the order. There are two possible approaches to, or models for, the determination of operating rules for the management and control of inventories. The general way in which the two alternative models will operate is shown conceptually in Fig. 4.1. The first is known as the reorder level reorder quantity system, alternatively as the continuous review system or the two bin system. This model requires that the stocks are reviewed continuously. When the level of the stock falls to some predetermined fixed point (the reorder level) then an order should be placed for some fixed predetermined quantity (the reorder quantity or the economic order quantity). Although the orders are fixed in advance, the times when orders are placed on the supplier are determined by the actual usage of the material in production. They are the days when the stock falls to the reorder level and below. Since demand is varying, these days cannot be accurately predicted in advance.

The alternative form of model is known as the cyclical or periodic review system. In this system the level of stock is reviewed only at predetermined regular intervals of time. The actual stock at the review is then compared with a predetermined quantity (the replenishment level). On the day of the review an order is placed with the supplier equal in size to the difference between the replenishment level and the actual stock. The days that orders will be placed are known in advance, but the size of each order is only determined on the days that the reviews are made.

In most practical situations, delivery of new supplies from outside sources does not take place instantaneously following the placing of the order. This delay is the delivery lead time. Providing protection against shortages, caused by fluctuations in demand over the lead time, is an important component of stock control. The fluctuations arise from both variations in demand over time and also from variations in the length of the lead time itself. Although the supplier will specify a lead time in accepting an order, his actual lead time can vary significantly above and below his specified level. Safety stocks must thus be carried to guard against both

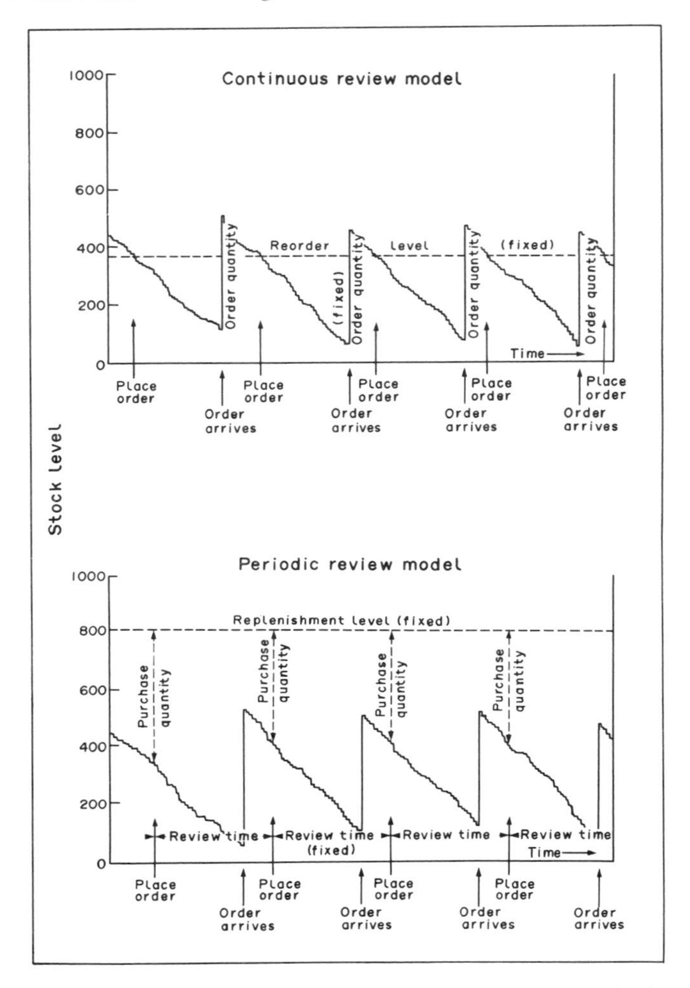

FIG. 4.1 Alternative classical statistical inventory control methods

types of fluctuation. These are an important component in determining the reorder level in the first type of stock control model and the replenishment level in the second type.

As mentioned in the case situations of Chapter 3, such variation is often built into the supplier's delivery promise. Linseed oil and maize imported into the U.K. must be purchased under the condition that delivery may take place on any day, at the supplier's free choice, within a specified month or half month. Guimarães (1981) has shown that even this agreed variation is often exceeded for international trading in raw materials. He analysed the delivery performance of ships carrying maize from the U.S. to Portugal. In this case delivery was specified as occurring within a 15 day period, usually the first half or the second half of a month. An analysis of 52 maize shipments over 1976, showed that on average ships arrived 2 days later than the end of the agreed period. A histogram showing the arrival times of the ships relative to the end of the agreed delivery period is shown in Fig. 4.2. This histogram shows that ships arrived up to one month after the end of the agreed period, and on some occasions even before the start of the agreed period. In fact this histogram fits closely to a well-known commonly occurring statistical distribution, the normal distribution. In this case the time that ships arrived after the end of the agreed period was found to be normally distributed about the mean of 2 days, with a standard deviation of 11 days.

In many practical applications, because of the multiplicity of items purchased, the classical statistical inventory control models are somewhat simplified. The order quantities are obtained by merely balancing the costs of placing orders and the costs of holding stock to meet the forecast mean demand only. The reorder levels are calculated separately on the basis of providing some service level that ensures that shortages, i.e. delays to the production programme, will never occur more frequently than once every 4 years, say, or alternatively that only some specified percentage of the annual production should be delayed. If the cost incurred whenever a

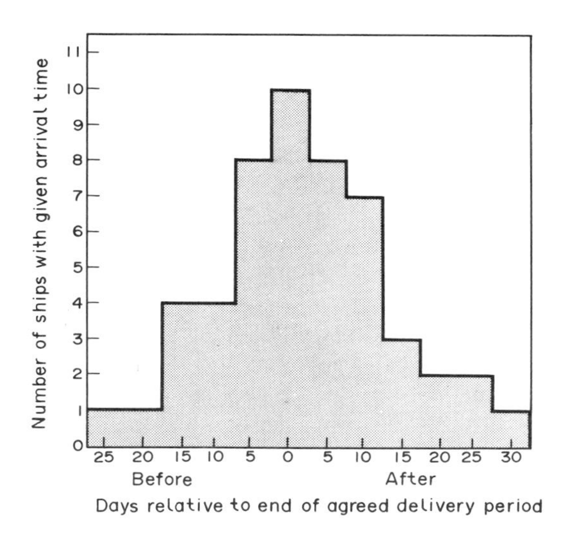

FIG. 4.2 Histogram of ship arrival times in days relative to the end of agreed delivery period (grouped data)

shortage happens can be calculated (for example the expected profit foregone on the production that is lost because of the delay in the arrival of material), then the best balance between the costs of shortages and the costs of holding safety stocks, to guard against delay and shortages arising from fluctuations in demand over the lead time, can be used to set the reorder levels.

The *Materials Requirements Planning* approach assumes that the company produces a master production schedule for the quantities of the final products to be made for several months into the future. The stochastic demand for the final product is smoothed out over time through a deterministic production plan. This details the size and timing of the production runs for the individual products. 'Parts Explosion' is then used to break down the final products into their component parts and raw materials. This break down, together with the master production schedule, specifies what amounts of raw material and components are required at what points in time. The production demands that purchasing must meet are irregular but specified exactly for a number of periods into the future. In these circumstances, it is unnecessary to have stocks of materials on hand all of the time. They are only required prior to the start of a scheduled production run using that material.

In several industries the final product is assembled from many component parts. These may be machined from a basic casting or made from different materials. In the first case the different component parts will probably have different economic production run lengths. In both situations a regular demand for the final product can transform into a very irregular and intermittent demand for the castings and the raw materials. The assumption of a roughly constant demand rate, as must be assumed to apply the statistical inventory models, is clearly inappropriate, say the advocates of this approach, for developing policies for the purchasing function. The ordering policies can be 'tailored' more closely to the demand pattern than is given by the statistical inventory models.

This approach is well-suited to the manufacture of large items assembled from many different individual component parts with lengthy manufacturing lead times. It suits also these cases where production is made to a customer's individual order rather than to stock, and also where customers must place their orders well in advance of delivery.

The need for some stability in the production levels over time for the successful application of this approach is obvious. Plans once laid down, must be adhered to, at least for the medium-term future. If frequent changes are made to the master production schedule, each month say, then the situation soon reverts to that of a stochastic demand for the basic components and raw materials.

Taking account of all these assumptions a formal mathematical model can be devised, whose solution will give the best operational and control procedures. In general the problem can be formulated in dynamic programming form. For those unfamiliar with this term it is described in more detail later in the book, particularly in Chapter 7. For the case of constant prices for the bought-in materials and components, the dynamic programming model can be solved generally to give a relatively simple procedure known as the Wagner–Whitin algorithm (1958), from which to work out the detailed operating rules. The calculation procedure is somewhat lengthy, however. For a small number of important items it is perfectly feasible to use this algorithm in the form of a computer programme. The advent of the new microcomputers makes this a less expensive possibility than before on a large main frame machine and the computer can be operated within the purchasing department. For large numbers of items though, the calculations may become rather excessive and expensive in computer time. For the more general problem where the Wagner–Whitin algorithm is inappropriate the computer time requirements become even longer. Various heuristic rules can be devised, that are appropriate to particular situations, which will give some reduction in purchasing and stockholding costs, although less than the maximum saving that could be made. The dynamic programming model is then best used as the yardstick for comparing the value of alternative simpler heuristic rules.

The problem of covering against variation in the lead time, so avoiding delays to the planned production, is frequently overlooked in the Materials Requirements Planning approach. An implicit assumption that appears to be made usually, is that, with orders programmed well in advance, suppliers' lead times become less variable. As we have seen earlier in this section, constant lead times do not exist for buying raw materials. In general some level of safety stocks will have to be kept.

In some industries, particularly those producing consumer goods rather than industrial products, such a rigid planning system may be inappropriate or impossible. The general marketing situation is often that customers require immediate delivery, or within a very short time period, otherwise the order is withdrawn. Such orders must be met from the existing stocks of the final product. In these circumstances the major part of production is planned so as to maintain the final goods stocks at some chosen level. Production overall may be 'planned' ahead on the basis of sales forecasts. However with many different products produced for sale, particularly if there are colour variations, the sales of the individual products will soon differ by large amounts from their forecasts. If production is planned for stock, then it will generally follow the variations in sales from month-to-month rather than smooth them out to give a deterministic schedule.

Purchasing has to regard 'production needs' as an 'outside customer', whose stochastic demands must be forecast. The forecast errors that inevitably will occur must be explicitly taken into account. Such circumstances inevitably lead to the use of some form of the statistical inventory control models rather than the materials planning requirements approach in the purchasing area.

4.3.2 Ordering policies for raw materials

Following the brief review above, the relevance of the various alternative methods described, for the control of bought-in supplies and components, to the purchasing and stocking of raw materials is now examined. It should be remembered that these methods were devised for items whose prices are fixed for a substantial period into the future. As seen in earlier chapters the essential characteristic of raw materials is that their prices are continually changing from day-to-day. Initially the classical statistical inventory control models are considered.

If a reorder level reorder quantity model is used, then a new order will only be placed with the suppliers on the day the stocks fall below the reorder level. Delivery then takes place one delivery lead time later. Although with this system the size of the order is known well in advance, since it is generally kept constant for a substantial period of time, the days on which orders will be placed cannot be known in advance. They depend entirely on how production uses up the raw material day-by-day. The price paid for raw material depends upon both the day when a purchase is made and the future time ahead that delivery of the purchase takes place. Remember that on the same day different prices are quoted for different future delivery times. Since neither the day of the next purchase nor its time of delivery can be scheduled in advance, the use of this stock control model denies the buyer any freedom to choose the price he will pay for his raw material. The stock control model itself automatically fixes the prices that will be paid for future raw materials. Thus the reorder level reorder quantity stock control model removes completely the major aspect of the commodity purchasing problem, the freedom of choice of the buyer to purchase advantageously at one of the many alternative price offers made to him.

The reorder level concept in statistical inventory control models is based on the likely usage of material over the delivery lead time. The delivery lead time is assumed to be outside the buyer's control, mainly determined by the supplier, and is taken as constant or as fluctuating randomly around some fixed mean value. In the raw material area the supplier does not quote a delivery lead time. He quotes the times when deliveries will be made. His price offers reflect this by quoting different prices for different delivery times as mentioned earlier. The whole commodity market structure via the mechanism of futures prices is organised so that things happen this way. Suppliers usually quote deliveries for successive half monthly or monthly periods. The futures markets where prices are quoted for deliveries up to 9 or 12 or 18 months ahead, depending on the particular raw materials, means that, subject possibly to a minimum value on the time ahead of delivery that an order must be made, the buyer himself can choose his delivery lead time. He can choose to purchase some raw material for delivery 9 months ahead, 6, 3, 2 or even one month ahead of delivery. Thus the reorder level concept itself in classical statistical stock control models is not applicable to the raw material purchasing situation. It is the time of delivery of material that is important in ensuring that production requirements are satisfied. Placing orders on the basis that stocks of materials held plus the forward orders have fallen below some chosen reorder level is entirely the wrong approach to use. Thus for several reasons this type of stock control model will not help those purchasing fluctuating priced raw materials.

If the buyer uses the other system, the periodic review model, he places his orders at regular intervals of time. He knows in advance on which days he will need to place an order. Knowing the days orders will be placed means that the periods in which orders will be delivered can be specified. Hence he has the freedom to choose amongst earlier days, prior to the regular review, at which to make a purchase for a particular delivery position at a lower price than on the review day. However it still ignores the situation just discussed that, unlike the case of manufactured and semimanufactured items, the buyer can himself decide his delivery lead time. This decision choice is taken away from him in this system also. More important is that unfortunately he cannot know in advance what size of order he should buy. The order size will be the difference between the replenishment level and his actual stocks on the day of the review and the consequent placing of an order. The order size can be quite variable and depends on how demand varies. A buyer has to place an order for a specific amount of material for delivery at some specific period. He cannot say that he would now like to buy 1000 tons at a particular price, but he may want to increase this by 300 tons, at the same price, or decrease it by 300 tons, in a couple of weeks time when the periodic review model tells him what the order size should be exactly. Thus for a different reason this other statistical inventory control model cannot help him either.

The classical statistical inventory models were developed for the case of fixed-price materials, so that the price paid did not depend on the day the order was placed nor on the period for which delivery was scheduled. In the commodity purchasing problem, choosing the day to place an order and the time of that order's delivery is automatically choosing the price to be paid. The very basic assumptions of the statistical inventory models are not satisfied. Furthermore, as discussed for maize, minimum constraints are often placed on the size of the order, whilst for linseed the constraints are even more restricted, to a choice of only two order sizes. Including these constraints would make either of the statistical models even less 'optimal' than normal.

In order to give the flexibility to buy at lower prices it is necessary for purchasing to place its orders and deliveries well in advance. The operating procedures for the statistical inventory control models are such that it is impossible to do this. Hence they cannot be applied to the purchasing and stockholding of raw materials whose prices fluctuate over time. Inevitably for the purchasing of raw materials with fluctuating prices, orders will have to be planned and scheduled using some form of the 'Materials Planning Requirements' approach. In general it will have to be based on the mean forecast usages for each future month or half month. As the forecasts of future usages are modified over time the buyer can adjust his orders to some extent.

As a separate exercise, safety stocks will need to be calculated to cope with the variabilities in the actual delivery time beyond the scheduled time period of delivery. Generally the safety stocks should be calculated so as to ensure that such delays, leading to lost production, should only occur relatively infrequently, once every 5 years for example, or should only cause a very small part of the total expected annual production using that raw material to be lost or delayed, 2% for example. If the demand from production is stochastically varying over time then this must also be taken into account in calculating the right level for safety stocks. One safeguard, that exists in most commodity markets, is that it is possible to buy materials, at least in small quantities, for immediate delivery, although a premium price may have to be paid. This provides a fallback position for those rare occasions when the safety stocks are exhausted.

In some instances, as in the copper buying situation of BMP, planning orders and their delivery is a relatively simple exercise. The consumption requirements are a known constant 20 tons per day. The supplier will deliver copper daily if BMP request it. The maximum amount that the purchasing manager can buy on any day is such that it, plus his stocks and any undelivered orders, cannot exceed 400 tons, 20 days consecutive usages. The buyer must work out his current physical stocks of copper and add to this the copper he has bought but which has not yet been delivered. Dividing this total by 20 determines how many future days' consumption requirements are currently covered. Let this be 12 days. Then his schedule for future delivery quantities is 20 tons to be delivered on the thirteenth day and on the fourteenth day, and fifteenth day hence etc. Because of the company policy on the maximum stock cover at any time, he can today only consider purchases for up to the twentieth day hence. The stock constraint covers the copper that has not entered into production. Each day 20 tons will be consumed by BMP reducing the stock cover by 20 tons. Thus tomorrow, because it is a rolling horizon constraint, he will be able to cover up to the twentyfirst day hence, and the day after up to the twentysecond day hence etc. Each day's requirements should be scheduled for delivery 2 or 3 days prior to usage, to ensure it is definitely there in the warehouse ready for production to use.

In both the linseed oil example for CFC and the maize buying example for MDL, orders must be planned for the whole of the longest possible buying period. A purchase could be made at the beginning of the buying period to cover all the usages over the period, if prices are forecast to increase strongly. Usages for each month of the buying period were first forecast. A materials planning requirements procedure was then used to convert these 'planned' irregular usages into purchase orders. Since reasonable assumptions or forecasts of the differences between prices for delivery in adjacent months over the future could be made, a relatively simple model was formulated to generate purchase orders. In both examples the model was in dynamic programming form, as discussed in the previous section. The complexities and difficulties that can arise are illustrated for the linseed oil situation in Chapter 7, where the model and solution developed for that case study are described in some detail. Because of the severe limitations on the delivery order sizes, the standard simple solution procedure of tailoring the purchase orders to the forecast usages, using the Wagner–Whitin algorithm, was not possible. The dynamic programming model had to be solved in its entirety for each set of forecast future usages.

4.4 THE STRATEGIC BUYING PERIOD

The determination of the strategic buying period entails consideration of the accuracy with which price forecasts over differing time periods can be made and also of the stock investment and financial implications. These two differing aspects of the problem are discussed separately. Forecasting is discussed first, followed by the stock investment and financial considerations.

4.4.1 Forecasting considerations

The intention is to clarify the particular requirements and properties for a price forecasting system that are needed for the determination of the buying period. This is carried out by analysing specific buying situations which are typical examples of the general problem. The initial situation

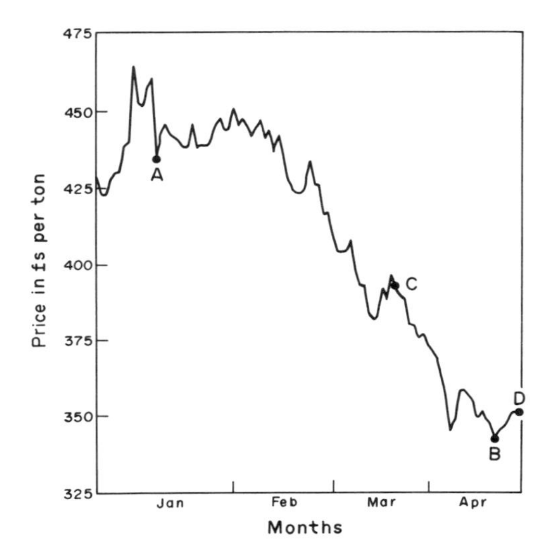

FIG. 4.3 Company buying price January to April 1967

covers general linear forecasting models. This is then extended to consider the modifications necessary when there are consistent seasonal price patterns, occurring because of the manner in which production takes place.

Figure 4.3 shows copper prices over 1967, taken from the BMP case study. The point A marked on the left-hand side of the figure is a day when prices fell by nearly £27 after a 12 day period of increasing prices. Any linear statistical forecasting system would almost certainly predict an upward trend to prices prior to the day's price offer at point A. A large fall in price in a situation where prices are on an upward trend would suggest a large purchase is indicated, or rather a purchase to raise the stock cover to a large value. As can be seen prices subsequently rose above the price of £433 at point A and stayed there for some time before falling even lower. Thus a large purchase at point A would be a good decision. Assuming that a regular 20 tons of copper is consumed per day as in the BMP case, a purchase to cover the next 5 day's consumption at point A would have saved £1090 compared to a back-to-back policy. Raising the stock cover to 10 days at point A would have saved even more, £1790. As the stock cover is increased then so are the savings, up a maximum of £6460 at a stock cover of 27 days. Thereafter, increasing the stock cover brings losses compared to a back-to-back policy. Prices fell eventually to below the level at point A. The subsequent continuous losses from increasing the stock cover further, start to counterbalance the earlier profits. At a stock cover of 35 days the cumulative profits have fallen back to under £5300. At a

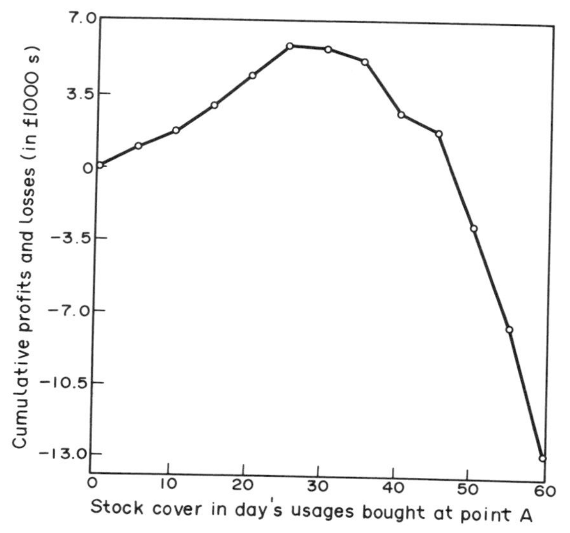

FIG. 4.4 Effect of increasing purchase and stock cover on savings relative to a back-to-back policy

stock cover of 50 days the losses more than outweigh the earlier profits, so the cumulative situation is a net loss of £2750. If the stock cover were raised to 60 days, then the net losses would exceed £13,700. (The net profits as a function of the stock cover are illustrated in Fig. 4.4).

The positive increasing trend predicted at point A persists for just over 20 days. Prices then begin to fall and a negative trend appears. The errors from the earlier price prediction at point A become larger and larger through late February, March and April, as do the losses made from buying the requirements for these periods at the earlier point A. The largest net profits appear when the purchase stock cover is roughly equal to the time the positive trend at A persists. On the other hand, buying heavily at point B in Fig. 4.3 would have brought continuous increases in the profits as the stock cover is increased without limit. This is because point B, 25 April 1967, was the lowest price for copper from 1966 to the present day. Point B marks the start of a long upward trend to prices. However any linear statistical forecasting system would have a negative trend at point B. Prices would be predicted to go even lower. Thus it is most unlikely that a large purchase would be recommended at point B.

The above discussion illustrates that the most appropriate length of the buying period, in respect of the forecasting problem, depends on both determining the length of time positive upward price trends will persist and whether it is possible, at a very early stage, to detect that a new period of persistent upwards price movement has begun. The difficulties in the latter problem are illustrated by points C and D in Fig. 4.3. Both points follow a 5 day rally of rising prices contrary to the overall long general downward decline of prices through February, March and April. Point C is a false herald of a reversal in the overall price trend and thus a very bad time at which to make a purchase. Point D turns out to be a very good time at which to make a purchase, see Fig. 4.5 which shows the price movements over the next 6 months of 1967. Yet in Fig. 4.3 there is apparently little difference between point C and point D.

For agricultural commodities with well-defined planting, growing and harvesting seasons the time at which the new harvest begins to affect prices may be a constraint on the buying period. The transitional period between the old crop and the new crop is a particularly difficult period for which to forecast prices in advance. Prices for the new crop will depend on so many events which have yet to occur, many of which are stochastic in nature. A good policy might be to limit purchases to deliveries from the old crop until well into the new crop growing period. This restriction ends the buying periods for different consumption periods at some specific calendar time. So different consumption/delivery periods will have different maximum buying periods.

The world maize market is a good example of this situation since it is virtually synonymous with the U.S. maize situation. In the U.S. maize is planted during April and May and harvesting occurs from late September to the end of November. The size of the crop depends on the land acreage planted and the yield. Particularly important factors in determining the

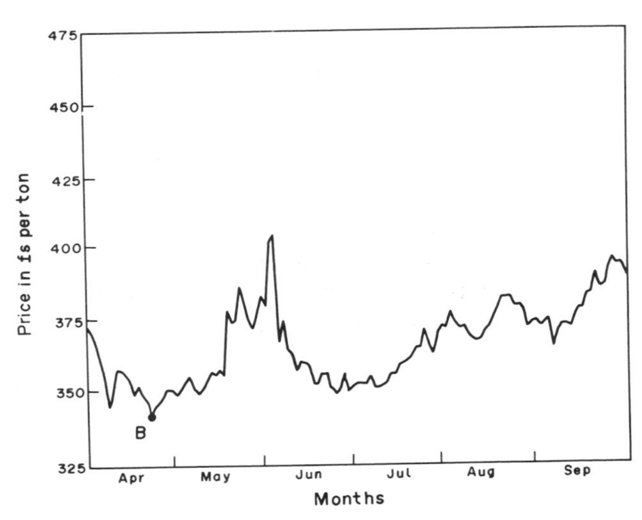

FIG. 4.5 Company buying price April to September 1967

yield are the moisture levels in the land at planting time, the amounts of fertiliser used, the weather over the critical growing months of July and August, when droughts may seriously affect the crop, and the incidence of crop diseases such as maize blight which again tend to occur in August. In January and February it may be possible to derive models to make reasonable forecasts of prices up to mid July, the end of the old crop year, since the supply of maize for that period is known with reasonable certainty. Prices for the period from August onwards critically depend on the size of the new crop. A good crop could bring very low prices, whilst a crop failure would bring very high prices indeed, up to three times those of an abundant crop situation.

In January and February the crop has not yet even been planted, so the acreage is not known. Weather forecasting for many months ahead is currently impossible, so forecasts of likely yields are not feasible. To the errors in any price forecasting model, based on 'known' supply and demand factors, must be added the large price differences that will result from the very large differences that are possible in the estimates of supply factors for the new crop next August, September etc. A discontinuous change in the level and nature of the forecasting errors for future prices occurs once one forecasts beyond mid July, the end of the old crop year.

Because of this discontinuity it may be appropriate to have different values for the length of the strategic buying period for the different delivery months of the year for some agricultural commodities. For most of the year purchases can be made to cover requirements up to the end of the old crop year, mid July. However, purchases for August, September should perhaps be delayed until after mid June, say, when the size of the acreage planted with maize is known, whilst October, November and December should be left until after July or August when some reasonable indication of the likely yield can be obtained.

4.4.2 Stock investment and financial considerations

In all the case studies described in Chapter 3, payment for the raw material was made by the consumer after delivery and not at the time of the purchase order. Thus the 'purchase and stock cover' can be increased at any time wtihout increasing the actual capital investment tied up in stocks at the same time. However, the consuming company is committed to those purchases and must have the money available to meet them as they are delivered. The capital loss, because higher prices were paid than would have been the case with a later purchase decision, still occurs. A company's attitude to the risks of possible losses versus levels of potential savings is thus an important factor. The higher the risk, i.e. the probability that the price forecasts are wrong and lead to a loss, the higher the size of the gain required. Since it is a real, albeit delayed, capital investment, the purchase and stock cover may be restricted by company constraints on the amount of its capital it is willing to have tied up in stocks of raw materials, part processed and finished goods.

When the material has to be bought for immediate delivery and physically stored by the company prior to its usage, or if payment must be made when the order is placed, costs of at least the interest on the capital investment in the raw materials inventory are incurred. If the cumulative storage costs, over 2 months say, exceed the maximum price change that can reasonably be expected to occur over the 2 months, then there is no advantage in having a maximum cover greater than 2 months. So the size of the storage costs relative to the purchase price affect the maximum buying period. Any limitations on the physical storage facilities will also set maximum values on the purchase and stock cover. Alternatively the commodity may be perishable e.g. fresh fruit, and have a limited shelf life. In this situation, provided a first-in first-out inventory usage system is employed, this limitation gives the maximum value for the purchase and stock cover, and the buying period.

A further consideration is the consistency of the purchasing performance over time. In one instance, it was found by evaluating the performance of a proposed new purchasing policy over several past years of prices, that it gave an average saving below a back-to-back policy of about 5%, if the maximum stock cover were 2 months consumption. This average obscured quite substantial variations in performance over successive short time periods of 2 months, ranging from a maximum saving of 12% in one 2-month period to a maximum loss of 6% in another, compared to a back-to-back policy. With a smaller maximum stock cover of only one month's consumption, the evaluation over the same period showed that the proposed policy would only save about 3% over the back-to-back policy. However the variations in performance over 2 monthly periods were much less, ranging from a maximum savings of 8% to a maximum loss of only 2%. Initially the company preferred the second policy with its overall smaller, but more consistent, gains to that of the larger maximum stock cover, with a higher gain but more variable performance. Such a choice reflects the 'risk aversion' nature of most decision makers with regard to purchasing for manufacturing industry. Since raw materials provided such a major part of the costs of the final product in the particular case study, variations in purchasing performance of the levels mentioned could possibly cause large variations in company profits from one quarter to the next. The company was concerned that this might cause unwelcome variations in the company's share price and hence its ability to raise new capital in the market.

The minimum stock cover, which will give the latest time at which any period's consumption order should be bought, depends on the company's attitude to guaranteeing security of supply and the possibility of having to pay premium prices, if the purchase is left so late that physically available supplies for immediate shipment become scarce in the market. It should be remembered that the prime responsibility of a purchasing department is to obtain supplies of material to keep production going. The price paid is a secondary consideration, unless company policies indicate that production is stopped if raw material prices exceed some specified high level. This is an eventuality that rarely occurs in practice. Companies prefer to avoid any possibility of not having material available for production needs, by biasing the minimum stock cover at a higher level than may be absolutely necessary. The actual value will depend on the particular circumstances, such as the level of activity in the market, the number of alternative suppliers and whether a shortage of supply is likely to occur in the near future or not. It must also take account of the lead time, that in practice, occurs between placing an order for spot shipment and its actual delivery to the factory. It some instances this can be very long, for example 3 to 4 months is typical for cocoa and rubber. This time must be added on to the time considered earlier, to avoid premium prices and 'stock-outs', to give the minimum time ahead of its usage that an order for raw materials must be placed. In the linseed oil case not only did the buyer have to ensure that the linseed oil was available in Rotterdam but, in addition, he had to ensure he had booked sufficient shipping space, among the limited number of special ships able to bring the linseed oil across the North Sea to his factory. He set his minimum stock cover at 2 months. On the other hand in BMPs case it was maintained at only 2 days. The difference was that in BMPs case the supplier was quite close to their factory.

4.5 THE ACTIVE BUYING PERIOD

4.5.1 The general concept

As discussed in the case studies of Chapter 3, we must distinguish between the strategic buying period and the active buying period. The strategic buying period is determined by overall company policies and strategies and sets the absolute limits on the freedom of action allowed to the buyer. The active buying period is the actual period of time within which the purchaser manager operationally decides to make a purchase for some particular delivery order of material. This may not be the same as the strategic period. Whilst the strategic buying period is fixed and constant his active buying period may be more adaptive to changing circumstances. The factors to be considered are how confident the purchasing manager is in his own price forecasts, or those of his buyers, at differing times and how well he believes his buyers can make good effective purchases leading to lower average purchasing costs, over different times ahead of the scheduled deliveries. There is obviously some interaction between the determination of the active and the strategic buying periods since the company may take into account an assessment of the buyers' actual performance in setting the absolute freedom of action given to the purchasing function. If they perform well the company may extend the length of the buying period, whilst if they perform badly the company may reduce it. Thus it is possible that the assessment of some new proposed purchasing policy may result in the extension of the strategic buying period as well as the active buying period.

There are two alternatives that can be adopted for determining the appropriate length of the active buying period. As will be shown in this section the choice depends on the type of forecasts that can be made in that buying situation. If the price forecasts are generated by a linear statistical forecasting system or by any method that basically simply extrapolates the past price movements into future time, then the best choice is to use a constant length for the active buying period for many months or several years at a time.

Alternatively, if price forecasts can be made that can anticipate changes in trend and reversals in the trend well in advance, then the active buying period can be made adaptive to changing circumstances and price forecasts.

A more fundamental and important point is that the nature of the active buying period is rather different in the two situations. In the first situation, where a fixed constant active buying period is used, the active buying period is essentially a modification of the strategic buying period to take account of the practical realities of how well the purchasing function can buy over differing lengths of time and how confident the purchasing manager is in his own or his buyers' abilities to achieve a consistent level of performance over time. The purchase and stock cover should never cover consumption further ahead than the length of the fixed active buying period whatever his price forecasts.

In the second situation it is asserted that much more confidence can be placed in the future price forecasts since they are based on a model that explains how prices are determined and can anticipate changes in level and trend well in advance. If such models can be determined, then confident forecasts that the prices over the next month or so will be the lowest for at least 4 or 6 or 9 months can be made. For example let the strategic buying period have a maximum value of 9 months. The forecasts may state that the consumption requirements for every period over the next 9 months will have their lowest price over the next month. In this case it is then worthwhile buying every future period's requirements over the next month or so. The strategic buying period gives the maximum purchase and stock cover, hence the total amounts that should be bought. However all of this total amount should be purchased over the active buying period of the next month or so. Hence the direct link between the value of the purchase and stock cover to be held and the length of the buying period is broken. The forecasts are used together with the strategic maximum and minimum purchase and stock cover (the minimum applies to the cover if prices are forecast to fall), to give the total requirements to be bought. The forecasts alone are used to determine the active buying period, when the lowest prices for each of those total requirements are expected to occur and hence the period over which to make purchases.

4.5.2 Statistical price forecasting methods

Any linear statistical forecasting system merely extrapolates its forecasts forward in time at a constant rate. If at the present time the forecasting system gives an upward trend of £1 per day, then its forecasts merely extrapolate the £1 per day continuously into the future without limit. It is not possible with such systems to anticipate a levelling off or a reversal of prices in 3 month's time, say, except when a consistent strong seasonal or periodic price pattern exists year-by-year. Thus such systems cannot provide any limitations on the amount of future consumption that should be covered. With an upward linear trend a larger saving is forecast for a purchase made now for the consumption in the fortieth day hence than a purchase for the twentieth day hence. Obviously trends do not continue for ever. The market circumstances change and some new trend is established in the market. Furthermore, a trend of £1 per day established on 31 January is the same as a trend of £1 per day established on 15 May. The implications, both for the future price forecasting and the size of the purchase and stock cover, will be the same on the two occasions. Trends of £1 per day, £0.30 per day or £5 per day will all suggest that large purchases to cover many future periods consumption requirements will bring savings in purchasing cost. Although the size of the savings in each case will be different, in all three cases the larger the purchase the larger the saving *pro rata*. On the other hand it is worth waiting to make a purchase whenever the daily price trend is negative, irrespective of whether it is only −£0.12 or −£12.0 per day. It is apparent that with a linear statistical price forecasting system the active buying period will take only two values, zero days or an 'infinity' of days. What should be used in practice as the numerical value for 'infinity' cannot be determined from the forecasting system itself? In general the value for 'infinity' should be less than the maximum value of the strategic buying period.

The only reasonable approach in these circumstances is to evaluate the performance of a tactical purchasing policy, using a linear statistical forecasting system as input, with differing values for the length of the active buying period (or the maximum purchase and stock cover). The average purchasing costs for each different value can then be calculated and compared. Provided that the average unit purchasing costs are relatively stable around their minimum value with respect to small changes in the buying period, this should be chosen as the appropriate value to use in practice. It is important to ensure that the evaluation is carried out over a period that is sufficiently long to include most of the many alternative patterns of price movement that do occur. This active buying period will then have the same constant value for all time. It will not change as the price forecasts change. The choice amongst the alternatives should generally only consider those values less than the strategic buying period. If a value larger than the current strategic buying period is indicated as best, it should be closely examined with respect to the financial investment in stocks, the likely size and frequency of losses due to unexpected falls in prices etc., so that if appropriate the company's strategy can be modified.

The linear statistical forecasting system may also contain regular seasonal or periodic components. If these are based on a 'natural cycle' as for an annual crop, then as discussed earlier in section 4.4.1, they may be used to set different values for the maximum buying period over the year to avoid those periods when considerable uncertainty as to the medium-term future price movements always occurs, such as the transition between the old crop year and the new crop year. However such a procedure does not give an active buying period that is adaptive to changing circumstances. Although the active buying period changes over the year the values will be consistent from one year to the next.

It would be unwise, as some have recommended, see Chapter 11, to go further than the suggestion of section 4.4.1 and try to determine different active buying periods over the year based entirely on an average annual seasonal pattern of prices. Two cautionary examples illustrate this warning. The first example is for U.S. maize prices over the late sixties and early seventies. For the period May to December of the year, two completely different seasonal patterns occurred. In the first, October was the lowest priced month whilst in the other September was the highest priced month over the period. The second example concerns cocoa prices.

Figure 4.6 shows a cocoa seasonal price index, produced by Shishko (1965), based on the New York exchange spot prices for Ghana Cocoa over the period 1947–63. The figure shows a strong seasonal pattern with a price peak in July/August, another brief upward movement in November

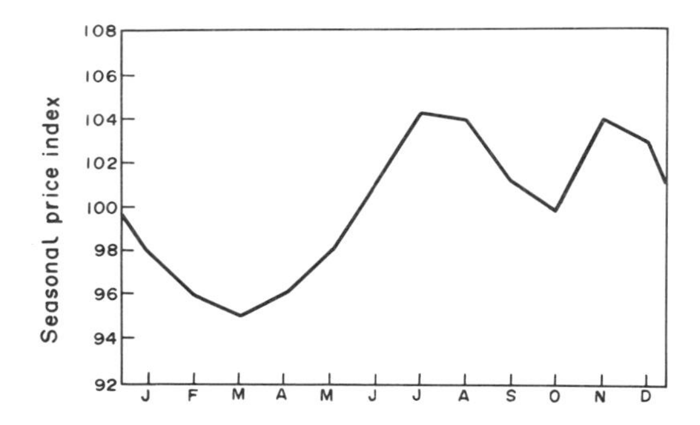

FIG. 4.6 Annual seasonal price pattern for cocoa due to Shishko, basis Spot Ghana New York 1947–63 equals 100

and a decline to the lowest price over January/March. The dominant physical factor is the very heavy shipment of the main cocoa crop from the African producers from late November through to March/April. Good stocks are built up in the importing countries and the manufacturer consumers' interest ebbs in anticipation of lower summer needs. In June and July demand improves in anticipation of the heavier cocoa grindings in the autumn. Supplies will be limited because the main crops of the African producers will have been sold out and only Brazil has some cocoa still to sell.

However in the 1960s the seasonal pattern tended to flatten out because of the larger stocks carried over from season-to-season, the trend towards longer harvesting periods and the improvement in the mid season crop in Ghana which affected prices over June, July and August. Thus concludes Shishko, a seasonal price index cannot be regarded as invariant and must be treated with caution.

This is confirmed by the price movements of cocoa illustrated in Fig. 4.7. This shows the Friday closing price for spot cocoa on the London terminal market for all of 1965 and 1966, as published in the *Financial Times*. It is seen that two completely different seasonal patterns occurred. In 1965 the seasonal pattern of price movements is completely inconsistent with Fig. 4.6, whilst in 1966 the seasonal pattern is quite similar to the seasonal index of Fig. 4.6. Particularly important is that the lowest price in 1965 occurs at the same time as the highest price in 1966. Furthermore in the 1968/69 cocoa marketing season the annual price pattern was again completely different to either 1965 or 1966, in that the highest prices were achieved in January and February. Note this is the time of the lowest prices of the year according to Shishko's index in Fig. 4.6.

Thus it must be concluded that, determining fixed different active buying periods for each monthly delivery of the year on the basis of an average

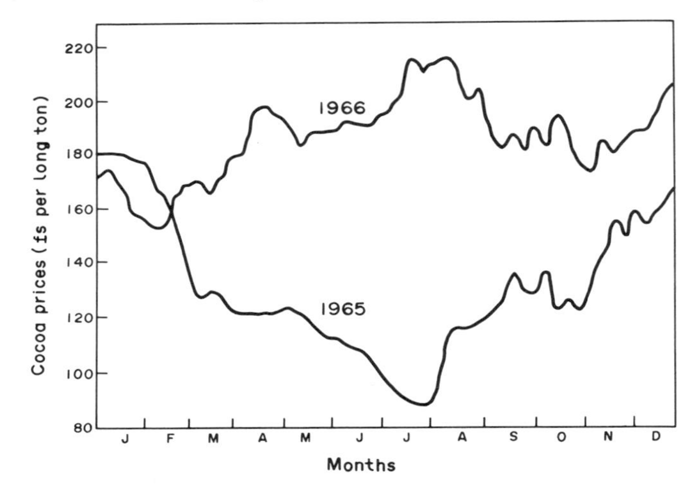

FIG. 4.7 End of week London cocoa prices over 1965 and 1966

seasonal index of past prices only, would not generally give a good purchasing performance and hence should not form the basis of any purchasing policy.

4.5.3 Econometric/behavioural price forecasting models

If the active buying period is to be made truly adaptive to the changing market and price circumstances, a price forecasting system is needed that can predict in advance that a change in trend will occur, including a complete reversal in the trend direction, and what will be the likely future high and low price levels. A model to achieve this must be an econometric behavioural model, that appears to explain how prices relate to, and are generated by, the actual values or 'expectations' of the values of the various supply and demand factors in the market. We are evaluating whether the buyers and sellers in the market etc., as a collectivity, have implicitly (perhaps in an unconscious manner) or explicitly determined a set of prices that they consistently find acceptable for various sets of market circumstances. Alternatively these prices may relate to the different alternative end uses of the commodity. As the prices increase certain uses, for different end products or in different geographical locations, become uneconomic because alternative materials become less expensive or because consumers' disposable resources are insufficient to buy the commodity at the new price. That end use disappears, perhaps permanently but usually in the short-term only temporarily. Different end uses fall away at different increasing prices. Similarly reducing prices may encourage the growth of other end uses.

Proving that a model completely explains market behaviour in all circumstances is a very time consuming task, by its very nature and theoretically and practically impossible. One can only hope to derive a model reasonably close to reality. For forecasting we need to show that consistent relationships between prices and market factors exist and persist for some time. Furthermore to provide the necessary confidence they must also be *a-priori* plausible causal relationships. Such models may require further models to forecast supply and demand factors in the future. Alternatively it may be possible to relate future prices to estimates of supply and demand factors at the present time or 'expectations' of these factors based directly on current values. A general approach to deriving such forecasting models is discussed in Chapter 8. The methods that should be used are described in detail in succeeding chapters for the case of forecasting prices of maize in Chicago and linseed oil at Rotterdam.

The importance of such a price forecasting model is that confident assertions can be made about precisely when the lowest prices for each scheduled delivery order will occur. This means that forecasts of prices must be made for the whole time period from now to the end of the strategic buying period. The forecasts from such a model will usually take the form of a mean forecast level for each future month together with some probability distribution for the likely fluctuations about this mean level, taking account of both the forecast errors and the day-to-day fluctuations about the average monthly level.

Let us assume that orders are scheduled for delivery each month and that the minimum strategic buying period, the safety level against delivery delays etc., is such that any month's delivery must have been purchased by the first day of the month preceding delivery. The maximum strategic buying period is assumed to be nine months. At the start of November say, the delivery requirements for December must have already been bought. Purchases to cover the delivery requirements for January, February, March, ... up to and including July can be made during November. The requirements for January must be bought over November, those for February over November and December, those for March over November, December and January, ... so that finally July requirements can be bought over the period November to May inclusive.

For each month's scheduled delivery the econometric/behavioural model price forecasts are used to determine which month over its buying period is predicted to contain the lowest price offers of the whole buying period. This can be done by a simple comparison of the forecast monthly mean values. Alternatively some consideration can be paid to the price fluctuations and forecast errors in the following way. Over the buying period for each scheduled delivery find the month with the lowest forecast mean price level. Use the probability distribution of the price fluctuations above and below the forecast mean price to calculate the particular price level for that month which has a probability that 75% of the prices quoted will be less than or equal to it. The 75% level is merely based on the intuitive notion that making profitable purchases on 3 out of 4 occasions will lead to lower purchasing costs. Other levels such as 2 out of 3 etc., could be used if desired. Next determine which months in its buying period have forecast mean levels above this 75% probability price. If this is the case for all other months in the buying period then buy that scheduled delivery in the lowest price month. Otherwise the active buying period will be the sum of the month with the lowest forecast mean price plus those other months whose forecast mean price levels are below the 75% probability price. In general these months will be consecutive ones. It is however possible that the months chosen by the above process will be December and May for example. In such situations the active buying period is 2 months in length, December and May, and a tactical day-to-day decision policy should be applied to buying over these 2 months, using the best possible price forecasts.

If, for example, prices are predicted to fall to a low in February followed by large price increases over May to August, then the lowest price month will be February. Hence it follows that January delivery requirements should be bought over November, February's as late as possible over the month of December, March's similarly over January. The requirements for the deliveries in April, May, June and July and August should all be brought over February. Since it is a rolling 9 month strategic buying period, if after updating the model at the end of January the forecast predicts that prices will rise strongly over the rest of the year, then in February purchases should also be made for the deliveries in August, September and October.

It is seen clearly that the period ahead at any time that reliable price forecasts can be made using the econometric/behavioural model will affect the maximum strategic buying period. If in the event reliable forecasts can only be made for 6 months ahead, the maximum strategic policy should be reduced to six months rather than the 9 months specified by the company objectives. For agricultural commodities the same comments made earlier concerning the old crop/new crop transition period, as regards different lengths for the strategic and active buying periods over the year, also apply.

A practical illustration of this approach is given in an appendix to this chapter for the case of cocoa. An econometric/behavioural model for forecasting cocoa prices is briefly outlined. The major part of the appendix then illustrates how the price forecasts from such a model can be used to continuously adapt the active buying period. The example covers cocoa over the period December 1969 to July 1972.

4.6 THE TACTICAL BUYING PROBLEM

As discussed in the introduction to this chapter, the case studies of Chapter 3 demonstrated that for the final stage, the tactical buying problem, there are two broad types of problem, one for the nonferrous metals and one for the agricultural materials. The differences between the two problems are inherent and arise essentially from the differing supply situations, as will be shown.

The supplier of copper to BMP is himself within the U.K. and does some processing work in his own factory to produce the refined copper wire that BMP actually use. His production of refined copper wire is continuous so that it can be delivered (and bought) every day if the consumer so wishes. The raw copper ingots used by BMPs supplier may however be delivered to him on a discontinuous basis since they are imported from overseas. The same situation occurs with nonferrous metal scrap, an important source of copper for U.K. consumers. Scrap metal dealers collect scrap on a relatively continuous basis and contact potential buyers on most days. Alternatively, scrap copper or alloys become available on a continuous basis from the production processes of manufacturers of nonferrous metal articles.

Metals are derived from ores mined out of the ground. The metal content of the ore deposits in each mine is relatively constant although it may differ significantly from one mine to the next. Mines are run on a continuous basis, producing the metal ore at a regular rate and smelting it to produce the metal at a regular rate. In both cases the production rate is under the control of the operators of the mine. The metal is very valuable so mining companies do not maintain large stocks. It is shipped out of the mines at regular intervals. It is less bulky than agricultural materials and is used and transported in smaller quantities. World mine production of copper is around 8 million tons compared to 400 million tons for maize for example.

The pricing basis for copper and other nonferrous metals on the London commodity markets reflect the continuous nature of their production and supply. Prices are quoted as two continuous price series, a daily spot price, for immediate delivery, and a 3 months (rolling) futures price. The futures price is always for delivery of copper 3 months hence and not for some fixed particular specified time in the future. Deliveries for specific dates in the future are negotiated as some weighted average of the daily cash and futures prices on the day of the agreement.

The supply of most agricultural commodities for the U.K. is on a discontinuous basis since they are imported by sea. As we have seen the economics of sea transport mean that the materials are shipped to the U.K. in very large quantities, of several thousand tons usually. The shippers of such materials have organised the market so that the buyers and consumers accept responsibility for physically storing them at their own expense until they are used in their production processes. Each delivery period for the materials is usually quoted as a separate individual price series. Although the differences between the prices for different delivery periods vary, the different price series are very strongly correlated over time.

Agricultural materials can only be harvested from each piece of land at definite times of the year, which are determined by the nature of the crop and the geographical location of the land. The farmer has little control over the harvesting time. He must gather the crop when it is ripe. Furthermore, the amount of the crop that can be harvested each year from each piece of land is highly variable and stochastic, depending critically on the weather conditions over the entire planting, growing and ripening period, and the absence or outbreak of crop disease and pest infestation, all of which affect the yield, and the weather over the harvesting period which can prevent the physical gathering of the crop. The use of fertilisers and insecticides, etc., can lead to a general increase in the yields per acre achieved, so giving the farmer some control. He can of course control the amount of land he plants. Even so, variations of the crop size from one year to the next of 10–25% are common. The harvest period is generally short. There is thus one intense peak of supply activity after which the crop must be stored over the rest of the year for usage by consumers. With temperate crops such as grains and soyabeans, crops are grown in both the Northern and Southern hemispheres so there are two short supply periods per year approximately 6 months apart.

The differences between the two types of purchasing situation are thus real and inherent in the nature of the commodities. It is interesting that these differences persist through the whole of the production, distribution and consumption chain.

From the buyer's point of view, after extracting the particular specific details of the nature of the commodity and its market, the nonferrous metal tactical purchasing situation reduces in its basic form to the following *Type 1* problem:

'Purchasing to meet a continuous daily requirement for consumption of a known amount of raw material with no limitation on the size of the daily purchase and with a single price offer each day.'

The situations for the agricultural materials reduce to the following *Type 2* tactical buying situation, although on any one day the buyer will generally have to cope with several such concurrent situations with differing deadlines:

'Purchasing a known amount of raw material before a known deadline with many alternative opportunities to make a purchase. At each opportunity there is a single price offer at which the buyer can purchase all or only part of his total requirement or wait until the next buying opportunity and price offer.'

Type 1 situation

When offered a particular price quotation, the buyer must make a decision as to whether to make a purchase at the present time or await the next opportunity to buy. When the price is low, the buyer would like to buy to build up his purchase and stock cover to meet at least part of the requirements for the periods when the price is high. When the price is high, he will wish to avoid purchasing and use his stocks to meet his current needs. Of course the problem is to decide what is a high or what is a low price, at any particular time. In deciding what to purchase on any particular day, the buyer must take account of the continuous nature of his purchasing problem. He will have the opportunity of making purchases tomorrow, the day after, or the next day and so on. He has to balance buying today, when lower prices may occur tomorrow or on other future days, against not buying today and leaving the purchase until later when he might find prices have increased. Obviously the buyer tries to guess or predict the lowest prices over future buying opportunities at which he can expect to buy each future period's requirements, if he does not buy them today. He tries to assess, in the light of the future purchasing decisions he will be able to make, the number of days' requirements which he should cover with a purchase at today's prices. For example, if he thinks it will be 5 days but not more until he will again be offered a price equal to today's or lower, then an obvious policy is to buy sufficient today to raise his stock level to cover the requirements for the next 5 days and no more.

It is important to recognise that the buyer must compare today's price offer with the prices he expects in the future and not compare them with the prices previously offered to him in the past. This would be misleading and can result in wrong purchasing decisions. Thus, if today's price is higher than prices have been in the recent past, this may mean that prices are on an increasing trend, so that a heavy purchase is appropriate. However if the price is compared with past prices, the buyer sees what appears to be a high price, which indicates that a very low purchase or no purchase at all should be made. The reverse situation occurs if today's price is low compared to the recent past.

One major constraint in this sitution is that if at any time the buyer only has sufficient in stock to cover consumption during the current buying opportunity, and no delivery ordered for the current opportunity, he *must* make a purchase whatever the price for immediate delivery. This is to meet consumption during the next buying opportunity. Otherwise there would be no material in stock at the start of the next buying opportunity and so production would have to be stopped. The buyer has thus failed to meet his primary duty of guaranteeing continuity of production. In most practical situations immediate daily delivery of new raw material, as in BMPs case, does not occur. The minimum physical stock level that a buyer should maintain will then be whatever is needed to cover consumption over the lead time for the delivery of new raw material from the supplier.

At some buying opportunity the buyer may purchase sufficient material to raise his stock level plus the total of the undelivered orders to the maximum purchase and stock cover level allowed by the company strategic policy. During the current buying opportunity some of the stock will be consumed by production. Hence at the start of the next buying opportunity the stock plus undelivered orders will be less than the maximum purchase and stock cover. Thus if a further very low price is offered a purchase could be made at that opportunity. It follows that in this Type 1 situation the buyer must keep a continual watch on prices since he might wish to make a purchase at any time.

Type 2 situation

In this second situation the buyer need not necessarily buy anything at all at any particular price offer, except at the very last one. There is only this one constraint on his freedom of action. At the very last offer prior to the deadline, the buyer must purchase an amount equal to the difference between the known requirement and his cumulative purchases to date. In the linseed oil and the first maize situation the buyer could only make one single purchase over the whole potential buying period. This was because suppliers did not allow several small purchases to be made and the minimum purchase quantity that the suppliers would accept was usually roughly equal to the known amount that had to be bought by the deadline. The problem here is that of picking out the lowest price offer over the whole period up to the deadline. As he moves forward through time the buyer has to determine whether the current price offer is lower than all of the unknown prices that will be offered over the remaining buying opportunities. As he moves forward through time the number of remaining opportunities in which to make a purchase continually decreases, since the deadline is a fixed point in time. Again for the same reasons as for the Type 1 situation the buyer must look forward into the unknown future for his price comparison not to the past. He can never go back to change an earlier decision. The past is past always. The past prices may assist him to make forecasts of the prices likely to be offered over the remaining opportunities to the deadline. The comparison must be with the forecasts of the future prices not with the past prices themselves. If he decides that he does not expect to receive a lower price in the future than the current price offer, then he should buy the whole amount that he requires. He then no longer has any real interest in that particular price series. He does not need to continually watch those prices. He should however continue to record the remaining prices right up to the deadline, as he will need them for the purpose of evaluating his purchasing performance.

In the second maize example of Chapter 3 the constraint on the minimum purchase size is less severe. There the buyer typically bought his known requirement in two or three subpurchases at different times over the buying period up to the deadline. In some purchasing situations it is theoretically possible to purchase the amount required in many small purchases or at least the equivalent of this by using the 'futures' markets. For example the U.K. soyabean market deals in lots of 100 tons on the SOMFA futures market. Thus 2000 tons of soyabean meal could be bought in 20 lots over a 2 or 3 month buying period. At the end, when the total amount required has been reached the soyabean meal 'futures' could be sold and a purchase of 2000 tons of meal made in the actuals or cash market.

However, as will be shown later, even if there are no constraints on the minimum purchase quantity the best strategy to minimise the expected cost of purchasing the known requirement is always the single purchase strategy. In practice in these Type 2 situations most buyers will purchase the amount required in two or three roughly equal subpurchases, to some extent equally spread out over the whole of the active buying period.

Such a policy is used by some animal feed manufacturers in a formal way known as the 'wedge' buying policy. If the active buying period is specified as three months then the 'wedge policy' states that one sixth of the required amount must be bought by the end of the first month, a further one quarter during the second month and the remaining amount during the final month of the buying period. Obviously differing percentages of the requirements to be bought in each month to those given above can be used. They all must have the property that the amounts purchased increase month-by-month over the active buying period and that at least 50% is bought over the final month. This approach recognises the validity of the motive underlying the single purchase strategy but says, that *in practice*, averaging over two or three prices will give lower purchasing costs in the long run and is also inherently less risky. The major element is to avoid really bad purchases and achieve less variability in purchasing performance from one month or quarter to the next. Thus some freedom is allowed to the buyers to make good purchases but the losses resulting from an incorrect reading of the market and future prices is minimised. The maxim used is 'don't put all of your eggs in one basket'. Companies and their buyers are often prepared to pay higher prices on average in the long run, than might be the case with another policy, to ensure that very bad purchases are avoided. This attitude really reflects buyers' difficulty in forecasting future prices and in making good purchasing decisions in the fluctuating uncertain priced commodity situations. It is a recognition, implicitly or explicitly, that they find great difficulty in systematically picking out the low prices. It is a reflection also of the target of an 'average' performance for purchasing similar to that achieved by their competitors.

Consider a Type 1 situation where there is one buying opportunity and price offer each day, and the buyer has 20 days' supply of material in stock. All consumption needs must be met as they occur. The buyer can make purchases of material to meet the consumption requirements of the twenty-first day hence either today, tomorrow, the day after, or on any day, or combination of days, up to the twentieth day hence. From an alternative viewpoint he has a known requirement to meet for the deadline of usage on the twenty-first day. He has 21 days with a price offer each day in which to buy that known requirement for the twenty-first day hence. This is precisely the Type 2 problem. Thus the Type 1 problem is a set of many concurrent Type 2 problems.

Consider a Type 2 problem where 100 tons must be purchased over a buying period of 20 days to meet a deadline on the twenty-first day. We can view this as a Type 1 problem with a consumption requirement of zero in every day except the twenty-first. If the current price is such that the stock should be raised to cover 10 days future requirements the purchase quantity will be zero. The cumulative demand over the next 10 days is zero, since it is zero in each period. If it is 17 days the purchase quantity is still zero. On the other hand, if it is 25 days, the purchase quantity will be 100 tons, since the cumulative demand over the next 25 days includes the 100 tons for the twenty-first day hence. Thus if the price offer is such that that stock cover should be raised to 21 days or more the 100 tons will be purchased otherwise it will be left until a later buying opportunity. The Type 2 problem is thus a very special near-degenerate case of the Type 1 problem.

Thus we see that the two types of problem are not really distinct. The solution to either one can be derived from the solution of the other. There is really only one tactical commodity purchasing problem. It is to derive some quantitative measure in terms of how many future offers will occur before an equal or lower price to the current price offer is quoted. Various alternative procedures have been put forward to operationally solve this problem. Some approaches directly consider the problem of measuring whether today's price is high or low compared to the prices that can be expected over the remainder of the active buying period, whilst others approach the problem in an indirect manner. These alternatives are discussed in detail in Chapter 11, together with their benefits and limitations to give the 'best buy' in particular circumstances.

4.7 ASSESSING BUYERS' PERFORMANCE

The structure of the commodity purchasing problem has now been finally determined as specifying the order deliveries required to meet the consumption requirements and also determining the active buying periods for each specified delivery order. A tactical buying problem exists within each active buying period. A great advantage in showing that this is the conceptual structure for making purchasing decisions, is that we are now in a position to set up formal measures of a buyer's purchasing performance. Such measures can enable standards or targets to be set for the purchasing performance for all the raw materials purchased. By comparing the actual purchasing performance with particular levels that the company might wish to achieve we can assess the likely benefit directly of trying to improve the purchasing policy. If there are several raw materials being bought, then this procedure will show on which raw materials it will be worth spending money, time and effort in trying to improve the purchasing policy. Up to now no acceptable and realistic ways of measuring purchasing performance have been proposed. Having seen precisely how the purchasing decision is structured we can now attempt to do this. Procedures for measuring purchasing performance are discussed and illustrated in the next chapter before going on, in the remainder of the book, to discuss the derivation of practical methods for specific situations to solve the various subproblems identified in each component of the purchasing decision in this chapter.

APPENDIX — Case Study

Adapting the Active Buying Period for Cocoa 1969–72

Provided an appropriate price forecasting model has been developed, the active buying period can be changed over time as the actual price forecasts given by the model change. An econometric/behavioural model is required that shows how the market determines prices over the medium-term future. The general principles behind and methods for deriving such models are discussed later in detail in Chapter 8. Such a model was developed by the author for explaining cocoa price movements on the London Market over the sixties, during late 1969 and early 1970. This section describes how this model could have been used for determining the appropriate values for the active buying period for a cocoa buyer over the three crop years 1969/70 to 1971/72. The actual forecasting model is only described briefly since the main emphasis here is on its usage for planning cocoa purchases.

As would be expected the model relates cocoa prices to supply and demand factors; more particularly to the relative levels of demand and supply. An important aspect of this model is that market estimates as published at particular points in time are used to give the values of the supply/demand factors. Furthermore the price at some *future* time is related to the state of the market as measured at an *earlier time*.

The relative abundance or scarcity of cocoa in the world can be measured simply as the ratio of the demand for cocoa to the supply of cocoa available. Demand is measured as the forecast of the yearly amount of the 'grindings' of cocoa and supply as the forecast annual production plus the stocks from the previous year. When this ratio is high there is a shortage and hence a sellers' market with pressure to generate high prices. When the ratio is low with abundant supplies, there is a buyers' market leading to low prices. The value of the ratio was calculated for the estimates circulating in the market for the various factors at the beginning of a quarter of the year. This is plotted against the price of cocoa on the market at the end of that quarter in Fig. 4.8. Each point represents one of the quarters over the crop years 1963/64 to 1968/69. Fig. 4.8 shows a model to explain cocoa prices over the sixties. The best fit relationship, the continuous line on the figure, shows that a small change in production or consumption in times of scarcity has more effect on price than the same change in times of abundance. The outliers of the relationship appear at high prices and situations of cocoa scarcity. This suggests some other factor comes into play at such times. In fact, the outliers of Fig. 4.8 could be explained reasonably well by a further relationship concerned with the rate of sales marketing of the crop.

The relationships as illustrated in Fig. 4.8 with unchanged parameter values, have been used to forecast prices over the 1969/70, 1970/71 and 1971/72 crop years. Forecasts are made at approximately quarter intervals, when the market generally receives new information. Each quarter, prices are forecast for the beginning of all succeeding quarters up to the end of the crop year. These forecasts together with the actual prices are shown in Fig. 4.9. The manner in which the active buying period can be adapted to changing circumstances, using the output of an econometric/behavioural price forecasting model, can be illustrated using these cocoa price forecasts.

The overall strategic company policies, reflecting the attitude to risk taking and investment in raw materials, might specify that forward cover should never exceed 9 months nor fall below 3 months. Thus the buyers are allowed a flexibility of 6 months, in which to buy requirements for any

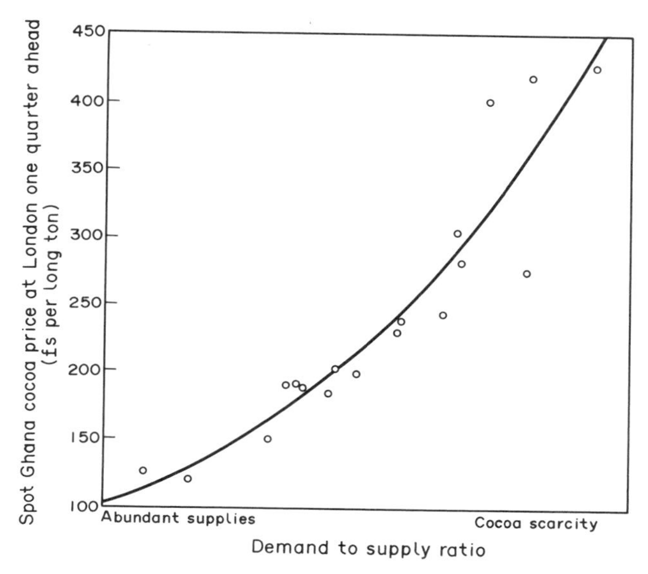

FIG. 4.8 A demand to supply price determination and forecasting model for cocoa

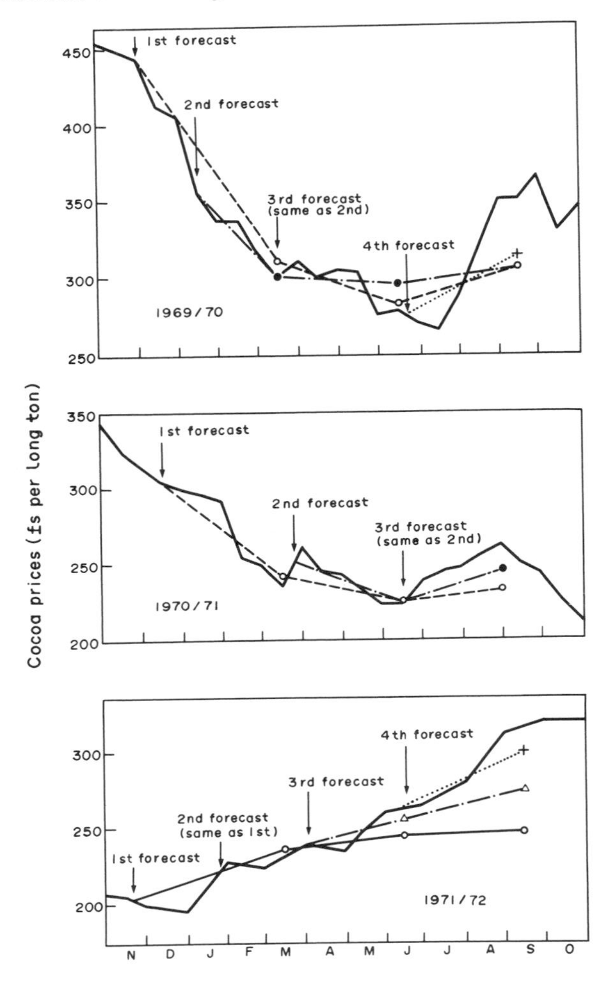

FIG. 4.9 Cocoa price forecasts based on Fig. 4.8 for the crop years 1969/70 to 1971/72. (The arrows mark the time the forecasts are made)

particular month. The market structure is such that supplies become available in monthly deliveries, rather than weekly or daily. The 3 months minimum cover, including both physical stocks and contracts placed but not yet delivered, is to ensure that supplies are guaranteed for production. It will include the lead time between placing an order and receiving physical deliveries from the marketing companies of the cocoa producing countries. In addition, it will ensure that the company does not run short and so be forced to pay high premiums for immediately available physical supplies.

The first forecast for the 1969/70 crop was made in late November when market views first crystallised on crop and consumption prospects. The prices were then at a level of £450 per ton. The first forecast indicated that prices would fall to £310 by early March, fall even further to £280 in June and then recover to £305 in September. Thus with falling prices, the maximum purchase and stock cover should be kept as low as possible. Requirements for intermediate periods not already covered should be bought as late as possible to restrict the minimum cover to 3 months. Each month's consumption will thus have an active buying period of one calendar month. No purchases to build up large stocks in advance of usage should be considered until the beginning of May. Revised market estimates of the state of supply and demand published in January were changed little. The second forecast confirmed the first, so that cover should be maintained at its minimum level and purchases delayed as long as possible. New market estimates in March were virtually unchanged; the third forecast made then was merely a repetition of the second.

At the beginning of May the active buying period has arrived. Purchasing should be delayed still, until prices fall below the £300 per ton level predicted, or new information modifies this forecast. The purchase and stock cover should be increased to its maximum value of 9 months over the period May to July, since prices are predicted to reach a low in June and then increase through to September. Over this period a tactical buying policy should be used to buy the consumption requirements for August (by the end of May), September (by the end of June) and October to March inclusive (by the end of July).

In June, with prices at £275 per ton, revised market estimates resulted in the fourth forecast predicting that prices would increase to £312 in September. This strongly confirms that June or July will contain the lowest price of the year. It reinforces the decision that the tactical buying policy should be actively used to ensure the requirements for the months mentioned are bought by the end of July.

The speculative surge through August and September, to £350 per ton, well above the June forecast was not supported by market fears of a forthcoming shortage. No purchases should be made until the speculative activity has subsided. The apparent bad forecasting performance for September is misleading; it is in fact a very good forecast since the high values of £350 are due to unsuccessful attempts at speculation. Speculators attempt to push prices up over August and September. The relationship of Fig. 4.8 indicated that there was no justification for such large increases in prices. The attempt failed and prices fell to £325 by the end of September. Shortly afterwards a U.S. bank subsidiary in Switzerland went bankrupt, allegedly due to losing nearly $30m. from cocoa speculation at this time. This period is in contrast to the successful speculation in 1968 when the relationships showed there were indeed real economic reasons for the very large price increases.

In late October, with cocoa prices in the range £330–£340 per ton, early estimates of supply and demand prospects for the forthcoming crop year became available from FAO and other sources. These suggest an average price level over the crop year of £320. In these circumstances purchasing should be delayed until market views have crystallised in November and December, since prices are currently above the forecast level. Cover will gradually be reduced by factory consumption. Note at the beginning of October the purchase and stock cover will have fallen to 6 months, covering consumption up to the end of March.

In early December with prices at £305 per ton, the forecast based on the prevailing market views was that prices would fall to £240 in early March, reach a low of around £225 in June and then increase slightly to £232 by September. The cover should be allowed to fall to its minimum level by factory consumption. Thus the active buying period, when the purchase and stock cover should be significantly increased voluntarily to cover forward requirements, will again be May to July as in 1969/70. By the end of December the cover will have fallen to the minimum 3 months level, covering consumption over January, February and March. The cover should be maintained at this minimum level until the next forecast, due in March, so that consumption requirements for April are bought over January, for May over February and June over March.

During March prices rallied from £235 to £260. The predictions based on the revised supply and demand estimate circulating on the market in March were unchanged from the earlier forecasts made in December. Thus this price rally is forecast to be only temporary. Prices will soon fall, so the cover should still be kept at its minimum level and purchasing, except for essential on-going requirements, delayed until mid May.

At the end of May, day-to-day buying should be started to extend the purchase and stock cover over June to its maximum of 9 months. The third forecast based on the prevailing views in June indicated that prices would increase by £20 up to September, thus confirming the earlier forecast. By the end of July the cover should have been raised to cover consumption through to the end of April 1972.

In late September of 1971 with prices at the levels of £240 as predicted earlier in the year, the early indications are that the 1971/72 crop will only be slightly better than that of the previous year. Assuming consumption grows at the recent average rate of 4% leads to an initial forecast, using Fig. 4.8, that prices over the year will be around £225 per ton. Since prices are currently above this level no further purchases should be made until the first reliable crop forecasts in November/December. In late October a leading London cocoa merchant forecast an even bigger cocoa surplus, which leads to a forecast price level of £210 per ton. So again no forward purchases should be made. As can be seen in Fig. 4.9, following such comments and views in the market, prices gradually drifted lower.

In late November with prices at £205 per ton, the first reliable market estimates are available. These lead to a first forecast for the 1971/72 crop year that prices will increase to £234 by early March, continue rising to £245 by early June and then level off for the remainder of the crop year. Since prices are forecast to rise well above the current levels, the purchasing stock cover should be raised to the maximum value of 9 months. At the beginning of December the cover will be 5 months, the 9 months achieved at the end of July less the actual consumption since then. A tactical day-to-day buying policy should be used to purchase the requirements for May–August over the current and following month, December and January.

In January 1972 when prices have risen to £225 the revised market views confirm the initial forecasts. The cover should be maintained at 9 months by buying September's consumption requirements over January and October's over February, etc.

Towards the end of April the general market assessment is of a reduction in the crop size compared with earlier views. The third revised forecast gives even higher prices for June and September, £255 and £266 respectively. In June the crop size estimate is further reduced. The fourth forecast is that prices will increase from the current value of £260 to reach £284 by September. By this time the U.K. government had decided to allow the pound sterling to float in value relative to other currencies. The pound had fallen in value by about 5%, so this should be added on to the forecast September price to give a forecast of £300 per ton, a large increase on the current levels.

Throughout the whole of 1972 the forecasts had been of continuously increasing prices over the year. The day-to-day tactical buying system will thus be in continual use, to ensure that the forward purchase and stock cover is maintained at the maximum 9 months allowed by company policy. From January onwards a purchase must be made each month to cover the consumption in the period 9 months hence.

In this example it has been assumed that the old crop/new crop transition can be ignored. In the case of cocoa this is essentially November when the first reliable crop estimates become available. In practice a company might decide on a compromise approach of only allowing a maximum stock cover of 6 months, midway between the minimum and maximum values, at the beginning of November. In this way it gets part of the benefit if prices rise for the new crop after November and mitigates its losses if the new crop prices fall, compared to allowing the cover to reach 9 months. In this example a 9 months maximum stock in November was always beneficial but it would not have been in November 1969, just prior to the start of the exercise.

5

Measuring Tactical Purchasing Performance

5.1 POSSIBLE MEASURES FOR PURCHASING EFFICIENCY

The lack of formal procedures by which a buyer of raw material commodities can evaluate his purchasing performance has been an impediment to research on improved commodity buying methods. A major contributor to the difficulties in measuring performance has been the sparsity of analysis to identify the various aspects of the purchasing decision. Variations in purchasing performance over time can be due to changes or inadequacies in any of the five stages of the purchasing decision as formulated in Chapter 4.

The discussion of the previous two chapters has shown the wide range of skills that the buyer needs to make good purchasing decisions. However, traditionally buyers have usually regarded the tactical buying problem, the final stage of the purchasing decision as discussed in Chapter 4, as their commodity purchasing problem. The buyer always has a choice of several days and different prices at which he can make purchases to meet a particular consumption requirement. Hence the major skill that buyers have exercised is the ability to choose the price or day, at which it is best to make purchases. Thus, at least initially, it is surely this skill in choosing the right price that should be measured.

Often the impact of the proper scheduling of the arrival of orders on total purchasing costs has not been appreciated. Company policy often treats this as the responsibility of production rather than purchasing. So it has been planned on the basis of production's needs without adequate consideration of the implications for buying. Alternatively because of the complexity of the problem, buyers have operated with relatively simple policies of constant maximum and minimum levels on the stock of materials plus forward purchases to be held at any time. The buyers must also operate within the company's overall strategic policy, which generally will impose limitations on how far ahead of its usage a purchase can be made.

As we have seen earlier the general answer to stages one to four of the purchasing decision is to prescribe a specific buying period for each and every purchase order, covering many alternative days and buying opportunities. We must only compare the price the buyer actually paid, or the average price if he makes several small subpurchases, with the prices offered at the alternative buying opportunities within the prescribed buying period for that purchase order. If the buying period is increased or decreased by company policies outside the buyer's control then this must be taken into account when setting up measures for the purchasing performance. For example, if there is a sudden shortage, because of poor production planning, which will require an emergency purchase order to be placed within 2 or 3 days, we must compare the price paid with the alternatives over those 2 or 3 days only, not those over the usual buying period of several weeks. Such an imposed order may significantly increase the average purchasing costs over a year and bring unfair comments on the buyer's performance. On the other hand if the answers to stages one to four of the purchasing decision give a 3 month buying period, but the buyer chooses to buy only in the last month, his purchase price must be compared with the prices over the whole 3 months and not just with the one month when he was actively buying.

The prices offered to him may be the market price or a price from a broker based on a 'world market' centred in a different country. We should ensure that only 'real' prices are considered that allow him to buy freely the quantity he needs. The buyer chooses a day or a price offer at which to make a purchase or in an informal market regards the price offer as a basis for negotiating discounts. The choice of day and the actual discount he achieves are two interrelated aspects of choosing a good price at which to make a purchase. We must therefore use the price offers made by the broker as the basis for setting standards for measuring purchasing performance, not the actual price paid after the discount. Buyers regard the discount achieved as part of their skill and an alternative to waiting a day or two for a possible lower price. The average 'discount' might also be a further measure of performance. However this should be analysed over time to see how it depends on market circumstances or whether a 'standard discount' is generally available from the broker.

In some companies, particularly in the nonferrous metals industry where buying can occur each day, a 'back-to-back' policy has been adopted, which requires the buyer to purchase each day exactly the same quantity of material that has been ordered by customers or used in production on that day, irrespective of the price level. This means that over the long run, the average purchasing cost will be the average market price. In many other situations, particularly for the agricultural commodities because of the way the supply market is organised, such a back-to-back policy is not possible. However the 'back-to-back' policy or the average market price over the buying period is surely a standard for measuring performance. If, for example, a day within the buying period were chosen at random, then over the long run the average purchase cost would be the average market price, the same as would be achieved in a back-to-back policy. Any purposeful purchasing policy should surely perform better over the long run than merely randomly choosing a day on which to make a purchase. Thus one possible measure, for each purchase order is to calculate the 'savings', defined as:

$$\text{Savings} = \text{Average market price} - \text{Actual price paid}$$

In general whilst the actual price paid will be the market price less whatever discount the buyer has negotiated, the average market price should exclude the discount. The average market price must be taken over the appropriate buying period for that order. The average value of the 'savings' each year, over all purchase orders in that year, shows the value of a purchasing policy.

However this measure alone is an inadequate measure of the buyer's performance, since greater 'savings' can be made when there are large price fluctuations. Small 'savings' may be a consequence of small fluctuations in price over the 'buying period' rather than the inadequacies of the policies used. It is necessary to take into account the size of the maximum potential savings. These can be estimated by calculating the market 'hindsight' price, which is the lowest price the buyer could have paid if he had predicted precisely how the prices offered by the broker or the market would move. Thus

$$\text{Maximum potential savings} = \text{Average market price} - \text{Market hindsight price.}$$

Whilst a good purchasing system should give a lower unit purchase cost over time than the average market price (equivalent to a back-to-back policy), it cannot be expected to be as low as the hindsight unit price, because in practice the future cannot be predicted with complete certainty. The efficiency of a purchasing policy can be measured by calculating the position between the average market price and the hindsight price, so that in percentage terms

$$\text{Purchasing efficiency} = \frac{100 \times \text{Actual savings}}{\text{Maximum potential savings}}$$

or

$$100 \times \frac{(\text{average market price} - \text{actual price paid})}{(\text{average market price} - \text{market hindsight price})}$$

An average purchasing efficiency of zero implies that over the long term, the purchase costs will be the same as the market average price. A 100%

efficiency represents the goal of buying at the lowest possible price. If the efficiency is negative, then the price paid is higher than the market average. Although this will happen on occasions, it ought not to occur consistently over long periods of time. Note that because the size of the price fluctuations above and below the average level over the buying period may be different, a nonsymmetric price distribution, purchasing efficiencies of less than −100% may occur on occasions.

Examples of the values of the purchasing efficiency achieved in practice for several branch buyers in one large animal feeds manufacturer for different raw materials are given in Table 5.1. Remember that negative values show where the buyer has failed to meet or beat the average price.

An interesting point about the values in the table is the lack of consistency of performance of the various buyers over the whole range of raw materials shown, except in Area 5. The buyer in Area 5 consistently

TABLE 5.1 Measures of buyers' performance

Raw materials	Area 1	Area 2	Area 3	Area 4	Area 5
Maize	9	10	2	13	13
Milo	39	10	−4	28	19
Barley	−45	41	27	7	20
Wheat	46	23	31	22	25
Sunflower	17	−5	−14	19	13
Soya	—	−14	17	49	28

beats the average but only by very small amounts whilst the buyer in Area 1 gives two good performances allied with one very bad one.

The requirements for each delivery position, hence the amounts bought, vary over time depending on total sales and seasonal variations in sales. With changes in consumer preferences over time the variations in demand for raw materials can be very large. Thus management is faced with the problem of deciding whether it prefers a buyer who consistently achieves small savings for all raw materials for all delivery positions or a more erratic buyer who frequently gives good results but less often gives very bad results. The risk is that his bad result may occur for a very large purchase order.

A table such as 5.1 may also assist a company in deciding how to organise its purchasing department. In the company of this example purchasing was decentralised to each separate factory. However, there is some evidence that organisation by commodity and some specialisation by the buyers could be advantageous. For example the buyer in Area 1 was good at Milo and Wheat but poor at Barley, whilst the buyer of Area 2 was good at Barley.

In comparing a buyer's performance from one year to the next we must ensure we compare like with like. The efficiency of a purchase is measured relative to the alternative opportunities at which purchases would have been made. Thus the buying periods for each order should have been the same for the 2 years in question otherwise we cannot conclude that the buyer has improved or worsened his performance etc. The changes may be due to changes to the other stages of the purchasing decision. In the same purchasing situation two buyers in different companies, with different lengths for their buying periods imposed by company strategies, would have different purchasing efficiencies even if they made purchases at the same prices. Since the buyer may not have much freedom to choose the length of the buying period, any comparisons and conclusions on his relative purchasing performance must reflect this fact. If target or standard levels for purchasing efficiencies that are realistic can be derived, then the purchasing efficiency measure can be used to estimate the potential value of increasing the length of the buying period. Alternatively, assuming that the same purchasing efficiency as achieved in the past by a buyer can be maintained for an increased length for the buying period, can allow the potential benefits of giving more freedom of action to buyer to be estimated.

5.2 AN ILLUSTRATIVE EXAMPLE OF THE PURCHASING EFFICIENCY CALCULATIONS

Table 5.2 gives the highest, lowest and average monthly prices quoted for cash sugar on the London Sugar Terminal over September 1972 to October 1973, in £s per ton. Consider a situation where the buyer needs to purchase sugar to meet a requirement of 1000 tons of consumption each month from December 1972 to November 1973. Each month's consumption could have been purchased on the cash market at any time over the preceding 3 months. Thus the 1000 tons needed for use in December 1972 could have been bought at any time over the period September, October or November. It must have been bought by the end of November. Let the actual average prices paid for each month's sugar consumption requirements for December 1972 to November 1973 be the values given in the second column of Table 5.3.

Since the sugar for usage in December can be bought at any time over the preceding 3 months, its buying period is September 1st to November 30th. The average price over its buying period is thus the average price over the months of September, October and November. From Table 5.2 this will be

$$\tfrac{1}{3}(72.02 + 76.16 + 75.66) = £74.6$$

TABLE 5.2 Monthly prices of sugar

Month		Highest	Lowest	Average
Sept. 1972		76.0	67.0	72.02
Oct.	"	78.5	72.5	76.16
Nov.	"	77.0	73.0	75.66
Dec.	"	100.0	77.5	93.05
Jan. 1973		105.0	93.0	98.25
Feb.	"	100.0	87.0	92.20
Mar.	"	93.0	87.0	93.41
Apr.	"	95.0	89.0	91.89
May	"	105.0	91.0	95.86
June	"	102.0	91.0	96.83
July	"	102.0	96.0	98.11
Aug.	"	99.0	89.0	94.27
Sept.	"	101.0	90.0	95.15
Oct.	"	106.0	96.5	101.63

The lowest price at which a purchase could have been made over the buying period will be the minimum of the lowest prices for September, October and November. From Table 5.2 these are respectively £67.0, £72.5 and £73.0. The minimum price that could have been paid for

TABLE 5.3 Measuring sugar purchasing performances

Month of usage		Price paid	Average price over buying period	Lowest price over buying period	'Savings'	Maximum savings	Purchasing efficiency
Dec. 1972		74.7	74.6	67.0	−0.1	7.6	−2.0
Jan. 1973		84.6	81.6	72.5	−3.0	9.1	−33.0
Feb.	"	93.7	89.0	73.0	−4.7	16.0	−29.4
Mar.	"	87.7	94.5	77.5	6.8	17.0	40.0
Apr.	"	89.0	94.6	87.0	5.6	7.6	73.7
May	"	92.1	92.5	87.0	0.4	5.5	7.3
June	"	95.1	93.7	87.0	−1.4	6.7	−20.9
July	"	93.7	94.9	89.0	1.2	5.9	20.3
Aug.	"	96.6	96.9	91.0	0.3	5.9	5.1
Sept.	"	93.1	96.4	89.0	3.3	7.4	44.6
Oct.	"	93.9	95.8	89.0	1.9	6.8	27.9
Nov.	"	97.0	97.0	89.0	0.0	8.0	0.0
Ave. values		90.93	91.79	83.16	0.858	8.62	11.1

December's sugar requirements is thus £67 per ton. The purchasing efficiency for the purchase of December's sugar requirements over its buying period is thus:

$$100 \times \frac{(74.6 - 74.7)}{(74.6 - 67.0)}$$

$$= 100 \times \frac{(-0.1)}{7.6}$$

$$= -2.0\%$$

The calculations to obtain the purchasing efficiencies for this and all other months are given in Table 5.3. The final column shows that the purchasing efficiency varies between a high of 73.7% and a low of −33.0%. The 'savings' achieved below a nominal back-to-back policy are about £10,000, approximately 1% of the total purchase bill over the year.

5.3 THE OVERALL AVERAGE PURCHASING EFFICIENCY

The average overall purchasing efficiency over some period is best taken as the simple arithmetic average of the efficiencies of the separate purchase orders, rather than weighting the average by the actual amounts bought for each order. The size of the orders and their buying period are set by company policies that do not depend usually on the magnitude of the daily price fluctuation about the underlying trends. The purchasing efficiency measures how well the buyer chooses the day and the price at which to make a purchase for each scheduled order. The size of the actual order was determined much earlier, prior to the start of the buying period. One would hope that the buyer will be as skilful with the large as with the small orders. If by chance in one year a very good (or very bad) purchase occurs for one particularly large order, whilst all the other purchases are poor (or good), then a biased impression of a good (bad) overall performance is given if the individual efficiencies are weighted by the size of the individual orders in the averaging process. In the sugar example the average purchasing efficiency for the monthly orders over the year shown is 11.1%.

An alternative method to measure the overall performance over some period of several orders is to work out the overall average values for each component of the Purchasing Efficiency formula separately. These are shown in the bottom row of Table 5.3. The average price paid for the sugar consumed over the twelve months was £90.93 per ton. The average of the average prices over the individual buying periods was £91.79. The average of the hindsight lowest prices was £83.16. The 'average' purchasing efficiency for the 12 months shown would then be

$$100 \times \frac{(91.79 - 90.93)}{(91.79 - 83.16)}$$

$$= 100 \times \frac{0.86}{8.63}$$

$$= 10.0\%$$

Note that this gives a slightly different value to the arithmetic average of the individual monthly purchasing efficiencies. Such small differences between the two ways of deriving an annual purchasing efficiency figure are an inevitable result of different averaging procedures and can arise from the small number of purchase orders included in the sample. As the number of purchase orders increase then the differences between them fall to zero. The differences between these two measures are discussed further in an appendix to this chapter, where it is shown that this alternative method is slightly less reliable than the original one.

For a small number of purchase order deliveries per year, as in the linseed oil and maize buying situations, it is relatively simple to calculate the individual purchasing efficiencies for each order separately and then average the efficiencies. This has the advantage that the buyer can consider his past performance to see whether he was relatively consistent or whether any good or bad purchases can be ascribed to some single factor which he could take into account in his future buying decisions. This is not possible directly if the alternative averaging method is used. For a few purchase orders the extra effort involved in the first averaging method for calculating an average purchasing efficiency for the year is very small.

However in the continuous consumption situation, as for the BMP copper buyer, there is an effective purchase order for each day of the year. So this would mean calculating at least 225 individual efficiencies per year. Furthermore the buyer is more likely to compare his performance over the quarters of the year rather than for each of the individual days. In this situation it might seem better to use the alternative averaging method with its saving in calculation time. With this method it is relatively easy to calculate the average price paid per ton of copper consumed during the year and the average hindsight minimum price that could have been paid for the copper consumed. Care must be taken however in calculating the yearly average of the average prices over the buying periods for each day's consumption. This is not the same as the average price over those days when consumption occurred, i.e. the average market price over the year. For a 20 day buying period for example, the buying period for the first day's consumption of the year will be the last 20 days of the previous year. For the second day's consumption it will be the last 19 days of the previous year plus the first day of this year etc. Using the average price of the year of consumption as a short cut procedure is wrong and will give a biased view. For the 20 day buying horizon, over a 12 month period the error is approximately 5% of the price change between the average price of the final 20 days of the previous year and the average price over the final 20 days of the consumption year. With large price changes such as can occur for copper (see Fig. 2.4) this can be a significant amount. For shorter time periods such as quarters of the year, the error increases to 20% of the price change over the 3 months. If prices are falling over a period then the average daily price over the consumption period will underestimate the average prices over the buying periods and indicate a worse performance than was the case. Conversely in an increasing price situation the average price over the consumption period will indicate a better performance than was the case. In the sugar example of Table 5.2 the average price over the consumption periods was over £94 compared to the average over the buying periods of £91.79. This would increase the purchasing efficiency to 28%, giving an impression of a much better performance than was the case at only 11.1%.

In view of the temptation to take a short cut and wrongly use the average daily price over the consumption period if the alternative averaging method is used, it is recommended that the purchasing efficiencies be calculated separately for each day's usage requirement in the continuous consumption situation of the copper buyer, as well as the other situations with only intermittent small numbers of purchase orders per year. The overall average purchasing performance over a time period should then be calculated as the average of the various individual orders' purchasing efficiencies.

The average purchasing efficiency of the 12 months illustrated in Table 5.3 is small and there are large variations from one month to the next. The question naturally arises whether the overall performance demonstrates the existence of a purposeful policy which generally buys well, or whether this same performance would be a chance result from some fortunate decisions of a policy effectively making random purchases. This can be tested by assuming that the individual monthly purchases are random and determining what range of values for the 12 monthly average performance could be expected.

Let E_i be the purchasing efficiency in month i and $\bar{E}$ the overall average efficiency. The calculated variance of the monthly purchasing efficiencies for N months, S^2, is given by

$$S^2 = \frac{1}{N-1} \sum_{i=1}^{N} (E_i - \bar{E})^2$$

From the data of Table 5.3, where N is 12, the value of S^2 is 1018. If the E_i are in fact random values then the standard error of the average value, $\bar{E}$ is given as $S/N^{\frac{1}{2}}$. Hence the standard error for our example is 9.2. The average value $\bar{E}$ divided by its standard error should follow approximately a Student-t statistical distribution. The value $\bar{E}$ is an estimate of the real underlying average performance that would be achieved over the long term. Statistical analysis shows that there is a 95% probability that this real average value lies within 2.2 standard errors of the calculated average. Hence the long-term average purchasing efficiency lies within the range −9.2 to 31.4. A policy of purchasing at random would over the long-term buy at the average market price and hence would have a long-term purchasing efficiency of zero. The statistical argument is that since the range of possible values for the long-term average value of the purchasing efficiency contains the zero value well within it, the results given in Table 5.3 could easily come from a random purchasing policy. There is no real evidence of a purposeful purchasing policy being applied to the sugar buying example.

The statistical test used here is subject to several conditions which are never exactly satisfied in practice. It assumes that the monthly purchasing efficiencies are independently distributed. This is unlikely to be true in practice as one would expect a buyer or a purchasing policy to perform well over certain sustained periods of price movements and badly over others. Furthermore in April 1973 for example, purchases could be made for the May, June and July sugar requirements. If a period of increasing prices is anticipated to occur for several months a good low priced purchase could be made for the requirements for the 3 months, May, June or July over April. If the forecast of prices is wrong then bad purchases are likely to be made for all 3 months. This factor is likely to bias the test somewhat to reject randomness in favour of a purposeful policy when this is incorrect.

The distribution of purchasing efficiencies in general is likely to be nonsymmetric, since the largest positive value for the efficiency is 100 whilst negative values below −100 can occur. Large negative efficiencies below −100 can arise if the price fluctuations above the mean value over the buying period are greater in magnitude than the fluctuations below the mean value. If this occurs quite often then, although the average price paid over the long-term will be the average market price, the average long-term purchasing efficiency from a policy of randomly choosing a price over the buying period could be slightly negative rather than zero. The test is thus biased to some extent against the existence of a purposeful policy.

An alternative way of expressing the test is that if the policy was really one of randomly choosing a price then there is a 1 in 8 chance of obtaining an average purchasing efficiency of 11.1% over a 12 month period. Conventionally, statistical tests adopt the criterion that the result should have only a 1 in 20 chance or less of occurring before the assumption of a random buying policy can be confidently rejected. It is a somewhat arbitrary criteria with little real rationale behind it. Whether it is too severe a criterion depends upon the circumstances and what changes might be made to company policies and operations. If it is a decision to criticise severely a buyer or replace him, then requiring a 1 in 20 chance or better that his performance is not the result of a random choice of prices is rather excessive. All that can be really said is that the policy used is not significantly better than a clerk placing random orders and that the policy can be improved. If it is a question of giving the buyer more freedom of action, by increasing the strategic or active buying period, to have a greater impact on the company's economic viability, then his policy ought to have a probability of merely being the results of random buying rather closer to the conventional statistics test level of 1 in 20 before such a change is made. Comparison with the 1 in 20 level should also be used if it is a question of deciding whether or not to invest money and effort into research to try to improve the purchasing performance. Although other factors as well must be considered in this latter decision.

The significance of the results achieved depends critically on the number of months evaluated. For example, if the same overall average purchasing efficiency and the same variance for the monthly efficiency were obtained as for the data of Table 5.3 over a period of 24 months, then the test shows that the 95% range of values would not contain the zero value. Hence it is strong evidence that a purposeful purchasing policy is at work buying generally at low prices.

If the variation of the monthly purchasing efficiencies shown in Table 5.3 is typical of what generally happens then from the above discussion a rough rule of thumb would be that if over one year the average efficiency is larger than 20% it demonstrates a good purposeful policy exists. If it lies between 10% and 20% there is probably a purposeful policy existing but which is not particularly good. If the average efficiency lies between 10% and −20% the policy effectively makes purchases at random. If the average efficiency is less than −20% there is a purposeful policy in existence but it is based on a completely mistaken premise or a disastrously wrong method of price forecasting. Over a 2-year period an average overall efficiency of over 10% indicates a purposeful policy at work.

5.4 POSSIBLE 'MISTAKES' IN CALCULATING EFFICIENCY MEASURES

The critical initial step in calculating the purchasing efficiency is to define the buying period for each purchase order. The next step is to calculate the average price offered and the minimum price from all the prices offered over the buying period. If these two steps are not performed correctly then the purchasing efficiency value obtained does not give a meaningful measure of the buyer's past performance. The buyer may give himself a false impression of how he is doing and hence wrongly, perhaps give less attention to that particular commodity in the future or make the wrong decision as to why he achieved such a performance. One 'such mistake' that did occur in a practical situation, for a real sugar buyer in fact, can be illustrated with the same situation as our original problem of Table 5.2 etc. This second buyer had exactly the same freedom of action as our first buyer, with a buying period of 3 months for each month's 1000 tons sugar consumption requirement. This buyer liked to record statistics on prices and performance on a monthly basis. Unlike many buyers he did try to examine and learn from his past successes and mistakes. The buyer noted that he had to buy the 1000 tons of sugar for usage during December by the end of November. He therefore recorded his information on the basis of the last month in which purchases could be made. He produced his data for the measurement of the purchasing performance for December 1972 to November 1973 consumption requirements in the form shown in Table 5.4, firstly filling in the first and second columns. He now made his fundamental mistake, in that, for the 1000 tons of sugar that must be purchased by the deadline of the end of November, he compared the price of £74.7 per ton that he paid with the average and lowest prices over November only, and similarly for the other months. [Remember that the actual purchase could have been made at any time over September, October or November.] His average price standard for the December usage requirements is thus the average price over November of £75.66 and the minimum price was the minimum over November of £73.00. He then calculated his 'savings' as £0.96 whilst his 'maximum savings' were £2.66 (75.66−73.0). Hence his purchasing efficiency was 0.96/2.66 = 26%. The various steps for each month in this incorrect calculation of his Purchasing Efficiency are shown in Table 5.4.

The average of the monthly purchasing efficiencies in Table 5.4 is 50.2%. This indicates a very good purchasing performance indeed. The standard error of the average purchasing efficiency from Table 5.4 works out to be 9.2, the same value as obtained for the proper calculation procedure. The underlying long term average efficiency will thus lie in the range 30.0 to 71.4. Since the lower end of the range is well above the value of zero, the purchasing efficiencies of Table 5.4 are not the results of a random buying policy. They apparently demonstrate the existence of a purposeful and very successful buying policy. It would be unreasonable to expect a buyer to do much better than the level of performance shown in Table 5.4.

The apparent performance achieved contrasts strongly with the real Purchasing Efficiency achieved of only 11.1% (see Table 5.3). There is a

TABLE 5.4 Incorrect measurement of purchasing performance

Month	Price paid	Average price	Lowest price	'Savings'	Maximum savings	Purchasing efficiency
Nov. 1972	74.7	75.66	73.0	0.96	2.66	36.1
Dec. "	84.6	93.05	77.5	8.45	15.55	54.3
Jan. 1973	93.7	98.25	93.0	4.55	5.25	86.7
Feb. "	87.7	92.20	87.0	4.5	5.20	86.5
Mar. "	89.0	93.41	87.0	4.41	6.41	68.8
Apr. "	92.1	91.89	89.0	−0.21	2.89	−7.3
May "	95.1	95.86	91.0	0.76	4.86	15.6
June "	93.7	96.83	91.0	3.13	5.83	53.7
July "	96.6	98.11	96.0	1.51	2.11	71.6
Aug. "	93.1	94.27	89.0	1.17	5.27	22.2
Sept. "	93.9	95.15	90.0	1.25	5.15	24.3
Oct. "	97.0	101.63	96.5	4.63	5.13	90.3
Ave. values	90.93	93.86	88.3	2.93	5.53	50.2

large difference between the two results. Calculating his performance incorrectly gives him a false sense that he chooses the particular days on which to make purchases very well, whereas in practice his performance is modest. This example shows the vital importance of measuring purchasing performance correctly by taking strict account of the buying period, the amount of time over which the buyer could have bought each purchase order.

The buyer here is confusing two aspects of the purchasing decision, the choice of the right length for the buying period for each order and then the choice of the right day within the buying period on which to make a purchase. He has set his standard of performance for comparison purposes as a monthly buying period, i.e. buying only over the month prior to usage. This monthly buying period gives him his average and minimum price standards for use in the efficiency formula. He has then given himself the possibility of buying sugar for each month's consumption up to 3 months ahead of the purchasing deadline. He went ahead and made purchases over various times within the 3 month buying periods for the various months' sugar requirements.

What he has shown is that it is clearly a good decision, for the period of Table 5.2, to go for a 3 month buying period rather than a one month buying period. The value of this decision could have been shown more directly. The average value of the average prices over one month buying periods for the sugar requirements was £93.86, whilst the average value for 3 months buying periods was £91.79, (see the third columns of Table 5.4

and 5.3). A 'random' purchasing policy would expect to buy at the average prices over the buying periods. Since the average for the 3 month buying period is so much lower than the one month buying period he would expect to achieve lower average purchasing costs with the longer buying period. His exercise has shown conclusively that he can take advantage of this potential for lower purchasing costs.

His 'incorrect' calculation of his Purchasing Efficiency has shown that his policy over a 3 month period is 'equivalent' to a 50.2% Purchasing Efficiency for a one month buying period. To endorse the decision to change to the longer buying period we must consider whether he could have achieved a more than 50% efficiency if he had kept to a one month buying period. Practical experience elsewhere would suggest that this was unlikely. It appears that 30 to 35% is the best efficiency that one can expect a good buyer to achieve using his skill and experience only, without formal quantitative methods. His decision has thus probably gained about £1 per ton on what he could have achieved with a one month period.

The decision to move to a 3 month buying period and give himself more freedom of action in which to make purchases in this instance was correct. The correct calculation of his Purchasing Efficiency gave an overall 11.1%. This means that he has not taken as much advantage as he might have done of the extra freedom of choice of prices from the longer buying period that he has given himself. He should thus give more attention to the choice of the prices within the 3 month buying period at which to make his purchase.

However these results have occurred for a long period of rapidly increasing prices. In this situation, being able to buy further ahead of usage than previously would inevitably lead to lower purchasing costs. In the opposite falling price situation this would not be so. Buying further ahead of usage in this case would lead to increased purchasing costs. Before he permanently changes to the longer 3 months buying period the buyer should wait to evaluate his performance over a period of falling prices to see what overall is best. If he can generally distinguish well in advance between a rising and a falling price situation, and hence accordingly buy earlier or later in the buying period he should change to the 3 month buying period. If he cannot do so then, with only such a moderate efficiency over the 3 month buying period, he might find it better overall to use the shorter previous value of one month.

5.5 A CASE STUDY — PURCHASING EFFICIENCY FOR MAIZE 1969–74

This section describes the application of the measures of purchasing performance just described to assess the maize buying policies for a company over a 6-year period. Following this assessment the derivation of standards or target buying performances is discussed and hence the potential savings from research into improved purchasing policies calculated.

As discussed in Chapter 3, the U.K. maize buying situation is such that deliveries are made on a monthly basis. Because of constraints on the minimum order size and the relative usage rates of the company, orders were placed for maize shipments to be made in particular months of the year. Because of varying demands for the company's products these need not necessarily be the same months from one year to the next year. The study covered 6 years from 1969-74. The date of purchase and the price paid had been recorded for every order placed over this period and was available for the study. The general policy of the company implied that each shipment month ordered had a buying period of around 12 weeks. The company had generally kept records on a graphical basis of the prices offered over the period when the buying department considered making purchases. In actual fact the buying periods could usually vary between 10 and 15 weeks. In 1972 company policy was changed for one year to reduce the buying period to around 8 weeks. Some of the records, as is often the case, were incomplete and only covered 4 or 5 weeks. The orders for these shipment months were excluded from the analysis since it was known that the purchase was made on a date outside the period for which price data was recorded. Furthermore all the price records for 1971 had been lost. The analysis of performance thus initially covered orders for 25 shipment months spread over 5 calendar years.

The general price situation in which the maize had to be purchased over the 6 years 1969-74 is illustrated in Fig. 5.1. This shows the market price quoted at the beginning of each month for maize to be shipped in the following month, on a delivered U.K. Port basis in £s per ton. It is seen that maize experienced an almost threefold rise in price from 1972 to early 1974 in common with many other commodities. Prices were relatively stable over the second half of 1969, increased slowly over 1970 to reach a peak in October 1970. Prices remained near to £30 per ton over the next 9 months but then fell rapidly over the next quarter to a low of between £23 to £24 per ton. Prices remained about this low level for the first half of 1972, then began a continuous increase to reach £32 per ton by December 1972, £52 per ton by June 1973, £63 per ton by December 1973 and finally a peak at almost £70 per ton in March 1974. As seen in Fig. 5.1 variations about the overall trends exist. In addition there are substantial fluctuations from day-to-day, about the beginning of the month prices shown in the figure.

A better picture of the problem, the potential opportunities for good purchases and the dangers of bad purchases is shown in Table 5.5. This illustrates for each purchase order actually made by the company, the price levels over their appropriate 'buying periods'. The second column of the

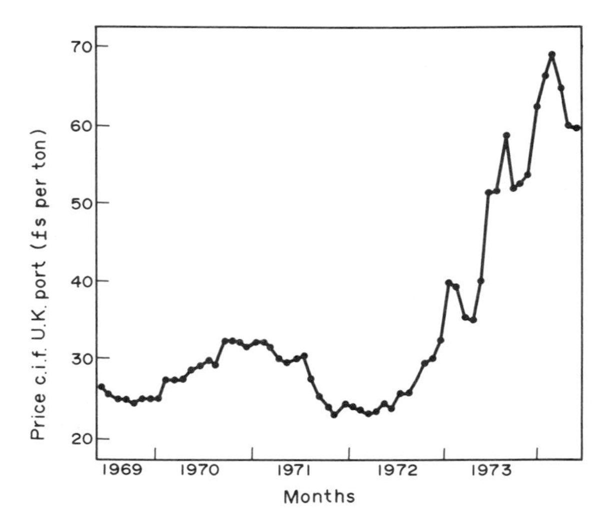

FIG. 5.1 U.K. prices for imported maize at beginning of month over 1969 to 1974

table shows the average price offer over the buying period, the third, the highest market price offered and the fourth the lowest price offered. As discussed earlier, the average price is a norm which should generally not be exceeded by the prices actually paid. The final column, the price range, is the difference between the highest and lowest prices over the buying period. It therefore indicates the price variability and the potential for making savings or losses against the average market price. If the price range is small, as in the example 1969, there is little possibility of making large savings in purchasing costs, whilst if the price range is high, as for 1973, there are numerous opportunities for making substantial savings in purchasing costs.

The company concerned in this case study had only stored the prices offered in the middle of each week, on the Wednesday, for all the various shipment months over the past, together with the actual prices paid and the day of the purchase. This meant that the average price and the minimum price over each buying period could only be calculated on the basis of a single weekly price. The purchases were usually made on days other than Wednesday. The actual prices paid for each shipment month are shown in the second column of Table 5.6. The third column shows the price recorded for the Wednesday of the week in which the purchase was made. The final two columns show the absolute and percentage 'discounts' on the price actually paid relative to its Wednesday price. A problem with such

TABLE 5.5 Price levels over each buying period*

Year	Purchase order shipment period	Average price	Highest price	Lowest price	Price range
1969	September	25.60	26.50	24.90	1.6
	October	25.00	25.50	24.10	1.4
	November	24.50	25.60	23.70	1.9
	December	24.90	25.60	24.20	1.4
1970	March	26.15	27.30	25.60	1.7
	April	26.35	27.20	25.80	1.4
	July	27.40	28.80	26.70	2.1
	August	28.20	30.10	26.70	3.4
	September	28.80	30.00	27.40	2.6
	October	29.65	33.20	27.90	5.3
	November	30.85	33.20	27.90	5.1
	December	32.20	33.60	28.70	4.9
1972	January	24.05	24.40	23.75	0.65
	July	24.45	25.00	23.75	1.25
	September	26.35	27.75	25.00	2.25
	November	28.50	30.75	27.50	3.25
1973	July	39.00	51.50	35.10	16.4
	August	41.85	51.80	35.70	16.1
	September	47.35	59.00	36.50	22.5
	October	55.65	63.25	48.50	14.75
	November	56.50	65.75	52.20	13.55
	December	54.75	57.00	53.00	4.0
1974	January	56.65	61.30	53.80	7.5
	February	59.55	64.80	55.00	9.8
	March	63.60	70.20	55.20	15.0

*The prices are for American No 3 Yellow Maize, except for July 1972, September 1972 and November 1972 which are for South African Yellow.

limited storage of past data, is that it is impossible to tell whether the 'discount' is due to the buyer's skill in negotiating a discount on the price originally offered by the broker supplier, or whether it is due to the normal fluctuations of the prices from one day to the next over a week. The latter effect exists, because the purchase for the November 1970 shipment is above the Wednesday price offer. All other prices are below their 'Wednesday' price, so either the buyer has exercised skill in choosing a reasonable day of the week in which to make a purchase or has been able to negotiate discounts from the supplier on a particular day.

Problems arise, because of this limited storage of data, in calculating the purchasing efficiency, most severely with regard to the minimum price offer. The first 2 shipment months in the tables illustrate the problem. For

TABLE 5.6 Prices paid for maize by company

Year	Shipment period	Price paid	Price offer that week	Absolute 'discount'	Percentage 'discount'
1969	September	24.75	24.90	0.15	0.60
	October	23.81	24.10	0.29	1.20
	November	24.00	24.25	0.25	1.03
	December	25.125	25.3	0.175	0.69
			Average	0.216	0.88
1970	March	26.43	26.8	0.37	1.38
	April	26.625	26.9	0.275	1.02
	July	27.625	27.7	0.075	0.27
	August	28.00	28.13	0.13	0.47
	September	28.875	29.1	0.225	0.77
	October	28.70	28.8	0.1	0.35
	November	33.60	33.2	−0.4	−1.20
	December	32.125	32.65	0.525	1.60
			Average	0.27	0.90
1972	January	23.90	24.20	0.3	1.24
	July	24.25	24.75	0.5	2.02
	September	27.50	27.75	0.25	0.90
	November	29.00	29.00	0	0
			Average	0.26	1.04
1973	July	49.5	50.6	1.1	2.18
	August	49.0	49.5	0.5	1.01
	September	52.75	53.7	0.95	1.77
	October	59.5	59.5	0	0
	November	52.25	53.5	0.75	1.40
	December	52.25	54.0	1.75	3.24
			Average	0.82	1.56
1974	January	62.35	64.8	1.55	2.39
	February	63.90	64.8	0.7	1.08
	March	64.5	65.2	0.7	1.07
			Average	0.98	1.51

both September and October 1969 shipments the price actually paid by the buyer was lower than any of the 'Wednesday' prices quoted over their buying periods. Use of the Wednesday price minimum and the actual prices would give a purchasing efficiency of more than 100%, which is contrary to its definition. On the other hand, it cannot be assumed that a similar sized reduction as occurred for the September and October 1969 shipments, should be applied to the Wednesday prices for all other months to give their minimum hindsight price. There is no published daily U.K. maize price series which is generally accepted as believable and accurate by consumers. Thus no assistance with the problem could be obtained from other published sources. The most sensible approach in the circumstances was agreed to be to use the actual price paid as the minimum price in the purchasing efficiency formula, whenever it was lower than the lowest 'Wednesday' price over the buying period for that shipment period, otherwise continue to use the lowest Wednesday price. It was recognised that this might give a slightly higher estimate of the purchasing performance, because the actual minimum price over each buying period would probably have been slightly lower than the minimum 'Wednesday' price. However, in view of the small variations in price from one week to the next that generally existed, this bias should only be small.

The savings, maximum savings and purchasing efficiency for each shipment order are shown in Table 5.7. At the beginning of the period illustrated some very good purchases were made. The first two, for example, were at the lowest possible prices recorded over their buying periods. The average purchasing efficiency for the 1969 purchases was 57%. This is a very good performance indeed. Thereafter the purchasing performance deteriorated. This is perhaps seen more clearly in Table 5.8 which shows the average values of the savings and efficiencies each year. The 1970 and 1972 efficiencies were remarkably similar at −20% and −16% respectively. These indicate small losses relative to a back-to-back policy. The situation worsened in 1973 and then again in 1974. What is somewhat curious are the very good purchases, at the minimum possible price, for the November and December shipments in 1973. These are preceded by a set of very bad purchases and also followed by some very bad purchases.

The purchasing efficiency was positive for only 10 out of the 25 orders purchased. It suggests that a random choice of the day of purchase within the buying period would lead to lower buying costs than were actually incurred. The average 'savings' achieved was £–1.27 per ton, showing the buyer paid £1.27 per ton more than the average market price, although most of this loss occurred in 1973 and 1974.

Eighty per cent of the purchase orders were bought in the second half of the Buying period, the exceptions being August and December 1970, July and November 1972, and March 1974. For those five purchases bought in the first half of their buying period, the average purchasing efficiency was −4%, much better than for the other purchases. On seven of the twenty-five occasions, purchases were left to the virtual end of the buying

TABLE 5.7 Measures of purchasing performance

Year	Purchase order shipment period	Savings	Maximum savings	Purchasing efficiency
1969	September	0.85	0.85	100
	October	1.19	1.19	100
	November	0.5	0.8	63
	December	−0.225	0.7	−32
1970	March	−0.28	0.55	−51
	April	−0.275	0.55	−50
	July	−0.225	0.7	−32
	August	0.2	1.5	13
	September	−0.075	1.4	−5
	October	0.95	1.75	54
	November	−2.75	2.95	−93
	December	0.075	3.5	2
1972	January	0.15	0.3	50
	July	0.2	0.7	29
	September	−1.25	1.35	−92
	November	−0.5	1.0	−50
1973	July	−10.5	3.9	−269
	August	−7.15	6.15	−116
	September	−5.4	8.85	−61
	October	−3.85	7.15	−54
	November	4.25	4.3	99
	December	2.5	2.5	100
1974	January	−5.7	2.8	−204
	February	−4.35	4.55	−96
	March	−0.9	7.4	−12

period. Generally purchases have been left until too late a date. This appears particularly true for the period of rapidly increasing prices of 1973 and 1974. It is possible that the strategy of a three-month buying period (12 weeks), as has been assumed to have existed over the past, was not followed for the whole period. The results quoted are thus measuring past

TABLE 5.8 Annual average purchasing performance

Year	Average savings (in £s per ton)	Average purchasing efficiency (% age)
1969	0.58	57
1970	−0.30	−20
1972	−0.35	−16
1973	−3.36	−50
1974	−2.74	−104

performance relative to a 3 month buying period strategy. Some of the very late purchases at the very end of the buying period, as for example July 1973 and January 1974, may have been caused by special circumstances, such as higher than expected usages of maize requiring special purchases with much shorter buying periods than normal.

The year 1973 marks a watershed in commodity prices generally and for maize in particular. The market circumstances for the 3 years 1969, 1970 and 1972 were radically different to those of 1973 and 1974. The 3 earlier years followed similar patterns of price movement to those that had been occurring since the mid fifties. The year 1973 and its successors were something new. The markets can perhaps be described as being in a stable situation for the earlier 3 years, but highly volatile and unstable after April 1973. The change in early 1973 was sudden and unexpected.

This change is seen clearly in Fig. 5.2 which shows the weekly 'Wednesday' prices over January to June 1973. Compared to the previous crop year*, 1972/73 was a year of a moderate shortage of U.S. maize due to increased domestic and overseas demands for maize. U.S. prices from September to April had very similar month-by-month movements to those of the 1970/71 crop year, an earlier year of a moderate maize shortage. As in 1971 prices over 1973 were generally expected to fall slowly from March through to August. Russia had had a serious grain crop failure with her 1972/73 crop, harvested in Autumn of 1972, which she had kept secret from the rest of the world. She now, in a most effective manner, bought millions of tonnes of wheat and maize from various U.S. grain merchants within one week. This 'grain raid', as it became known, was carried out so

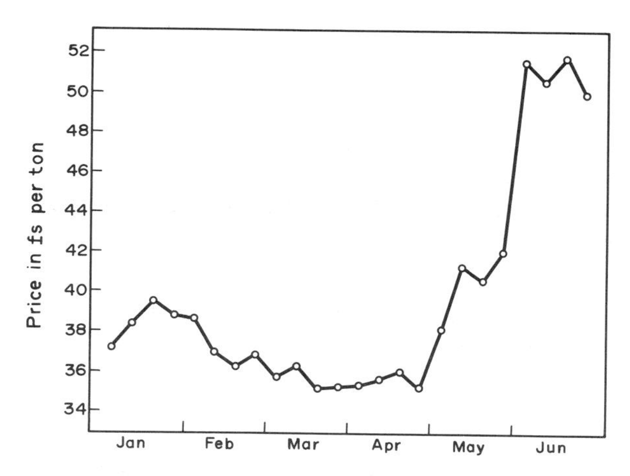

FIG. 5.2 Midweekly U.K. maize prices for July shipment over January to June 1973

*The maize marketing year is internationally accepted as the U.S. crop year and is on an Oct./Sept. basis.

skilfully under conditions of commercial secrecy, that no-one was aware of the need for the Russian purchases until they had been made. The purchases were made at low prices consistent with the moderate shortage prevailing prior to the Russian demands. The Russian purchases 'overnight' transformed the world maize situation to one of an extreme shortage. U.S. maize exports over 1972/73 were 12 million tonnes up on those of 1971/72, an increase of more than 50%. U.S. maize stocks were virtually depleted and in August 1973 there were real fears that the U.S. would need to impose export controls to ban some maize exports. U.S. farmers protested bitterly that they had been 'cheated and forced' to sell their grain at too low prices. Following such protests the U.S. government took steps to ensure Russia could no longer buy such extra large quantities of maize without giving public advance notice of their intention to do so. Over the next 3 years after 1972/73, crop shortages, high demand and the decline of the relative value of the U.S. dollar led to continued high prices and large price fluctuations.

Like most operators in the maize market the buyer of maize in our case study was taken by surprise by the events of early 1973 and its effects on prices. He was slow to react and understand the new situation. It was a new market situation, completely different to anything that had been experienced during the previous 20 years. Panic and fear of not obtaining maize requirements became the major cause of price movements. The July, August, September and October shipments were bought at late stages in their buying periods, presumably because our buyer thought that the initial panic would subside and that prices would fall to more normal levels. The very bad purchasing efficiencies for these four purchase orders (see Table 5.6) are thus very understandable.

It is thus perhaps more reasonable to assess the buyer's performance over the first 3 years initially and then from the November 1973 shipment onwards, when the buyer had had some months of experience of the new situation. This is certainly true if we are to be able to assess what purchasing standards should be set and what possible savings might result, since no forecasting system could have predicted the Russian 'grain raid' in advance.

The average purchasing efficiency achieved over the shipment months of 1969, 1970, and 1972 was in fact 0%, the same as a back-to-back policy would achieve if such a policy were feasible. Thus overall during these years the company neither gained nor lost by the purchasing policy followed by the buyer. The initial year 1969, as mentioned earlier showed a good performance. The years 1970 and 1972 were generally years of continuous small price increases whilst by contrast over 1969 prices generally fell. Generally, buying at too late a time in the buying period over these years would have led to the results that actually occurred, a high purchasing efficiency in the year of falling prices and low efficiencies in years of rising prices. It is usually accepted that most buyers find it easier to buy effectively in a rising market rather than a falling market. Interestingly our buyer appears to demonstrate the opposite characteristic.

5.6 A TARGET OR STANDARD FOR PURCHASING PERFORMANCE

The question naturally arises as to what ought to be a standard or target performance buyers should attempt to achieve. Over the first year of the period the buyer achieved an average efficiency of 57%. It is surely then not unreasonable to set something near this level of performance as a target to aim for over other periods, of rising markets for example. A reasonable target might be to achieve an average purchasing efficiency of 50% over long periods of time. This is half way between the average price offer over the buying periods, the back-to-back policy, and the lowest possible price over the buying period.

An exercise can be performed to estimate what savings in purchasing cost would have been made if this target standard had been consistently achieved over all shipment months for 1969, 1970 and 1972. However it is possible to have one or two prices within the buying period substantially less than all of the others. In such circumstances the target price giving a purchasing efficiency of 50% may only have those one or two price offers below it. It would be most unrealistic to expect a buyer to always purchase at the one price offer or even at one of the two price offers below the target price. A meaningful target price must give several opportunities at which the buyer can buy at or below the target price. Accordingly the target price performance is best defined as the larger of the following two values:

1. The price level giving a 50% purchasing efficiency

and

2. The price such that one quarter of all prices within the buying period are less than or equal to it. This may be called the 'lower quartile' price.

These two values together with the target performance price are shown in Table 5.9 for 1969, 1970 and 1972. Note on the majority of occasions there is little difference between the 50% purchasing efficiency and the lower quartile price. Thus this procedure gives a target performance price such that there are many opportunities to buy at a lower price quotation. The potential savings that could have been made with a purchase at that price, rather than the price that was actually paid, are shown in the final column of Table 5.9.

If purchases had been bought at the target price then about 1% extra in purchasing costs would have been paid in 1969 whilst savings of 3% in both

TABLE 5.9 Target purchasing performance pre 1972/73

Year	Purchase order shipment period	50% Purchasing efficiency price	Lower quartile price	Target performance price	Actual price paid—target price (£s)
1969	September	25.175	25.4	25.4	−0.65
	October	24.4	24.6	24.6	−0.80
	November	24.1	24.1	24.1	−0.10
	December	24.55	24.6	24.6	0.525
1970	March	25.9	25.6	25.9	0.53
	April	26.1	25.9	26.1	0.525
	July	27.1	26.8	27.1	0.525
	August	27.5	27.6	27.6	0.40
	September	28.1	28.4	28.4	0.475
	October	28.9	28.6	28.9	−0.20
	November	29.5	28.6	29.5	4.10
	December	31.15	31.6	31.6	0.525
1972	January	23.9	23.95	23.95	−0.05
	July	24.1	24.0	24.1	0.15
	September	25.7	25.5	25.7	1.80
	November	28.0	28.00	28.00	1.00

TABLE 5.10 Target purchasing performance post 1972/73

Year	Purchase order shipment period	50% Purchasing efficiency price	Lower quartile price	Target performance price	Actual price paid—target price (£s)
1973	November	54.35	53.2	54.35	−2.15
	December	53.5	53.6	53.6	−1.35
1974	January	55.25	55.0	55.25	7.10
	February	57.3	55.4	57.3	6.60
	March	59.4	59.4	59.4	5.10

1970 and 1972 would have been incurred. Over the 3 years, assuming an average 6000 tons bought for each month's shipment, the total net savings would amount to over £50,000. Thus if this target performance could have achieved, quite substantial reductions in purchasing cost would have been achieved.

The 50% purchasing efficiency represents a target to aim for. Over relatively stable prices as occurred in 1969–72 it is a very feasible target as will be seen later in Chapter 12. However as mentioned earlier, experience would suggest that a good buyer is unlikely to exceed an efficiency of 35%. This might be a better initial target. This level of performance, if achieved for each shipment month, would have given an extra 1.7% in costs over 1969 and a saving of 2.7% and 2.3% in 1970 and 1972 respectively. The financial implications would be a net savings over the 3 years of about £44,000.

In terms of setting standards for purchasing performance by which a buyer's performance must be judged, a lower efficiency of 25% is probably more reasonable. If this level had been achieved substantial net savings of around £35,000 over the 3 years could have been achieved. A 25% purchasing efficiency price means that approximately a proportion of 3 out of every 8 prices offered over the buying period will be below it. It is surely not unreasonable to expect a buyer to be able to purchase at these prices.

The buyer's performance over the period of more volatile price movements of 1973 and later was not as good as that of the earlier period. For the reasons previously discussed, the assessment is confined to the last 5 shipment months November 1973, December 1973, January 1974, February 1974 and March 1974. The purchases for the first 2 of these 5 months were very good, giving a virtually perfect purchasing performance. Prices, after increasing to a peak in August 1973 because of the fears of U.S. export controls, fell to a minimum level in October and November under the influence of a good new U.S. maize crop. Thus prices generally fell over the buying period for these two shipment months. The purchases for both these shipment months were left until close to the end of the buying periods, as was often the case with the earlier months of 1969, 1970 and 1972. It is difficult to avoid the conclusion that the buyer had good fortune on this occasion rather than that he exercised good judgement in delaying his purchase. However losses were made on the next three purchase orders. Over the whole 5 months the average purchasing efficiency was −24%, indicating that the buyer was worse than a back-to-back policy. [Unfortunately the two examples of maize purchasing discussed in Chapter 3 show that a back-to-back policy is not a feasible policy.]

The target prices, together with the 50% purchasing efficiency and lower quartile prices for these months, are shown in Table 5.10. The fact that the lower quartile prices for November 1973 and February 1974 are so far below the 50% efficiency prices gives some indication of the unusually large price fluctuations that occurred over the later months of 1973. The final column of the table shows the savings that would have been achieved on the actual price paid if this target performance had been achieved in all 5 months. On average, assuming equal amounts purchased for each month, the savings would have been £3.06 per ton, giving a 5.2% saving in purchasing cost. Such savings would have amounted to almost £100,000 for those 5 shipment months alone. Achieving the 35% purchasing efficiency price would have given net savings of around £70,000 whilst reaching the 25% efficiency purchasing standard would have meant net savings of over £55,000.

This analysis of the buyer's past performance suggests that it could be improved with great benefit to the company. The buyer appears not to be taking advantage of the flexibility given him with regard to the length of the buying period. He is generally leaving purchases until too late a date. In the past he appears to have been slow to react to situations of increasing prices. Perhaps because of other purchasing responsibilities he has been unable to devote sufficient time to keeping up-to-date with the world maize situation. The analysis has shown that such attention and effort is well worthwhile. If the performance he achieved over the first of the 5 years could be achieved consistently in other market circumstances, then very significant reductions in purchasing cost would be the result. The question of whether policies can be devised to help buyers achieve such a target performance in both stable and volatile market conditions is explored in succeeding chapters. The purchasing efficiency measure, together with any desired level of target performance, can be used in any circumstances to estimate the potential savings and thus whether an investment in a research study would be worthwhile.

5.7 ALTERNATIVE PURCHASING EFFICIENCY MEASURES

The purchasing efficiency measure has been criticised as inappropriate from the statistical viewpoint because of its nonsymmetric nature. As was seen in the case study on maize large negative purchasing efficiencies, almost as low as −300%, can occur, whilst the maximum value can never exceed 100%. The general approach that the savings achieved should be assessed relative to the variations in prices over the buying periods is accepted. It has been suggested that the maximum savings concept in the denominator of the definition of the purchasing efficiency should be replaced by some alternative measure of price volatility over the buying period, for example the standard deviation of the prices. For ease of calculations, the use of the average value of the daily price changes over the buying period, disregarding their positive or negative signs, has been advocated as a more suitable statistic. This value is roughly proportional to the standard deviation for symmetric distributions.

However applying this approach to some of the maize purchases just discussed in the case study shows that this modification does little to mitigate the nonsymmetric nature of the purchasing measure. The purchasing efficiencies for January 1972 and July 1973 shipments were 50% and −269% respectively. The purchasing efficiency values calculated using the absolute average daily price change, rather than the maximum savings, are 82.9% for January 1972 and −656% for July 1973. Hence it exaggerates the differences between the 2 months rather than reducing them. A slight improvement occurs if the standard deviation of the prices is used instead of the average absolute price changes. In this case the measure of performance has a value of 83.3% for January 1972 and −194.5% for July 1973.

It thus seems, that whatever factor one puts in the denominator, the same problem of very highly negative purchasing efficiency values will occur if there are a few prices very much higher than the rest. The purchasing efficiency measure is simpler to calculate than the two alternatives put forward. More importantly it measures the actual performance relative to what the buyer could have done at best. As was shown in the case study, the potential cash savings can easily be calculated for setting different target performance levels above the current performance to see if such an improvement is worthwhile. Thus the purchasing efficiency concept appears the most appropriate one to use in practice. Furthermore large negative values show that the buyer has completely misread the market situation.

One problem is that the purchasing efficiency measure is very sensitive to one unusually low price offer in the buying period. This will give the minimum price offer to be used to calculate the maximum potential savings, the denominator of the purchasing efficiency measure. It is rather excessive to expect any buyer consistently to pick out the lowest price offer over the buying period, particularly if this is much lower than all the other prices. Most people would expect such an unusually low price to herald the start of a new period of lower prices, so would not make a purchase at the price. Thus using an unusually low price to calculate the maximum savings is rather overestimating the maximum savings one could expect in practice. To avoid such distortions it may be better to use the third or fourth lowest price over the buying period to measure the maximum potential savings. It is not so unreasonable to expect a buyer to pick out one of the three or four lowest prices offered over the buying periods on a good number of occasions.

Appendix
Overall average purchasing efficiency analysis

For the ith purchase order the purchasing efficiency expressed as a proportion, m_i, rather than a percentage M_i, can be written as

$$m_i = \frac{\text{Average market price} - \text{Actual price paid}}{\text{Average market price} - \text{Minimum price}}$$

$$= \frac{\bar{p}_i - x_i}{\bar{p}_i - y_i} = \frac{u_i}{v_i} \text{ say,}$$

where $\bar{p}_i$ is the average of all the price offers made for the ith purchase order over its total buying period. If we consider a whole set of N consecutive orders covering some long period, then as we have seen there are two ways to calculate the annual average purchasing efficiency. The simplest approach is to calculate m_i for each individual order and then take the average value over the N orders that year as

$$\bar{m} = \frac{1}{N} \sum_{i=1}^{N} m_i.$$

An alternative approach is to work out the averages over the period for each component of the formula separately and define the average efficiency as

$$\hat{m} = \frac{\bar{p} - \bar{x}}{\bar{p} - \bar{y}}$$

where $\bar{x}$ is the average price paid per unit purchased over the year, i.e.

$$\bar{x} = \frac{1}{N} \sum_{i=1}^{N} x_i$$

Similarly

$$\bar{y} = \frac{1}{N} \sum_{i=1}^{N} y_i$$

and

$$\bar{p} = \frac{1}{N} \sum_{i=1}^{N} \bar{p}_i.$$

The average market price measure $\bar{p}$ should be taken as the average value of the various average market prices over the buying periods for each other. Thus it follows that

$$\hat{m} = \frac{\sum_{i=1}^{N} (\bar{p}_i - x_i)}{\sum_{i=1}^{N} (\bar{p}_i - y_i)} = \frac{\sum_{i=1}^{N} u_i}{\sum_{i=1}^{N} v_i}$$

Now

$$\bar{m} = \frac{1}{N} \sum_{i=1}^{N} \left(\frac{u_i}{v_i}\right) = \frac{1}{N} \sum_{i=1}^{N} m_i$$

But

$$\hat{m} = \frac{\sum_{i=1}^{N} u_i}{\sum_{i=1}^{N} v_i} = \frac{\sum_{i=1}^{N} m_i v_i}{\sum_{i=1}^{N} v_i}$$

$$= \sum_{i=1}^{N} m_i \left(\frac{v_i}{\sum_{i=1}^{N} v_i}\right)$$

The aim of any effective purchasing policy is surely to buy just as well when there are large potential savings as when there are only small ones and vice versa. Thus the purchasing efficiency, m_i, should be independent of the maximum savings, v_i, that can be achieved. We can thus assume that the m_i are random variables independently distributed about the real underlying value of the purchasing policy used, with the same variance σ^2. Let us assume that different values for each m_i are obtained by repeated purchases (samples) for the same N purchase orders. Then the same set of v_is are used in each sample to give different values for m_i. For ease of notation let

$$w_i = \frac{v_i}{\sum_{i=1}^{N} v_i}$$

Since the v_i may be treated as constants with respect to the sampling then so may the w_i. Thus

$$E(\hat{m}) = E\,(\sum_i w_i m_i) = \sum_i w_i E(m_i) = \sum_i w_i m$$

where $m = E(m_i)$.
Hence

$$E(\hat{m}) = m \sum_i w_i = \frac{m \sum_i v_i}{\sum_i v_i} = m.$$

Also

$$E(\bar{m}) = \mathrm{E}\left(\frac{1}{N} \sum_i^N m_i\right) = m.$$

Hence in the long term as the number of purchase orders included in the assessment of performance increases the two alternative averages tend to the same value. Since the m_i are statistically independent with the same variance, σ^2, then

$$V(\hat{m}) = V\left(\sum_i w_i m_i\right) = \sum_i w_i^2 \sigma_{m_i}{}^2 = \sigma^2 \sum_i w_i^2$$

$$\text{i.e. } V(\hat{m}) = \sigma^2 \sum_i \left\{ \frac{v_i}{\sum_i v_i} \right\}^2 = \sigma^2 \frac{\sum v_i^2}{\left(\sum_i v_i\right)^2}$$

and

$$V(\bar{m}) = V\left(\frac{1}{N} \sum_i m_i\right) = \sum_i \frac{1}{N^2} \sigma^2_{m_i} = \frac{\sigma^2}{N}.$$

Now

$$\sum_i v_i^2 - \frac{1}{N}\left(\sum_i v_i\right)^2 = \sum_i (v_i - \bar{v})^2 \geqslant 0$$

so that

$$\frac{\sum v_i^2}{(\sum v_i)^2} \geqslant \frac{1}{N}$$

hence $V(\bar{m}) \leqslant V(\hat{m})$.

Since $\bar{m}$ and $\hat{m}$ have the same expected value, but $\bar{m}$ has the smaller variance it is surely preferred as a measure of the average purchasing performance over a period of time.

The difference between these two measures will be

$$(\hat{m} - \bar{m}) = \sum_i w_i m_i - \frac{1}{N} \sum_i m_i$$

$$= \sum_i m_i \left\{ \frac{v_i}{\sum_i v_i} - \frac{1}{N} \right\}$$

$$= \frac{1}{N \sum_i v_i} \sum_i m_i (v_i - \bar{v}).$$

Hence on the same assumptions as above

$$V\,(\hat{m} - \bar{m}) = \frac{\sigma^2}{N^2\left(\sum_i v\right)^2} \sum_i (v_i - \bar{v})^2.$$

For the sugar example illustrated in Table 5.2

$$\sum_i v_i = 103.42 \qquad \sum v_i^2 = 1053.07 \qquad N = 12$$

$$\therefore \quad \sum_i (v_i - \bar{v})^2 = 161.762.$$

Hence $V(\hat{m} - \bar{m}) = (0.0125\ \sigma)^2$

The variance of m_i, σ^2, is 0.1018. So that

$$V(\hat{m} - \bar{m}) = (0.004)^2.$$

Since the expected value of $\hat{m} - \bar{m}$ is zero the difference between the two averages should usually be less than 0.01, or 1% for the percentage measure M. Hence the numerical differences are very small and insignificant.

6

Purchasing Partially Substitutable Basic Raw Materials—The Blending Problem

6.1 INTRODUCTION

In terms of the structure of the commodity purchasing decision presented in Chapter 4, this chapter describes some examples of the problems arising in stage 1, the determination of what quantities of which materials should be bought. In many instances the final product made by a company has to be manufactured from a mix or blend of different raw materials. The choice of which materials to use in what quantities is complicated by the situation that the materials are generally only partially substitutable for each other. If for some reason a decision is made not to use a particular material this cannot be replaced directly by the same quantity of an alternative material. The quantities of all the other materials used in the blend will have to be changed also. It is thus impossible to consider the best amount for each material separately. The total set of possible materials must be considered as a whole. With at least twenty-five alternative materials available in the simplest practical blending problem the number of alternative blends that satisfy the production requirements will be very large indeed. Such a problem cannot be solved by simple inspection or simple logical procedures.

Fortunately all such blending problems have a basic common structure which can be described by a particular mathematical model. This mathematical model can be solved by the use of a now well-established operational research technique called linear programming. In the form of a computer program this technique can be used to solve very large problems in a very short time.

The use of linear programming, and in recent years nonlinear programming, to determine the most economic mix of materials to purchase for the production of iron, alloys or blended products, represents one of the most successful uses of operational research methods and techniques in the purchasing area. Typical applications of these methods have been in the steel industry to determine the mix of ores, coal and limestone to purchase to make iron of a required quality at minimum cost, in the animal feeds industry to determine the mix of raw materials to make animal feeds satisfying the required nutritional qualities, in the food industry to make margarine and in more recent years in the nonferrous manufacturing industry to determine the mix of metals and nonferrous scraps to make metal alloys at minimum cost.

These applications have generally risen as an extension of work that has been carried out on the production processes. A mathematical model has been developed to fit the practical situation with regard to all possible production outputs and all possible raw materials. It has then been shown by running the model retrospectively against situations which had occurred in the past, and had been carefully measured, that the costs of producing the known outputs using linear programming methods were less than those that had been evaluated by the traditional rule of thumb methods.

The problem then arises to find a way of using this model in practice, since the materials that can be used on any particular day in any particular process can only be selected from those materials available in the stocks at that time. Over a period of time the actual overall pattern of materials to be consumed must be the same as the actual pattern purchased. Money can only be saved at the point where it is spent. Thus it is necessary to attempt to apply the mathematical model to the purchasing function, where the relative costs of any materials at any time have a direct immediate impact on the economics of the purchase and therefore the usage of materials, as well as to the production processes otherwise actual economies in cost will not be achieved.

The problems in applying mathematical programming methods to the 'General Blending Problem' will be initially illustrated via simple illustrative examples. The way practical operating systems have been devised will be illustrated by discussing in detail three industries with which the author has had direct involvement. These cases represent the range of problems and the work that has been carried out in this area. They also show the limitations of the methods and the necessary extensions that have been devised.

6.2 THE BLENDING PROBLEM IN THEORY

The problems that arise in deciding what quantities of each of the partially substitutable materials to purchase can be seen via the following highly simplified animal feeds example. The buyer has a choice of three possible materials that he can use to make a feed for poultry. The company's research department has identified two critical constraints on the nutritional make up of the feed. The first is that at least 18% of the blended product by weight must consist of crude protein. The second is that the metabolic energy of the feed must be at least 2600 kilocalories per kilogram of feed by weight. The make up of the three possible input materials in terms of the nutritional qualities together with their current prices are shown in the table below:

Material	Crude protein % age by weight	Metabolic energy Kcal. per kilo	Current price £s per ton
A	9	3400	170
B	15	1400	160
C	45	2400	180

There is no yield loss in the blending process. Essentially the materials are ground, mixed thoroughly and then extruded in the form of small cakes. Because of this the problem can be formulated in terms of producing 1 ton of blended product. The quantities of each material required to make 1 ton can be multiplied by 1000 if the company wish to produce 1000 tons of poultry feed, by 5000 if it wishes to produce 5000 tons of feed etc.

Let a tons of material A, b tons of material B and c tons of material C be the amounts blended together to produce exactly 1 ton of poultry feed.

Since there is no loss of material in the blending process the material input, in whatever form it is measured must equal the material output. Thus

Tons in = tons out

Hence by definition of a, b and c it follows that

$a + b + c =$ tons in $=$ tons out $= 1$.

Similarly, crude protein in = crude protein out.

From the table the crude protein in is

$$\frac{9}{100}a + \frac{15}{100}b + \frac{45}{100}c$$

From the research department's work the crude protein out must be greater than or equal to

$$\frac{18}{100} \times 1$$

(since there is 1 ton of total output). Hence we must choose values for a, b and c such that

$$9a + 15b + 45c \geqslant 18.$$

(The symbol '$\geqslant$' merely means 'greater than or equal to'. Similarly '$\leqslant$' means less than or equal to.)

Similarly metabolic energy in = metabolic energy out

Metabolic energy in is $(3400a + 1400b + 2400c) \times 1000$ (assuming 1000 kilograms equals 1 ton), and the metabolic energy out must be greater than $2600 \times 1 \times 1000$. Hence simplifying it follows that we must choose a, b and c such that

$$34a + 14b + 24c \geqslant 26.$$

Since we cannot use negative amounts of material in the blending process we must have

$$a \geqslant 0,\ b \geqslant 0,\ c \geqslant 0.$$

Furthermore, the cost of 1 ton of blended product will be in £s,

$$F = 170a + 160b + 180c.$$

The poultry feed purchasing problem has been transformed into the following mathematical problem. Find numerical values for the three variables a, b and c that satisfy the following constraints:

$$a + b + c = 1$$

$$9a + 15b + 45c \geqslant 18$$

$$34a + 14b + 24c \geqslant 26$$

$$a \geqslant 0,\ b \geqslant 0,\ c \geqslant 0$$

that will give the smallest possible value for the linear function

$$F = 170a + 160b + 180c.$$

This is a simple example of a linear programming problem which can easily be solved by a special mathematical technique called the simplex method. The solution for the given set of prices is

$$a = 0.5,\ b = 0.3 \text{ and } c = 0.2.$$

These values are the number of tons of material A, B and C respectively that should be used to produce exactly one ton of poultry feed. A simple derivation of this solution is given in an appendix to this chapter for those who are not familiar with linear programming methods.

All blending problems, at least initially, can be formulated in the same way as our simple poultry feed example. Obviously in practice they will contain many many alternative materials (i.e. variables) and the choice amongst them will be constrained by many many more constraints. The constraints will always be a mixture of equal to or less than inequalities, equal to or greater than inequalities and strict equalities. However in all cases the left-hand sides of the constraints will be linear combinations of the variables (the alternative materials). If the cost function is also a linear combination of the variables the simplex method always gives the solution to the larger practical problem, no matter how many variables or constrain equations it contains.

6.3 THE PROBLEMS AT THE FORMULATION STAGE

An initial major problem in applying this approach to practical problems is that of actually deriving the total set of constraints on the way the alternative materials can be blended together. This requires a major research effort, particularly if the blending process involves complex chemical and physical reactions only imperfectly understood. A good example of this arises in the iron and steel industry in determining what mix of the various alternative iron ores available should be blended with coke and limestone to make a specified quality and quantity of molten iron at the lowest possible cost. This is the first case study of a 'practical' blending problem described in this chapter. This problem has three aspects which in many ways make it the simplest example of a blending problem. Firstly only one quality of molten iron has to be produced. Secondly all of the constraints are technical. They are based on the processes that occur inside the blast furnace where the blending process actually takes place. Finally the prices of the materials that can be used are fixed for a substantial period of time ahead as are the maximum quantities of each one that can be bought.

As the physical and chemical processes that take place inside the blast furnace have become better understood changes have been made in the design of each new blast furnace. Each new blast furnace is slightly different, so that separate linear programming models, sets of constraints, have to be formulated for each individual blast furnace. The centralisation of the purchasing function for a set of blast furnaces in different geographical locations plus the development of a partial blending of the ores in a pretreatment process before putting them in the blast furnaces, giving an increased efficiency to the blast furnace process, enormously complicates the purchasing problem. The simple model of our poultry feed example requires large modifications to cover this problem. How this can be done is also described in the iron and steel case study.

The setting up of the constraints presents similar problems in the production of metal alloys in the nonferrous metals industry. Again they are basically determined by the technical processes taking place. In other industries however, such as animal feeds, margarine or tobacco for example, the constraints are necessarily less precise. They may be based on qualities such as the taste or the colour of the final blended product which cannot be assessed in a precise quantitative way. Some may be based on an assessment of the marketability of the product rather than a precise technical description. For example higher protein in the animal feed may not necessarily be an advantage in the efficient rearing of beef cattle. However, farmers may think it is an advantage and hence are prepared to pay higher prices or buy more of the product if they have higher protein values. These aspects are discussed in the second case study covering mainly animal feeds.

6.4 THE DIFFICULTIES ARISING FROM THE FLUCTUATING PRICES OF MATERIALS

A major difficulty in applying the linear programming model of the simple poultry feeds example in practical blending problems, with the exception of the iron and steel situation is the everchanging nature of the prices of the alternative materials that can be purchased and used. There is no longer a fixed price for each material which will remain fixed into the foreseeable future. In the buying situation discussed in Chapter 3, the animal feed manufacturer, LFF, adopted a monthly planning basis for his production of feed. The materials for making a feed in a particular month could have been bought over the previous three months. Thus the buyer will have many opportunities to buy each of the various raw materials, but different prices will be quoted to him at each buying opportunity. He will have to take into account the changes in price of the materials that will inevitably occur from the time when he begins to purchase them for a particular production mix to the time when that mix goes into production. The price changes that occur imply that the differences between the prices of the alternative materials will also change so that it may be necessary to completely revise the amounts of each material that he should use for the mix. If prices change by large amounts in the short-term, as we have seen in earlier chapters can happen, the buyer may have to be continually changing the amounts he should try to buy. Since most manufacturing companies adopt a policy of never reselling materials that they have bought previously, the problem becomes more complicated since the final amounts of material to be bought must take into account the amounts already bought.

These problems can be illustrated in our very simple poultry feeds example. It is perhaps clearer to see this problem graphically. Using the first constraint of the linear programming model we can convert it into a two dimensional problem by substituting the expression $(1-a-b)$ for the variable c in the other constraints, the nonnegativity conditions and the objective function. After some simplification of the algebra the problem reduces to:

Find the values of a and b satisfying the constraints

$$a + b \leqslant 1$$

$$12a + 10b \leqslant 9$$

$$5a - 5b \geqslant 1$$

that give the minimum value of the cost expression

$$F = 180 - 10a - 20b$$

Each point in Fig. 6.1a will have an a and b coordinate relative to the two axes shown. Choosing a value for a and b is the same as choosing a point on the graph. Each constraint can be represented as a line dividing the graph into two regions, only one of which has a and b values satisfying the constraint. The various constraints and the nonnegativity conditions together form a triangular region as shown in Fig. 6.1a. All the points inside the shaded triangular region or on its boundaries have a and b coordinates that satisfy the constraints. It can be shown that the solution minimising some linear cost functions must always be at one of the corner points of the triangle P, Q or R. Each corner point will be the solution for different ranges of prices of the three materials.

For the current set of prices £170, £160 and £180 for materials A, B and C respectively, the solution minimising the cost of producing 1 ton of poultry feed is 0.5, 0.3 and 0.2 tons of the three materials respectively (i.e. corner Q of Fig. 6.1a). These values, multiplied by the number of tons of poultry feed that will be produced, are the amounts of each material the buyer should seek to purchase.

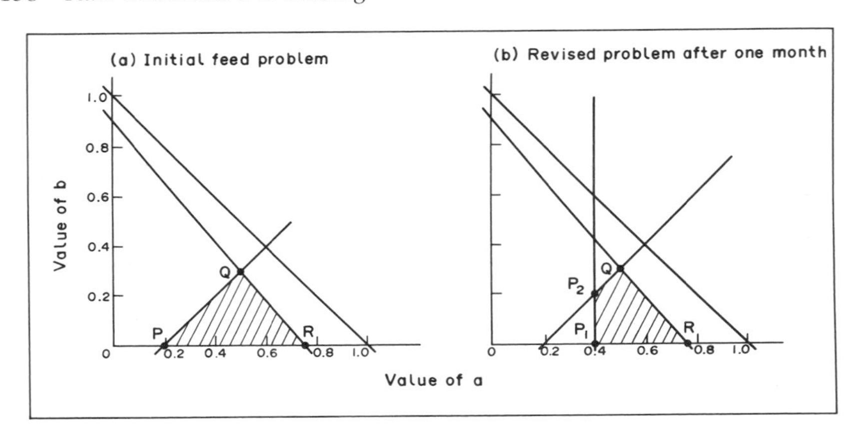

FIG. 6.1 Graphical representation of first simple feeds purchasing problem

Let us assume that as for the LFF situation the buyer has 3 months in which to buy the required amounts. Let us further assume that over the first month there are only very small fluctuations in prices about the initial levels and that the buyer purchases the equivalent of 0.4 tons of material A. Let us assume that now because of a possible drought affecting crops or some such event the price of material A increases strongly to £200 per ton, whilst that of material C falls slightly to £170 per ton and the price of material B remains unchanged.

At these new prices the solution minimising the total purchasing cost would be at corner P of the triangular region of Fig. 6.1a. This implies purchasing 0.2 tons of material A and 0.8 tons of material C. However, the company has already bought 0.4 tons of material A. The company policy, as already stated, is that material cannot be resold once it has been purchased. Since this 0.4 tons is already paid for, it does not make sense to only use 0.2 tons of it and buy a further 0.8 tons of material C. The extra 0.2 tons of material A would be wasted. We must obviously look for solutions which use up the 0.4 tons of A already bought. Our previous linear programming model must be extended by a further constraint that a, the amount of material A to be used in the mix, must be greater than or equal to 0.4.

In Fig. 6.1a this is the same as drawing a line parallel to the b axis through the point $a = 0.4$ and confining the choice of solution points to the region either on or to the right of this line. This is done in Fig. 6.1b. It is now seen that the feasible region of all possible solutions is the shaded polygon shown with corners at P_1, P_2, Q and R. The cost functions for the two-dimensional problem also changes to become

$$F = 200\,(a - 0.4) + 160b + 170\,(1-a-b-0.4) + 0.4 \times 170$$

$$= 158 + 30a - 10b.$$

As before, the solution minimising the cost expression must be at one of the four corners P_1, P_2, Q or R, Note that there are now four choices rather than three. In this case the solution is at the corner P_1. This means that at these new prices no more of material A should be bought but that 0.2 tons of B and 0.4 tons of C should be purchased.

For practical problems with many more variables and constraints there will be many more 'optimal' solutions for the differing sets of prices. If there is little correlation between the prices of the alternative materials then moderate fluctuations or trends in prices from day-to-day can lead to frequent changes in the target amounts of each material that should be purchased over the remaining buying period. The practical difficulties in operating a purchasing policy in these circumstances to buy the minimum cost mix over some buying period can easily be imagined.

6.5 THE INTERACTION OF THE FLUCTUATING PRICES AND THE MANY DIFFERENT FINAL PRODUCTS

A further complication arises in practice from the fluctuating prices for the various materials. There are usually many different blended products that need to be produced which can use the same raw materials. Each of the products will contain different chemical, nutritional or production constraints on the mixing of the partially substitutable raw materials. For example, animal feeds differ between those for beef cattle, dairy cattle, poultry and hogs, and for summer or winter use. The problem is that at the same price different amounts of the raw materials would be required for the different products. A change in prices may cause a greater change to the amounts of the different raw materials that should be used for some products than for others.

For example, consider a second feed product using the same materials A, B and C. Assume that the constraints on this product take the following form; where again a, b and c are the tons of material A, B and C to make 1 ton of blended product.

$$a + b + c = 1$$

$$16a + 16b + 6c \leqslant 14$$

$$20a + 19b + 18c \geqslant 19.$$

The purchase cost function will be exactly the same as for the first poultry feed example. Converted into its two-dimensional form in *a* and *b* only, the constraint space containing all possible solutions to the problem will take the triangular shape through the corners *S, T* and *V* shown in Fig. 6.2. In this case the optimal solution will always be at one of the corners *S, T* or *V*. For the original set of prices £170, £160 and £180 for materials A, B and C respectively the solution giving the minimum purchasing cost lies at corner *T*. The solution is to buy 0.2 tons of A, 0.6 tons of B and 0.2 tons of C for each ton of blended product made. This solution also remains the best for the new set of prices at the end of the first month when the price of A rose to £200, C fell to £170 and B remained unchanged at £160. Thus we see that the change in prices has a large effect on the amounts to be purchased for the first feed product but no change at all for the second.

Let us consider what the buyer should do if, over the first month of the buying period, he has bought the equivalent of 0.2 tons of material B for the first feed and 0.3 tons for the second, and nothing of materials A and C. Remember that prices fluctuated around the initial price levels of £170, £160 and £180 for materials A, B and C. The price change now occurs. If we continue to treat the two feed products separably we can calculate what materials to buy for the two feeds in the same way as we considered the price change for the first feed earlier. The new regions within which solutions must be are shown in Fig. 6.3a and 6.3b, the triangle Q, P_1, R_1 for the initial poultry feed problem and the triangle S_1, T, V_1 for the second problem. For the new set of prices, £200, £160 and £170 for materials A, B

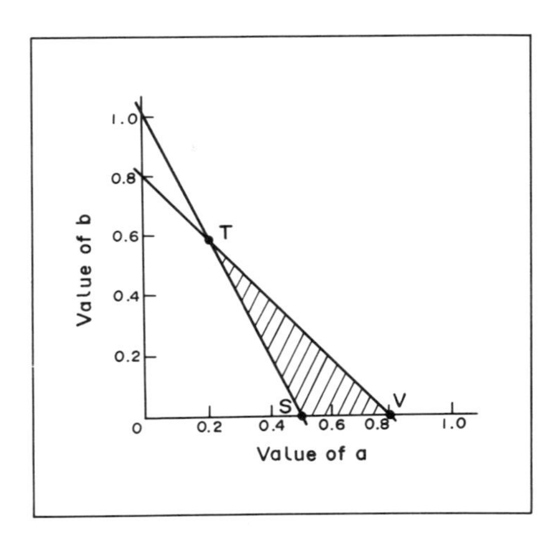

FIG. 6.2 Graphical representation of second feed purchasing problem

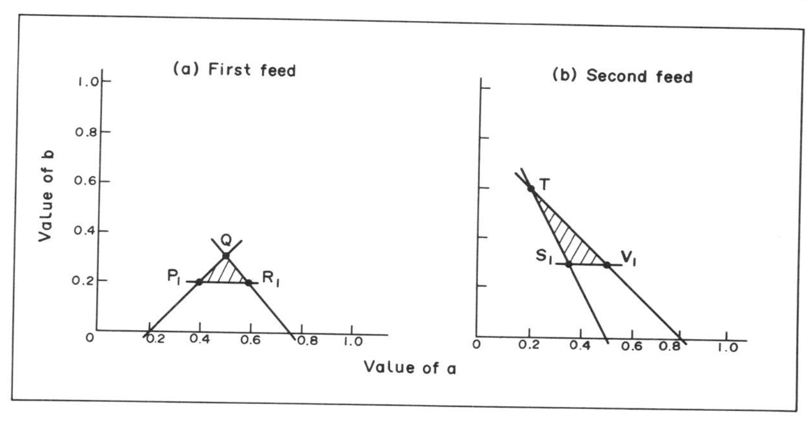

FIG. 6.3 The two feeds purchasing problem using up existing stocks of material B

and C respectively, the cost minimising solutions are at the corners P_1 for the first problem and the corner T for the second problem. This means that the buyer should buy a further 0.4 tons of A and 0.4 tons of C for the first feed and a further 0.2 tons of A, 0.3 tons of B and 0.2 tons of C for the second feed. Assuming no further price changes the cost of these purchases would be £148 for the first feed and £122 for the second feed, amounting to a total of £270.

Under the new prices the same amounts of each material should be bought for the second feed as for the initial set of prices. If no material B had been bought for the first feed during the first month of the buying period, the best solution (corner P of Fig. 6.1a) would be to buy only 0.2 tons of A and 0.8 tons of C, i.e. less of the now expensive material A than we require under the new conditions. The buyer has already bought 0.5 tons of B in total which were intended to be split into 0.2 tons for the first feed and 0.3 tons for the second feed. There is no particular reason why this original split should be maintained. One course of action is to reallocate this 0.5 tons of B between the two feeds in a different way to see if the further purchases to complete the mix of materials for each feed can be bought at a lower total cost. Since the second feed requires more of material B than we have bought in total so far, and the first feed would prefer to use no material B at all under the new prices, an alternative obvious policy is to allocate all of the 0.5 tons of B already bought to the second feed. If this is done then to minimise further purchase costs the buyer would need to buy 0.2 tons of A, only 0.1 tons of B and 0.2 tons of C for the second feed. He would then need to consider the first feed on the basis that he has bought no materials for it at all. His solution to minimise costs would then be to buy 0.2 tons of A and 0.8 tons of C. The cost of these further purchases would be £176 for the first feed and £90 for the second feed. In total this amounts to £266. This represents a saving of £4 per ton on the original policy of continuing with the initial allocations of the material B that was purchased over the first month of the buying period.

The total cost of producing one ton of the first feed has been reduced from £180 to £176 whilst that for the second feed has remained unchanged at £170. In practical problems with say ten–twelve different products it may be that whilst the final purchasing costs are reduced for most products they are increased for some. This may cause some minor problems for the accounting system in working out costs and profits achieved per product. Care is obviously needed in interpreting such figures and making decisions about changing product sales prices or the product range. However such clerical and recording difficulties should not be allowed to prevent changes being made to the purchasing policy which can reduce the total purchasing cost, the money the company actually spends.

In practice with several different products being made there will be many more alternatives to be considered concerning the reallocation of existing purchases than the two of our simple illustrative example. Fortunately this problem can also be formulated in linear programming form similar to our previous examples. However it becomes larger and more complex. In the case of our example it could take the following form.

Let a_1, b_1, c_1 be the amounts still to be bought for the first feed and a_2, b_2, c_2 be the amounts to be bought for the second feed. Let the amounts of each material already purchased be A, B and C. Let x_1, y_1, z_1 and x_2, y_2, z_2 be the amounts of the three materials already purchased that are allocated to the first and second feed respectively. The constraints will then take the form:

$$
\begin{array}{rcccccl}
a_1 + x_1 & + & b_1 + y_1 & + & c_1 + z_1 & & = 1 \\
9(a_1 + x_1) & + & 15(b_1 + y_1) & + & 45(c_1 + z_1) & & \geqslant 18 \\
34(a_1 + x_1) & + & 14(b_1 + y_1) & + & 24(c_1 + z_1) & & \geqslant 26 \\
a_2 + x_2 & + & b_2 + y_2 & + & c_2 + z_2 & & = 1 \\
16(a_2 + x_2) & + & 16(b_2 + y_2) & + & 6(c_2 + z_2) & & \leqslant 14 \\
20(a_2 + x_2) & + & 19(b_2 + y_2) & + & 18(c_2 + z_2) & & \geqslant 19 \\
x_1 & + & x_2 & & & & \leqslant A
\end{array}
$$

$$y_1 \quad + \quad y_2 \qquad\qquad \leqslant B$$

$$z_2 \quad + \quad z_2 \qquad\qquad \leqslant C$$

and the total further purchasing cost function to be minimised will be

$$200\,(a_1 + a_2) + 160(b_1 + b_2) + 170(c_1 + c_2).$$

Instead of two linear programming problems, each of three variables and three constraints, we now have one linear programming problem containing twelve variables and nine constraints. This will require substantially more computer time and expense to solve it. With ten products which individually would need a linear programming model of twenty-five to thirty variables and constraints, as for animal feeds for example, the enlarged problem would become very large indeed. Furthermore, as well as reallocating earlier purchases in a different way between products to that originally intended, it may be advantageous to reallocate them between the product production runs in different successive months also. Considering periods of three successive months will more than triple the number of variables and constraints. In animal feeds one could easily end up with a linear programming model of well over 2000 variables and about 1000 constraints. The cost of preparing and solving a problem of such size becomes very significant and time consuming. If such an approach is beneficial there are obvious practical questions to be answered, such as deciding whether an overall review should be made at regular intervals of time or only whenever price changes over time exceed some preselected level.

An important aspect of research into such blending problems is trying to find some appropriate operational system that gains a significant reduction in purchasing cost yet can be applied in practice in a convenient easy routine fashion. Possible solutions to this are discussed in the second case study which further examines these problems in the context mainly of the animal feeds problem.

6.6 THE CONTROL OF PURCHASES PROBLEM

A final area, often neglected, is the control and administrative procedures to be adopted to ensure the purchasing decisions do achieve the targets laid down by the purchasing policy. This is particularly important in the nonferrous metals industry. A substantial part of the raw materials used in the manufacture of nonferrous metal alloys is reclaimed scrap. The major source of this scrap is from scrap dealers. There are typically many scrap dealers who on any particular day can only provide relatively small amounts of scrap. Furthermore, each contract placed for the final nonferrous product produced is usually priced on an individual basis depending on the metal prices on the London Metal Exchange on the day the contract was placed. There are obvious control problems in matching the total metal requirements of the various contracts with the purchases of scraps and metals that should be made over time. The final case study describes a particular system devised for one manufacturer to cope with these problems. The result is to free the buyers from much of the chores of the routine 'housekeeping' problems and enables them to devote more of their time to decide at what prices to make purchases day-by-day. Setting up control procedures that give a day-to-day updating of the amounts bought, the average prices paid etc., is an essential step for achieving good purchasing results overall.

6.7 THE BLAST FURNACE BURDENING PROBLEM—CASE STUDY 1

The use of linear programming for selecting the most economical mix of iron ores and other input materials to purchase for use in a blast furnace to produce molten iron was the earliest successful use of this technique in the purchasing area. In many ways this is the simplest situation for the use of linear programming. Generally only one single product is produced from the blast furnace, i.e. one type of molten iron. Secondly the prices of the various raw materials used are stable and fixed for long periods of time. Neither of these two conditions are satisfied for the other two examples.

The operation of a blast furnace is semicontinuous. Each production run of the blast furnace produces several thousand tons of molten iron. The input raw materials are iron ores, which contain between 20% and 60% iron in the form of oxides, a variety of other metallic and nonmetallic oxides such as limestone, and coke. These materials are mixed and the blast furnace is then filled with the mix, the 'burden'. They are then burnt inside the furnace so that various chemical and physical reactions take place. The output is the molten iron, still containing some impurities notably carbon and phosphorous, which must be removed in the steelmaking process, slag containing most of the impurities, which is of little value, and blast furnace gas which provides a useful source of fuel for other parts of the steel works. The costs of the iron ore and coke are roughly similar. The object of the operating policy is to select the amounts of particular ores and other materials, the production rate and mode of operation which will minimise the cost of producing the required quantity and quality of molten iron. Further details of this case study are given in the paper by Lawrence and Flowerdew.

The coke was produced from coke ovens within the iron and steel company using coal purchased from the National Coal Board. Limestone was produced from the company's own quarries. Although some iron ore is mined within the U.K., the major proportion of iron ore used by the U.K. steel industry is imported from all parts of the world. This work was initially carried out in the early sixties when the U.K. steel industry was still organised as separate competitive private companies with only one of them being state owned. However co-operation between them took place for the importing of the iron ores. A joint company called BISC(Ore) Ltd., had been set up with the responsibility for obtaining all the supplies of iron ore for the U.K. This company arranged long-term contracts to take given amounts of ore from the different ore producing countries of the world. They also arranged the details of transportation of the ores to the U.K. using a mixture of long-term and short-term chartering of specialised ore carrying ships. Each year BISC(Ore) would have specified amounts of each type of ore available for use by the steel companies. Each ore was made available to the steel companies at a pre-set fixed price. An annual 'auction' took place to determine the allocation of the ores to the various steel companies. In this auction prices were fixed and companies bid against each other by asking for particular amounts of each ore. If the total amount requested by the steel companies was less than the amount BISC(Ore) had available for the year then all demands were met. If it exceeded the amount available then each company received only some proportion of its original request. In the 'first round' of bidding amounts of each ore were allocated to each company and each company informed of what amounts of what ores were still available. The companies then made revised requests in a 'second round' of bidding. The process continued until each company had sufficient ore for its planned production of iron. Naturally the amounts available of the 'best' ore were always oversubscribed. The problem was to determine what mix of 'best' ores and 'poorer' ores could be most economic. Taking account of the usual excess demand for the best ores so as to ask for a 'good' blend of the better of the 'poorer' ores as early as possible in the bidding process gave a particular company an advantage over its competitors. In addition, after each round of bidding, using the ores already allocated to the company and then determining what would be the best mix of the remaining available ores in a rapid systematic manner would also help in the various negotiations that occurred in each round of bidding. The end result of this bidding system was that each company would know what quantities of each ore it had available at fixed prices to determine the mix of ores to use to provide the molten iron it required at the lowest possible cost for the next year.

In order to apply linear programming techniques to this problem it is necessary to derive a model of the blast furnace that will relate the outputs to the inputs, i.e. to know the way in which the individual chemical components of all the different inputs will distribute themselves between the molten iron output and the slag output. Many of the relationships must be assumed. Some of them are well-established whilst others are less certain. At the original time of this application no one had produced a 'final and complete physico-chemical model' of the way a blast furnace works. These considerations however, apply whether the policy is produced by the use of linear programming or by the more conventional methods then currently used. The inadequacies and limitations of the work in this area arise from the lack of a complete understanding of these relationships. In this instance the major problem arises in developing an adequate model of the blast furnace process, the 'black box' in which the inputs are converted into the outputs. The types of relationships that have been generally used are as follows:

1. *Production rate.* The proposed production rate is normally fixed at some desired level unless any surplus iron that may be produced can be sold outside the company. It is usually assumed that a constant percentage (the yield) of all the iron in the raw materials emerges in the molten iron product.

2. *Iron ore.* There are limits on the availability of the various ores. These are determined from the annual negotiations that take place between the companies and BISC(Ore) Ltd. Different ores have different percentages of iron and different types and amounts of impurity. The proportion that finds its way into hot metal of the various impurities is assumed to have a fixed value.

3. *Coke.* The amount of coke consumed in producing each ton of iron depends on the particular furnace and also on the make up of the mix of materials charged to the furnace; for example, the percentage of slag, hard ores and soft ores, and the percentage of part refined ores etc. The relationship which determines the coke consumption rate is a matter of great controversy in blast furnace circles. Many studies have been carried out to attempt to determine this relationship more exactly, particularly by the British Iron and Steel Research Association. Most of the relationships rely to a very large extent on multiple regression analysis rather than on a complete understanding of the physical and chemical reactions that take place. Thus this is at present an inadequacy in the models.

4. *Ore quality.* The efficiency of the operation of the blast furnace in terms of the rate of output depends on the amount of 'hard' ores and also the amount of 'fine' ores in the burden which is charged into the

furnace. At present there does not exist an equation relating the way in which these properties influence the efficiency of the furnace. Because of the imperfect knowledge in this area these constraints are usually taken care of by placing upper bounds on the percentages of hard and fine ores that may be used in the mix.

5. *Slag Properties.* For various technical reasons the level and mix of impurities in the slag must be controlled within certain prescribed limits.

A detailed specification of those constraints is shown in Fig. 6.4 and the resulting linear programming model in Fig. 6.5.

FIG. 6.4 Constraints for a blast furnace burdening model

The following relationships are those used for a typical programme.

(i) Ore A has analysis:
50% Fe, 0.1% P, 0.1% Mn, 0.03% S, 3% CaO, 0.2% MgO, 3.0% SiO_2, 1.0% Al_2O_3.
This analysis is typical of the many ores from which we select.

(ii) 95% of all Fe in the raw materials emerges in the hot metal. The hot metal itself is made up of 93% Fe and 7% impurities.

(iii) Total hot metal required = 10,000 tons/week.

(iv) Coke consumption $\leqslant$7500 tons/week.
Coke contains 88% carbon and the carbon consumption is given by
cwt of carbon/ton of iron = 12+0.35 × cwt of slag/ton of iron
+0.1 × cwt of hard ore/ton of iron.
−0.125 × cwt of sinter/ton of iron.

(v) Slag 'volume' = 1.02 × total weight of CaO, MgO, SiO_2, Al_2O_3 from all ores, less the amounts known to get into the hot metal.

(vi)	% of ores which are defined as 'hard'	$\leqslant 20$
(vii)	Total % of fines from all ores	$\leqslant 20$
(viii)	Total P/total hot metal	$\leqslant 0.28\%$
(ix)	Total Mn/total hot metal	between 1.2% and 1.5%
(x)	Total S/slag volume	$\leqslant 1.6\%$
(xi)	Total MgO/slag volume	$\leqslant 7\%$
(xii)	Total SiO_2/slag volume	$\leqslant 33\%$
(xiii)	Total Al_2O_3/slag volume	between 14% and 19%
(xiv)	Basicity ratio $\frac{CaO + MgO}{SiO_2 + Al_2O_3}$	= 0.95

(xv) There are upper limits on the amount of some ores set by commercial restrictions.

The objective is to minimise the sum of the costs of the ores plus the cost of the coke. This gives the simplex tableau shown in Fig. 6.5.

Reproduced from the paper by Lawrence and Flowerdew (1963)

Fig. 6.5 Simplex tableau for blast furnace linear programme

Restriction no.	Nature of restriction		Hot metal	Coke	Sinter	Limestone	Ore A	Ore B (hard)	Ore C (fine)	Slag vol.	Total SiO_2 + Al_2O_2	Total burden
1	Define production per week	10,000=	1									
2	Yield of Fe from ores	0=	−.98	.005	.6	0	.5	.6	.55	0	0	0
3	Carbon balance in coke	0=	.6	−.88	−.125	0	0	.1	0	.35	0	0
4	Coke usage upper limit	7,500⩾		1								
5	Slag 'volume' definition	0=	−1.7	8	20	56	10	12	14	−98	0	0
6	Total burden definition	0=			1	1	1	1	1			−1
7	Hard ores ⩽20% of burden	0⩾						1				−.2
8	Fine ores ⩽20% of burden	0⩾							1			−.2
9	Phosphorus ⩽.28% of hot metal	0⩾	−.28	.06	.2	0	.1	.2	.3	0	0	0
10	Manganese ⩽1% of hot metal	0⩾	−1.5	0	.3	0	.1	.4	.5	0	0	0
11	Manganese ⩾.8% of hot metal	0⩾	1.2	0	−.3	0	−.1	−.4	−.5	0	0	0
12	MgO ⩽7% of slag vol.	0⩾		.16	3	.5	4	5	4	−7	0	0
13	S ⩽1.6% of slag vol.	0⩾		.8	.03	.02	.05	.01	.1	−1.6	0	0
14	SiO_2 ⩽33% of slag vol.	0⩾	1.5	5	4	1.5	3	2	5	−33	0	0
15	Al_2O_3 ⩽19% of slag vol.	0⩾		3	5	.5	2	3	4	−19	0	0
16	Al_2O_3 ⩾14% of slag vol.	0⩾		−3	−5	−.5	−2	−3	−4	14		
17	Define SiO_2 + Al_2O_3 in slag	0=	−1.7	8	9	2	5	5	9		−100	
18	Basicity ratio = 0.95	0=	0	0	11	54	5	7	5		−95	
19	Commercial restriction on sinter	10,000⩾			1							
20	Commercial restriction on ore A	3,000⩾					1					
	Objective function (to minimise cost)		0	8	6	1	6	5.5	5	0	0	0

N.B. (1) The absence of an entry implies a zero.
(2) All costs are fictitious, and chemical properties are approximate but add up correctly.
(3) Slack variables are not shown but are arranged so that equalities or inequalities arise as shown.

Reproduced from the paper by Lawrence and Flowerdew (1963)

The lack of technical knowledge at the time the case study was carried out, expressable in a linear form as required for the model, can be a serious drawback for several reasons.

(a) The equations used to predict process performance may not be accurate enough; this could affect the efficiency of the purchasing decisions that must be made. The use of sensitivity and parametric analysis on the solution to the model could, however, mitigate this particular problem.

(b) Some variables which might be usefully used in the model may have to be omitted because there is no reliable information on the effect of changing the variables.

(c) The restrictions used in the model may be weaker or stronger than is strictly necessary. Savings in purchasing cost would result if certain restrictions could be relaxed.

(d) Confining attention to the blast furnace operations may lead to a suboptimal approach to the problem. For subsequent processes there is an upper limit on the phosphorous content of the molten iron that is produced from the blast furnace. In general, it is cheaper to produce higher phosphorous iron and more expensive to refine it. It is obviously more economical to determine the purchase quantities from a total model that includes buying the materials, the blast furnace operation and also refining the iron into steel.

Not withstanding these inadequacies, the use of linear programming in the Iron and Steel industry led to significant reductions in the costs of purchasing raw materials and improvements to the iron making efficiency. In general this has been achieved by changing the mix of the middle range ores, in terms of price and iron composition, that are purchased, rather than a better use of the limited good ores that are available.

In order that the potential savings given by the annual purchasing policy model are achieved it is necessary to set up a system so that each burden charged into the blast furnace for each production run is the same as that given by the annual model. Using the recommended ores in incorrect quantities will produce a burden which is more costly than the optimum (otherwise it would have been recommended). Thus even if the total amounts of ore used by such burdens adds up to the amount ordered by the annual model the total annual cost of producing the molten iron will be higher. The work described above was carried out in the late fifties and early sixties. Since then substantial research effort has been devoted by many iron and steel companies all over the world, to obtaining a better understanding of the technical processes that occur inside a blast furnace. Great success has been achieved. It is fair to say that at the present time many of the inadequacies discussed under (a), (b) and (c) earlier have been overcome.

6.8 EXTENDING THE BURDENING MODEL TO COVER SEVERAL DIFFERENT BLAST FURNACES

The company where the above work was carried out, Richard Thomas and Baldwins Ltd., had two separate works, one at Ebbw Vale and Spencer Works at Newport. Spencer Works was a new plant being built during the early stages of the development of the work described above. Ebbw Vale had two blast furnaces which, having been built at different times and being of different sizes, had different operating characteristics, whilst Spencer Works had two blast furnaces which were virtually identical for burdening purposes. The model described above was for burdening a single blast furnace. The total company purchases of iron ore could be obtained by running three different linear programming models and then cumulating the individual ore requirements. Obviously some problems arise with this procedure in allocating limited ores between the different blast furnaces.

It had also been discovered that some, but not all of the iron ore, could be pretreated before being charged into the blast furnace, to give a material with better physical properties for iron making than the unprocessed ores. This process is known as sintering. Such a sinter plant was constructed as part of the Spencer Works development. Initially the sinter plant was operated so as to produce a sinter to satisfy the requirements of the Spencer Works blast furnaces. This same specified sinter was also made available for use in the Ebbw Vale furnaces. Experiments with the linear programmes for each of the separate Ebbw Vale furnaces showed that they would operate more profitably with a different sinter make up than that which was best for Spencer Works and also for the other blast furnace at Ebbw Vale. Three different types of sinter are thus required for the most economical furnace operation for the whole company. Unfortunately the sinter plant operating characteristics were such that only one type of sinter could be produced. This should obviously be some compromise between the best mixes for the three different blast furnaces.

To determine this it was necessary to set up a company ore purchasing model which integrated together the three different blast furnaces and the sinter plant as well as producing burdens for Ebbw Vale and Spencer Works. Such a model necessarily specifies the make up of the optimal sinter mix as one of its outputs. This implies that the three separate linear programming matrices for the burdening at the different furnaces must be combined into a single matrix with a term added to the objective function representing the cost of producing the sinter. In addition, in each blast furnace matrix the single variable giving the sinter quantity to be used, say S_1 for the first furnace, is replaced by several variables S_{i1} for the ith chemical element, e.g. iron, manganese, phosphorous etc., one for each of the chemical elements of concern in that furnace. Note in the initial work the sinter composition was known and fixed so that the amount of the ith chemical for the first blast furnace was merely some known constant times S_1. Variables S_{i2} and S_{i3} are defined similarly for the blast furnaces 2 and 3. A further set of equations will be required defining the total amount of ith chemical, $S_i = S_{i1} + S_{i2} + S_{i3}$, as a linear function of the amounts of ores used to make the sinter, i.e. input to the sinter plant. However, if no further constraints are added the linear programme could choose the proportions of any two elements, manganese and phosphorous for example, to be different for the sinter intended for use in either Ebbw Vale furnace or at Spencer Works. Thus a further constraint must be added that the chemical make up of the sinter going to the three furnaces must be the same. This can be most conveniently handled in the following way.

Let P_1, P_2 and $1 - P_1 - P_2$ be the proportions of the total sinter produced going to furnaces 1, 2 and 3. Using the same definitions of variables as before then for all chemicals components we require

$$S_{i1} = P_1 S_i$$

$$S_{i2} = P_2 S_i$$

$$S_{i3} = S_i - S_{i1} - S_{i2}.$$

This will ensure a constant sinter mix for all furnaces. (Note in addition we require $1 - P_1 - P_2 \geqslant 0$.)

In terms of the problem P_1, P_2 and S_i are all variables so that nonlinear constraints have now been added. The ordinary simplex method can no longer be used to solve the problem. Fortunately by one of those remarkable coincidences, beloved by novelists but all too rare in practice, a powerful general, but mathematically relatively simple method, called 'separable programming', had just been devised by E. M. Beale to solve exactly this type of problem. The company purchasing model could thus be solved.

Note that

$$P_1 S_i = \left(\frac{P_1 + S_i}{2}\right)^2 - \left(\frac{P_1 - S_i}{d}\right)^2$$

$$= z^2 - y^2$$

where

$$z = \left(\frac{P_1 + S_i}{2}\right) \text{ and } y = \left(\frac{P_1 - S_i}{2}\right)$$

In separable programming any general function of a single variable, i.e. z^2 and y^2 above, is replaced by a piecewise linear approximation based on a finite number of points. The problem then takes a linear form and a special procedure based on the simplex method can be used to solve it. The details of the method are given in the paper by Beale, Coen and Flowerdew (1965).

6.9 THE ANIMAL FEED MIX PROBLEM—CASE STUDY 2

In contrast to the blast furnace burdening problem the process technology of the animal feeds industry is relatively simple. Various materials are mixed together thoroughly and then transformed into a cake or pellet form for the animals to eat. The major potential input materials are the cereals (wheat, maize, barley, sorghum) and protein bearing vegetable oilseed extracts or meals, the residue after the oil has been removed. In addition small amounts of minerals, vitamins and synthetic proteins are also added. The animal food produced must satisfy certain nutritional requirements, for example minimum amounts of crude protein and metabolic energy, maximum amounts of crude fibre, minimum amounts of amino acids such as methionine, lysine etc. Such constraints were initially based on research into an animal's nutritional requirements for particular purposes, e.g. cows to provide meat or cows to provide milk. These constraints will vary from animal to animal. There will also be alternative sets of constraints for each animal representing alternative feeds that are produced for particular purposes, for example for summer and winter feeding. Table 6.1 shows some examples of how these constraints vary for different brands of animal feeds. Table 6.2 shows the make up in terms of these nutritional elements of various possible input materials.

Assuming fixed prices it requires only a simple linear programming formulation to find the minimum cost mix of materials for a single brand. However it was soon found that the science of nutrition was not as precise as first thought. Intangible qualities such as palatability to the animal had to be considered. Certain raw materials had to be included or the use of others restricted because they contributed some desired imprecise quality to the feed such as the texture of the final product, or its taste, or simply to give a nice colour. For example fishmeal is a valuable source of crude protein but too much of it in the feed gives too strong a taste for the animal or in the final meat for human consumption. The difficulties posed to the processing, handling and storage caused by the physical characteristics had to be taken into account. In addition the views of the marketing department of the effect on sales of nutrition qualities above the minimum necessary levels must be included. Methods of quantifying all of these factors using experience or deliberate experiments had to be devised. The specification of an individual brand of animal feeds soon becomes quite complex.

TABLE 6.1 Varying nutritional constraints for different animal feeds brands

Constraint		Feeds					
		1	2	3	4	5	6
Moisture	Max	13.00	13.00	13.00	13.00	13.00	13.00
Crude protein %	Min	12.00	14.00	18.00	16.00	16.00	12.00
Crude fibre %	Max	5.00	8.00	4.00	8.00	8.50	7.00
Crude fat %	Min	—	—	—	—	2.00	2.00
Calcium %	Min–Max	0.5–0.7	0.6–0.8	0.8–1.0	2.4–3.3	0.8–1.0	0.45–0.6
Total phos.	Min–Max	0.4–0.6	0.45–0.65	0.6–0.8	0.6–0.8	0.6–0.8	0.4 –0.6
Ash %	Max	8.0	8.0	10.0	13.00	10.00	9.00
Poultry M.E. (kcal/kg)	Min	—	—	3100	2600	—	—
Pigs N.E. (UF/100 kg)	Min	100	100	—	—	—	—
Cattle M.E. (UF/100 kg)	Min	—	—	—	—	95	100
Methionine %	Min	0.28	0.20	0.38	0.34	—	—
Meth + Cystine %	Min	0.56	0.40	0.79	0.59	—	—
Lysine %	Min	0.72	0.50	1.01	0.61	—	—

TABLE 6.2 Nutritional constituents of some input materials

Nutritional constraints	Maize	Sorghum	Fishmeal	Peanut meal	Soya meal	Sunflower meal	Maize gluten meal	Wheat bran	Cane molasses
Moisture %	13.0	13.0	8.0	10.0	10.0	1.00	10.00	12.0	26.0
Crude protein %	9.0	10.0	72.0	50.0	45.0	30.0	60.0	15.0	3.0
Crude fibre %	2.5	2.5	0	7.0	7.0	28.0	1.5	10.0	—
Crude fat %	4.0	3.0	8.5	1.0	1.0	1.5	2.5	4.5	—
Calcium %	0.02	0.04	4.0	0.15	0.25	0.25	0.01	0.1	0.8
Total phos. %	0.30	0.3	2.5	0.6	0.6	1.2	0.4	1.2	0.08
Ash %	1.5	2.0	11.0	5.5	6.0	6.0	1.5	5.5	8.0
Poultry M.E. (Kcal/kg)	3370	3300	3200	2750	2240	900	3600	1300	1950
Pigs N.E. (UF/100 kg)	115	116	115	92	94	38	106	71	86
Cattle N.E. (UF/100 kg)	115	95	—	91	102	61	—	70	80
Methionine %	0.17	0.17	2.0	0.55	0.64	0.64	1.54	0.16	—
Meth + Cystine %	0.39	0.35	2.7	1.28	1.38	1.10	2.52	0.37	—
Lysine %	0.26	0.22	5.4	1.73	2.90	1.08	0.98	0.42	—

However for each brand, all of the constraints of whatever type can be formulated as linear inequalities on the amounts or proportions of each raw material that can be used for each brand. Hence the basic blending for each brand is achieved by linear programming. The proportions of each raw material that should be used for each brand can then be multiplied by the brand sales. The total requirement for each raw material is simply the sum of the requirements for each brand and these are the amounts the buyers need to purchase at the lowest possible price.

One major problem is to determine at what price each raw material should be costed into the linear programme. The problems here arise from the market structure for the buying and selling of such raw materials, whose day-to-day fluctuations have already been illustrated in the opening chapters of this book. It is useful also to consider the way in which buying is generally done in the animal feeds industry.

Generally speaking few raw materials are bought on a 'spot' or hand-to-mouth basis. The normal method is to buy forward, that is, to place contracts today for materials to be delivered in one of several months' ahead, (see the LFF case situation in Chapter 3). The prices at which these materials are available fluctuate daily according to the various market forces at work. If a buyer judges that maize, for instance, is a rising market and will continue that way he may buy his maize requirements for several months ahead. Conversely, if he believes another material to be falling in price he may hold off making a purchase until the last possible moment. If we take the time horizon as, for instance, 6 months ahead he may be in a position of having bought part of his requirement for some materials to be used in the sixth month ahead, a little more for the fifth, likewise for the fourth and third months and he may have completed his requirement for the 2 months immediately ahead. In some cases it may be possible to sell these contracts and thus never see the goods. However in general animal feed manufacturers adopt a policy constraint that once they have bought some raw material they will never resell it, whatever happens to prices, and will thus always use it in their own production processes.

The prices to be used to cost raw materials into the linear programme, in order to decide what amounts of each material should be purchased, could be based on a stock valuation of materials already purchased, which is updated frequently. This is past history and will not give good guides to future price if they are on a falling or rising trend. Furthermore it repeats the pattern of past purchases and may include bad purchases at higher prices than necessary. Other alternatives are to use some notional price based upon the buyer's judgement and experience—the value of this approach obviously depends on his price forecasting record—or to use the first price quoted when buying for some future months product mix requirements can begin. A common approach is to use a replacement price, that is the cost per unit of replacing today all the materials that have been purchased which are in stock or still to be delivered.

A serious inadequacy exists however with this replacement approach. Using a price at the formulation stage taken from the offers at a particular point in time is not satisfactory. This implicitly assumes that the differences between the price of the different raw materials at this initial point in time will remain the same over all the feasible buying opportunities. This is rarely true. The prices of the materials change over time as, more importantly, do the differences between them. This approach is used so as to provide a buffer between the actual purchasing of materials and the formulation of the materials required to make the brand. Using these fixed prices, determined at an initial point in time, enables the model to give the amounts of each particular material that should be bought by the purchasing manager. The purchasing manager will then go ahead and attempt to buy these preselected amounts of material at minimum cost. In many instances, however, there is no feedback from the buyer to the formulation manager, once the requirements have been set, of the changes in price that do occur. Thus often if the price of one material increases whilst another decreases, no change will be made in the requirements to be bought. It often follows from this that if, once all materials required for any production run have been bought, the linear programme is rerun with the price of each material costed into the program at the average price actually paid, the amounts of each material that would have been used will often be quite different from the amounts that were actually purchased.

Treating each brand separately is also a misconceived approach. Purchases of raw materials are made for each brand and allocated to that particular brand. When the linear programme is rerun monthly or weekly with updated replacement prices, the amounts of each material already purchased for each brand must be included in the linear programme for that brand at zero cost in the objective function. At each updating and rerunning stage the replacement prices will be different to the average prices paid for the materials already purchased, for example, after half the initially specified requirements have been bought the price may have increased. So it is no longer a useful material to buy to minimise raw material costs. At each updating stage the problem is to buy the *remaining* materials required for each month's production of each brand. As already mentioned, the prices at which particular raw materials are used will differ from brand to brand and for the same price each brand will normally use different quantities of that raw material. It may thus be better to reallocate the total material already purchased between the brands in a different way to the present allocation to achieve this objective, for example using it in large proportions for some brands and zero for others instead of using some for all brands. Similarly in practice some materials may only be available in limited quantities over periods of time because of supply shortages. Treating each brand separately gives no guidance on how this limited amount should be shared between the different brands.

It is necessary to take account of these interactions between brands so as to ensure that overall company purchases are made at the lowest possible cost. This can be achieved by combining all of the linear programs for each single brand into one large program with a set of linkage equations. The first set of linkage equations ensure that the amounts of each material already purchased that will be used in each brand sum up to the total amount already purchased. This is an overall usage constraint and means that at each updating stage the amounts allocated to each brand can be modified to minimise the overall cost for all brands. The second set of linkage equations merely sum up the amounts of each material still to be purchased for each brand. Overall constraints on total purchases can then be imposed at maximum or minimum limits if necessary. The problem here is one of size, since over 100 separate brands of animal feeds are normally produced. However, many of these are only produced in very small quantities. Typically 10 brands may account for 80% of total production, whilst the next 5 brands account for a further 10% of production. The remaining 90 or so brands only account for 10% of total production. A multibrand linear programme would thus in practice only consider the 10 most popular brands. The remaining 90 plus brands would be covered by separate linear programmes. Since they make up such a small proportion of production the potential saving lost through ignoring the interactions between them is not worth the vast increase in computation cost that would result if they were included in the multibrand linear programme. An alternative approach would be to consider whether a single composite linear programming model could satisfactorily represent all of these brands, at least for the purposes of purchasing the major raw materials of interest.

The basic production planning and hence purchase planning time unit is one month. Many of the points with regard to the usage of limited materials between brands also apply to spreading the usage over 2 or 3 months of production. Hence it would be appropriate to consider extending the multibrand linear programme to a multibrand multitime period linear programme. This obviously increases the computational effort and cost. Obviously in practice experiments need to be made to see what level of complexity is required to achieve the best possible purchasing cost. It may be that a hierarchy of models is best with different models being run at the various updating stages.

All the factors discussed should be taken into account in formulating models to generate optimal purchasing policies in these situations. The purchasing problem is thus essentially a continuous decision process

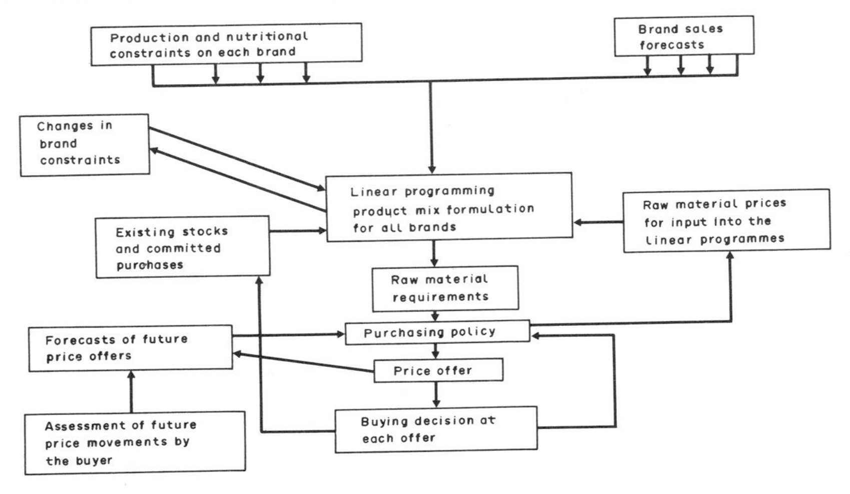

FIG. 6.6 General logic of purchasing problem

determining what materials to buy, how much of each material to buy and when or at what price to buy. In the general problem each of these three aspects of the situation are interrelated. The general logic of the situation will be as shown in Fig. 6.6. A general model would thus entail a multibrand linear programme, to take care of the blending problem and the interactions between brands, embedded within a dynamic processing model to take care of the price variability and uncertainty, and the spreading of stocks over time. (The term dynamic programming has been mentioned earlier and is described in more detail in Chapter 7). Problems arise here, however, since a model of this size and complexity will be impossible to solve analytically and expensive in terms of computer time to solve numerically, even if it could be formulated correctly. In order to achieve savings in practice, it is obviously necessary to derive a simplified model, perhaps to split up the various factors and consider each of them separately or formulate some 'average' mix at the purchasing stage. Separate combined linear programming models would then be formulated at the production stage for the differing products that need to be made. In animal feeds, for example, the problem could be divided so as to

1. Improve the brand formulation by introducing a feedback between price changes and the formulation stage by rerunning the linear programming models for each of the brands at successive stages, for example once per week with the latest prices that are available, or with revised forecasts of the lowest prices in the future at which one can expect to buy the required amounts of material.

2. Taking all brands into account and formulating a multibrand, multitime period linear programme to make the best use of existing stocks, committed purchases and the materials in short supply and running this, which is obviously more costly in computer time, at more infrequent intervals, say monthly.
3. Once material requirements have been set, to improve the buying process so as to obtain each material at the lowest possible price.

Chappell (1974), in an excellent paper well worth reading, described the system set up for a particular U.K. animal feed manufacturer. The emphasis is on the second phase above, of a multi-brand multi-time period linear programme as a basis for forward planning of raw material requirements. The Chappell paper is concerned with the problems of setting up and running a routine system in an animal feeds company. It was found that such a system could be set up at each factory with moderate computing costs more than exceeded by the ensuing reduction in purchasing cost. Williams and Redwood (1974) consider the blending and purchasing of vegetable oils to produce margarine, again on the basis of individual factories. Their problem was more complex than the animal feeds case so that much of the paper is concerned with computational considerations. They show that by use of a particular special method, the Dantzig-Wolfe decomposition algorithm, the model can be applied as a routine operating system with moderate running costs.

In both papers however the fluctuating price aspect of the problem is ignored. The basis on which materials are priced into the linear programmes is not discussed. Chappell incorporates into his system extensive use of sensitivity analysis to give some guidance to the buyers on how the requirements for raw materials should change as prices change. In general however this critical aspect of the problem appears left to the company buyer's own judgements. Only half of the problem has thus been covered. This leads naturally on to the tactical buying and price forecasting problems to be discussed in subsequent chapters.

Various approaches to the fluctuating price problem have been devised for such blending problem situations. The general approach is to try to avoid it or wish it away. In the nonferrous metal alloy example of Rule, described in the next section, the problem is avoided to some extent by buying metal each day to meet the orders for the alloy products received each day. Each order received is treated as a special order and is produced as a special order. The supply side is such that raw material can be purchased each day in small quantities of the same size as the final product order or smaller. Such a situation requires the development of extensive data processing systems. Much of the Rule paper is concerned with this problem. Rule copes with stages (1) and (2) of the problem by formulating an 'average mix' for the purchasing aspect which covers all of the various alloys produced. Separate linear programmes are then run at the production stage for each alloy.

In the animal feeds situation it is not possible to make individual purchases for each customer's order. Most of the raw materials used are imported. The materials are sold in large quantities of several thousand tons, being a major portion of a grain ship's carrying capacity for example. Sales of the final animal feed product are in small quantities to individual farmers. It is just not possible to buy an amount each day corresponding to the sales to farmers of the feed on that day. Only a very small number of intermittent purchases can be made. The fluctuating uncertain future price situation is the environment in which the animal feeds buyer has to operate. He can in no way avoid it. One approach is to delay purchases to as late a stage as possible, say the month before usage, in the vague hope that this reduces the time period in which buyers can make mistakes and achieve more stable prices. This, however, reduces the possibility of good purchases. It also suffers from the possibility that delaying too long can mean the material scheduled for delivery in the next month by the suppliers has already been bought by competitors and that there is in effect a 'local shortage' resulting in higher prices. A more sophisticated version of this approach to mitigating the situation is the 'Wedge' policy adopted by some companies. In this the buyer is given a 3 month time horizon in which to make the purchases for a particular month's usage. In the first month he has to buy 10% of the material requirement, in the second 30%, and the third and final month 60%. Obviously many alternative sets of numbers can be used, although in all cases the values are based on judgement rather than any formal assessment. They reflect the intuitive notion that the consequences of bad judgements should be minimised and that bad judgements (from a hindsight viewpoint) increase with the time span allowed for purchasing the materials for some particular month's production.

6.10 THE NONFERROUS METAL ALLOYS PURCHASING PROBLEM—CASE STUDY 3

This final section is taken from a paper by Rule (1970), presented in 1970 to an International Metals Conference, describing a system he applied within one of the major U.K. nonferrous metal manufacturing companies.

There is no doubt that in the copper and brass semimanufacturing industries, the central economic problem is to reduce to a minimum the investment in metal. The development of more advanced plant and control mechanisms, such as tight stock control and efficient production scheduling, all contribute to this objective. Perhaps the most obvious way of improving the situation is to pay less for basic materials in the first place. A system has been devised in the Rod Division of the Delta Metal Company to assist the Metal Purchasing function with the strategy and tactics of its operations. The system brings together three developments which will be discussed in turn. They are:

1. The use of a computer for routine data processing activities.
2. The use of online links which enable 'conversations' to take place.
3. Linear programming methods.

The Metal Buying Problem

The policy for purchasing metal in the Company, and in fact, throughout the copper and copper alloy industries, affects considerably the effectiveness of linear programming techniques for calculating charges for alloy blending.

The company, because of the rapidly fluctuating price of copper, adopts what is generally known as a back-to-back policy for metal buying, i.e. as orders for products are received, metal is bought on the same day to match those orders with a little extra added to allow for melting losses. The aim is to maintain the 'metal at risk'—that quantity of metal owned by the company which is not committed to customers' orders—at a constant level and thereby minimise to a large extent losses or profits due to changes in metal prices.

Thus, since metal is being purchased on a given day to match sales of products on that day, it is the mixture of materials required for those products sold that is important; the alloys might not be manufactured for some days or even weeks after receipt of the orders since they tend to be batched into economic lots. Calculations relating to alloys being melted on any day are therefore not very meaningful to the purchase requirement.

The problem of metal buying within the company is complicated by the fact that buying is carried out from two separate locations for five separate foundries and the computer system has been designed to minimise the difficulties created by this situation and to capitalise on the advantages that it can present.

In practice adhering to this policy with any degree of accuracy and, at the same time, satisfying foundry requirements, is more difficult than may be supposed for the following reasons:

(i) Sales orders are received continuously throughout the day via telephone as well as through the mail. It is therefore difficult to keep track of the total commitment that has to be covered.
(ii) When metal is sold it often happens that an agreement is entered into with the customer for the company to repurchase at a fixed

price the scrap that will be generated. The precise quantity involved and certainly the timing of its delivery cannot be known accurately.

(iii) It is rarely possible to delivery precisely the quantity of metal alloy that the customer has ordered. When overdelivery takes place on an order, metal is being sold which was never covered. On the other hand, when underdelivery takes place, metal remains in the company which had been bought to cover that part of the delivery that was never made. At the other end of the factory over and underdeliveries occur with the supply of raw material. In addition, allowances made for melting and other losses are only known statistically and may vary from time to time with the quality of material used and also with the pattern of products that are made. All these factors need to be fed back and summated as they occur, and the resulting figures used to adjust metal cover requirements. If this is not done there is a danger that the supply of material will slowly diverge from actual requirements.

(iv) Everything that has been said so far is true if only one material is under consideration. In the manufacture of brass the situation is further complicated by the fact that a large range of materials may be purchased which vary, not only in their metallurgical contents, but also in their physical characteristics.

(v) In general, material is purchased in relatively small quantities from a relatively large number of suppliers—mostly scrap merchants and industrial companies which generate scrap. It is therefore difficult to know what materials are available at any one time, and decisions to buy or not to buy have to be made with inadequate information on the overall supply position.

(vi) If the buyer is responsible for the provision of raw materials to a number of geographically separate foundries, he has the problem of scheduling deliveries to maintain a balance at each location. These foundries will have limited storage capacity and they may each have varying requirements both in quantity and makeup. Relative difficulties of delivery and transport costs should desirably be taken into account when purchases are made.

In the past, buyers attempted to balance all the above factors by applying judgement to historical information.

A data processing system

For simplicity, it is assumed in this section that there exists only one location, one Metal Buying Department, one Sales Department and one Foundry. It is further assumed that all deliveries of material to and from the company are made during the day and that normal commercial data processing activities are carried out on the computer overnight.

After the end of the normal working day the following sequence of events occurs

1. Full details of all new orders received during the day are fed into the computer and added to the existing order bank.
2. Full details of all new metal contracts entered into during the day are fed into the computer and added to the existing contract file. The difference between the total metal content of (1) and the total metal content of (2) is computed—call this quantity X.
3. Details of all despatches made including the exact weight and order number are fed into the computer. The computer matches each despatch note with the appropriate order in the order bank and an invoice is produced. The unfulfilled part of that order remains in the order bank unless the order is considered to be complete, in which case any discrepancy between the quantity ordered and that delivered is noted. When all delivery notes have been processed a total figure for over or underdelivery on that day is computed, Y.
4. Details of all raw materials arrivals are fed into the computer and compared with entries on the contract file. Procedures analogous to those for deliveries are carried out and the total net figure for over or underdelivery of raw material into the foundry is computed, Z.

(Further routines follow 3 and 4 to perform various accounting and statistical tasks.)

The overall figures $(X + Y + Z)$, which results from the above calculations, is used to adjust the metal buying requirements of the following day. In fact, a series of numbers is computed corresponding to the quantities of each element (copper, zinc, etc.) for which discrepancies arise.

To this extent we have a classical example of the use of a computer to carry out a rather lengthy but straightforward data processing job. Such a computer system is of value in keeping track of the various commercial transactions that take place and in ensuring that feedback occurs to keep a balance between supply and demand. However, it does not assist the metal buying function in its decision-making on what metal to buy and at what price. It is assumed, above, that it is somehow possible to purchase metal to cover the orders received on any particular day before the information arrives in the computer.

Until very recently it would not have been possible to consider anything more sophisticated than this and such a system would have been justified on the grounds of the increased accuracy of the information provided and the elimination of much tedious clerical effort.

Interactive computer system—stage 1

The next stage of development of the computer system is to keep track of sales orders and metal purchase contracts as they occur. The recent development of so-called online computer terminals makes it possible to place all the power of the computer at the finger tips of a metal buyer. The type of terminal used in this particular system is the Visual Display Unit (VDU) which consists of a cathode-ray screen, similar to a small television set, with a typewriter-type keyboard attachment.

Imagine a number of metal buyers, each with a telephone and a visual display unit on his desk, connecting him directly with a computer that is completely up-to-date with the position on all outstanding orders and all outstanding purchase contracts. He is speaking to a merchant and he is negotiating a purchase of material. He keys into the machine a code to describe the particular metal under discussion. Instantly on his screen appears a written description of the type of metal—to ensure that the wrong code number had not been quoted—together with the maximum quantity that may be purchased. If a bargain is struck, the remaining information—quantity, delivery date, supplier code—is entered into the computer which immediately re-evaluates its requirements. If, 2 seconds later, another buyer keys a request for information, he will be provided with guidance taking into account all the latest data. Whenever necessary, details of any existing contract can be called up on the screen for reference.

For the computer to keep the score accurately it is necessary for information on orders received to be fed in and, to achieve this, the Sales Department is also equipped with VDUs. No attempt is being made at present to enter individual orders but summaries are keyed in from time to time as orders are received, broken down into a number of alloy types and batched. It is envisaged that the time may well come when a larger number of terminals will be available in the Sales Offices, which will allow individual orders to be entered and thus eliminate the need to punch them all separately for a subsequent offline computer run.

The system does, however, remove the necessity for a separate computer run to read-in information on purchases. All documentation, the printing of purchase contracts, and the updating of data files for accounting and statistical purposes follow automatically with no further clerical intervention.

To sum up the system as described so far. A number of overnight data processing runs are carried out, including those numbered 1, 3 and 4. These runs, apart from being necessary for accounting and statistical purposes, compute corrections to be brought forward into the next day's purchasing programme. Details of purchase contracts are entered into the computer as they occur. The computer is used as a scorekeeper and calculating device as well as a data processing machine.

Before describing the next and perhaps the most significant development it is necessary to trace the evolution of thought that occurred elsewhere while the straightforward data processing system was being developed.

The foundry mix problem

The problem of producing nonferrous metal alloys is similar to that of producing molten iron in a blast furnace. Production occurs in foundries again on a semicontinuous basis, each production run being called a 'melt'. The production of one melt of a particular alloy in a particular foundry at a particular time can be formulated as a simple linear programming problem. The basic information required is as follows:

(a) the required output composition in terms of basic constituents—copper, zinc, etc.
(b) the composition of all available raw materials that can be melted, also in terms of basic constituents,
(c) the current stock levels of all raw material,
(d) all physical and technical constraints on the use of available materials—e.g. the amount of swarf (a particular form of light weight scrap) that can be melted is limited by volume.
(e) current replacement costs of all raw materials.

The linear programming technique calculates the mix of raw materials that will keep within all the constraints, and which minimises the cost. The concept can be extended to include a large number of melts taking into account the interactions between them. Interactions occur for two reasons:

(i) Different melts will compete for the same material,
(ii) The raw materials to be added to produce a given alloy will be constrained by the pool of metal remaining from the preceding melt.

A great deal of detailed work was carried out at one foundry. A mathematical model was developed to fit the practical situations with regard to all possible melts and all possible raw materials. It was shown, by running this model retrospectively against situations which had been carefully measured, that the costs developed by the computer were less than those which had been evaluated by the traditional rule-of-thumb methods.

Because a number of different alloys are produced the linear programming model for the foundry scheduling problem requires a matrix of approximately 300 constraints and 650 variables. However for the buying of the various raw materials it was found that a single composite alloy could adequately represent all the various different alloys made. Thus the linear programmes for each day's purchases could be reduced to a single programme with a size of only 70 constraints and 100 variables.

Interactive computer system—stage 2

The second stage of the system is the inclusion of the linear programming model and the presentation of information to help directly the buyers in making purchasing decisions. The 'overnight' data processing runs are carried out as before. First thing in the morning, all the sales orders received in that morning's post are analysed into alloy types and fed into the computer. Also, at this time, the expected London Metal Exchange opening copper price is fed in.

The computer then enters the linear programming model, which evaluates the minimum cost mix of raw materials to cover the metal requirements, assuming that the costs of the various materials will bear the same ratio one to another that occurred on the previous day. With this basic mix evaluated, extensive sensitivity and parametric analysis is carried out. A large number of other mixes are calculated by varying the individual prices of materials, and with each new set of costs, a minimum overall cost is indicated to cover the total requirement. From this lengthy sequence of calculations a matrix of quantities and costs is set up on a magnetic disc. The computer refers to this matrix of information each time it is interrogated throughout the day as the metal buyers carry out their business.

When the metal buyers are engaged upon negotiating contracts they can interrogate this file by means of their visual display units for advice on quantities that should be purchased and the appropriate maximum prices. If the buyer settles a deal he enters the relevant information directly on to the display screen and transmits it to the computer which automatically produces the purchase contract for the supplier, updates the files and re-evaluates the purchase requirements. The company has a manufacturing plant in London as well as Birmingham. The metal buyer in London also has a display unit. As well as enabling the same control system to be applied to operations at this other site, it is possible, making allowance for transport costs, to take advantage of the differences in material costs and availability that frequently arise between the two areas.

The system has been designed around the requirement of dealing with approximately 500 sales order items per day and up to 50 contracts might be raised for the purchase of metal to cover these orders.

At this stage of development, therefore, the system performs four functions for the metal buyer:

1. It removes all the tedious clerical effort.
2. It ensures that metal requirements are known accurately at all times.
3. It ensures that the mix of metal bought is within all the physical and technical constraints imposed by the foundry.
4. It keeps the total price paid for materials as low as possible, taking into account all available information.

None of these things could be achieved without a computer and none of these could be achieved to the same extent without online links.

Appendix
The Simplex Method for Linear Programming

The simple blending problem derived in section 6.2 was converted into the following mathematical problem.

Find numerical values for the three variables a, b and c that satisfy the following constraints

$$a + b + c = 1$$

$$9a + 15b + 45c \geqslant 18$$

$$34a + 14b + 24c \geqslant 26$$

$$a \geqslant 0,\ b \geqslant 0,\ c \geqslant 0$$

that will give the smallest possible value for the linear function

$$F = 170a + 160b + 180c.$$

Using the first equation the problem can be transformed into a two-variable problem by replacing c by the expression $(1-a-b)$ in the other constraints. This gives the problem as: find the nonnegative values of a and b that satisfy the constraints

$36a + 30b \leqslant 27$	or more simply	$12a + 10b \leqslant 9$
$10a - 10b \geqslant 2$		$5a - 5b \geqslant 1$
$a + b \leqslant 1$		$a + b \leqslant 1$
$a \geqslant 0,\ b \geqslant 0$		$a \geqslant 0,\ b \geqslant 0$

that give the minimum value of the linear function

$$F = 180 - 10a - 20b$$

The advantage of deriving a two-variable problem is that we can use a graphical approach. Any point on a two-dimensional graph will have an a coordinate measuring its distance in a horizontal direction from some origin, and a b coordinate measuring its distance in a vertical direction from the origin. Choosing a pair of values for a and b is thus the same as choosing a particular point on the graph. Furthermore, each constraint, such as the third restriction that $a + b$ must be less than or equal to one, can be represented as dividing all possible points on the graph, pairs of values for a and b, into a permissible and non-permissible region; see Fig. 6.7a. A similar result follows for the other constraints also to give us two more pictures like Fig. 6.7a. The problem is that we must confine our choice of values for a and b, that is points on the two-dimensional graph, to those that satisfy all three constraints at the same time and in addition the restriction that a and b cannot be negative. The total set of values of a and b that meet all of these constraints and restrictions will be the coordinate values of all the points lying either within the shaded triangular region of Fig. 6.7b and on the boundaries of the triangle. Any point in this region has values of its a and b coordinates that give a feasible practical solution to the problem. For this reason it is called the 'feasible region'. The value of the third variable, c, can be obtained by simple subtraction as $1 - a - b$.

As can be seen, there are a very large number of alternative feasible solutions to this purchasing problem. It is now necessary to find the particular values of a and b that minimise the purchasing cost, that is the point in the feasible region whose a and b coordinate values give the lowest value of the expressions $F = 180 - 10a - 20b$. All combinations of values of a and b which give the same value for F lie on a straight line on the graphs of Fig. 6.7a and 6.7b. For example, the point with an a value of 0.36

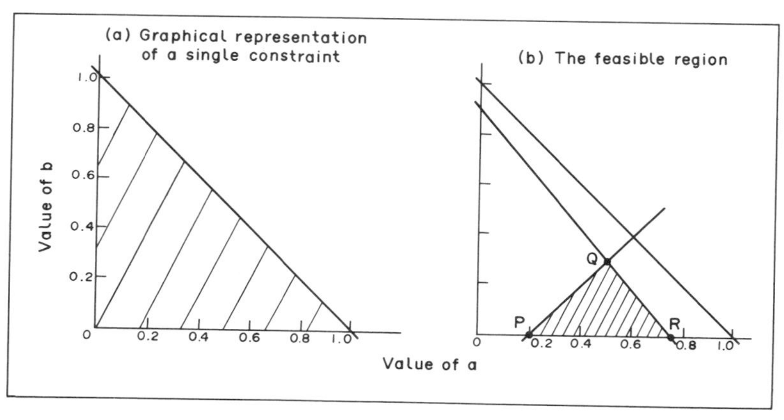

FIG. 6.7 Graphical illustration of the linear programming simplex method

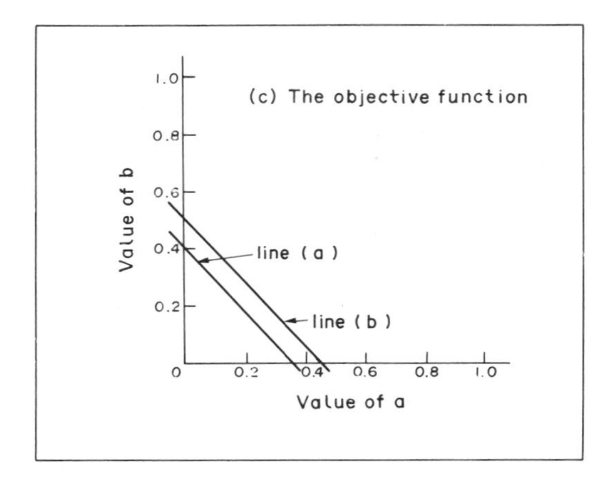

FIG. 6.7 contd.

and a b value of 0 gives a value for F of $200 - 32.4 = 167.6$, as does the point with an a value of 0 and a b value of 0.4. Indeed, all points on the line (a) in Fig. 6.7c drawn through these two points gives the same value for F of 167.6. A line drawn through the two points with a and b values of 0.45 and 0 respectively and 0 and 0.5 respectively gives a value for F of 159.5. This is line (b) drawn on Fig. 6.7c. The two lines shown on Fig. 6.7c are parallel. The one to the right, furthest from the origin point, contains points whose a and b coordinate values give a lower value for the cost expression, F. This shows that if we take line (b) and move it further to the right away from the origin, keeping it parallel to the first line (a), the points on it will give lower and lower values for the cost expression, F. The lowest value will be found by continuing to move line (b) to the right until only one point of the shaded feasible region lies on it. This last point will be one of the corners P, Q or R. In this case it is the point Q where a is 0.5 and b is 0.3. It follows that c is 0.2.

The solution to the practical purchasing problem is given by reinterpreting the mathematical variables. The poultry feed is produced at minimum cost by blending together 0.5 tons of material A, 0.3 tons of material B and 0.2 tons of material C for each ton of poultry feed that needs to be produced. This will give the lowest raw material purchasing cost for the set of prices given.

This is a simple example of a linear programming problem. It was shown that all the possible solutions to the problem must be within the feasible region bounded by the constraint inequalities. It was further demonstrated that the particular solution minimising the purchasing cost expression was at one of the corner points of this feasible region. In fact for different cost functions, F, i.e. different prices for each material, it can be shown that the best solution will always be at one of the corner points, P, Q or R. For problems containing more than two or three variables graphical methods of solution are not practicable. However, the above two properties are also true for similar problems with many variables and many constraints, provided all the constraints and the purchasing cost expression are linear combinations of the variables. These results provide the basis of a systematic method of solution to the linear programming problem. This gives a systematic sequential procedure moving from an initial 'corner' of the feasible region to adjacent 'corners' which give a lower purchasing cost, until the corner giving the minimum possible cost is reached. This solution procedure is called the 'simplex method'. The actual arithmetic involved is straightforward but extremely tedious and long. Fortunately programs exist for all modern computers which can easily and quickly give the solutions to very large problems involving hundreds of constraints and variables.

7

Scheduling the Delivery of Linseed Oil Requirements

7.1 DEFINITION OF THE PURCHASING PROBLEM

The linseed oil purchasing situation in the U.K. has been described earlier in terms of a fictitious company, Cartmel Floor Coverings Ltd. (CFC), in Chapter 3, section 3.2. The discussion showed that no practical policy for buying at the average market price existed. The company buyer had to make decisions in the face of future uncertain fluctuating prices as to whether a particular price offered to him is the lowest he can expect for some time to come, hence a good price at which to buy some part of his future requirements. He can in no way avoid such decisions daily. The tactical commodity purchasing decision must be faced in its most difficult form. In addition CFC's buyer is confronted with complex strategic problems concerning the scheduling of delivery months and determining the active buying period for each delivery month. A method for coping with the first of these strategic problems is developed in this chapter. The second problem is considered in Chapter 9, which covers the development of medium-term price forecasting models for linseed.

Planning linseed oil deliveries has to take place on a calendar monthly basis, since the suppliers only offer linseed oil on the basis that delivery will take place on some day of a specified month, at their own free choice. However detailed loading of the various machines used in the production process by CFC takes place on a weekly basis. Unfortunately calendar months are not an integral number of weeks. To convert production needs for raw materials into orders on suppliers, CFC has adopted a planning procedure that a year consists of 12 planning months each containing a complete number of weeks. If the first 2 days of a calendar month occur on a Thursday or Friday, then that week is put into the previous month, whilst if the first 3 days occur on Wednesday, Thursday and Friday that whole week is put into the current month. Public holidays and works holidays can lead to large variations in the working days from one planning month to the next. Since usage of linseed oil in production generally occurs at a relatively constant daily rate, there will be quite large variations in linseed oil consumption requirements from one planning month to the next.

The usual company policy is to update sales forecasts each quarter of the year. Monthly sales forecasts are made for the final product sales for the current year and at times for the first quarter of the following year. These are converted into a monthly production schedule for the final product. From this schedule the monthly linseed oil requirements are calculated. Annual sales forecasts are made for the following 2 years. These annual sales can be converted into linseed oil requirements. The planning months for each of those 2 years together with the public and works holidays are known. The annual linseed oil requirements are divided by the total number of factory working days of the year to give a working day's linseed usage. Multiplying by the number of working days in each month adjusted by any seasonal effects will give the monthly linseed oil consumption requirements for those further 2 years. Hence at any time forecasts of the monthly linseed oil requirements for the current year and the following 2 years are available. These are the requirements that the company buyer must meet at the lowest possible total purchasing and stockholding cost.

The manufacturing process from taking linseed oil out of the storage tanks to producing the final product ready for sale is of the order of 3 months. Allowing for the settling time of linseed oil in the tanks and for the stock turnover, sales of the final product that are going to be made in June of 1981, for example, will require linseed oil to be physically delivered to the company in January or February of 1981. The company policy is that linseed oil may be bought from brokers to cover future deliveries for usage as far ahead as 9 months. This implies that linseed oil sold by the company in its final product form in June 1981 can be purchased any time from June 1980 onwards. The purchasing of linseed oil must be considered 1 year ahead of the actual selling of that oil in the form of the final product to the outside world. At any time CFC must be making forecasts of sales for 1 year ahead of the present time in order to determine its future usage of linseed oil. The problem thus becomes one of sales planning as much as forecasting. It is necessary for the company to plan what level of sales it is trying to achieve in the forthcoming year rather than merely extrapolating sales figures from now over the next 12 months.

Presumably because of this planned nature the difference between the forecast and actual monthly usages, within the current year, had usually been less than 10%. Occasionally much larger errors than this occurred, due to some unexpected event. One example that we shall see of this was the U.K. coal miners strike in February 1972. The resulting U.K. power shortages virtually forced the shutdown of the factory for several weeks. February linseed oil usage was only a fraction of the level forecast several months earlier. Errors of less than 10% in forecasting monthly usages up to 12 months in advance is a good performance. It was felt that the pay off from investing any of the limited resources available for the study into linseed oil buying policies in the demand forecasting area was too small and uncertain to be a worthwhile activity. The current company sales forecasting and planning procedures were to be used as the basis for the monthly linseed oil usages that the buyer had to make purchases to meet.

Having specified raw material requirements the next stage is to convert these into purchase orders. As shown earlier in Chapter 3 purchase orders must be in the form of amounts to be delivered in particular months because of the market structure on the supply side. Problems here arise from two sources. The first of these is the delivery constraint. Order quantities must be either 400 or 600 tons because of the way oil has to be shipped over from Rotterdam. Furthermore, delivery cannot be specified by the linseed oil brokers more closely than that delivery will take place on any day of a particular month. The second is that linseed oil prices quoted on a particular day for deliveries in different future months are not the same.

The first constraint, together with the delivery conditions imposed by the linseed oil sellers, means that the buyer must choose particular months in which to have a shipment of 400 or 600 tons delivered. The second problem means that a 400 or 600 ton order for use in some future month can be bought in several different ways. Consider the example of the prices quoted by the linseed oil broker on 28 November 1972, already shown in Table 3.1. The broker quoted prices for oil to be delivered in December, January, February, March, April and May 1973 as given below in £'s per ton.

Delivery Month

Immediately	*Dec.*	*Jan.*	*Feb.*	*Mar.*	*Apr.*	*May*
111.25	112.0	114.0	115.0	116.25	117.5	118.75

Let us assume the buyer needs 600 tons of oil to be available for use in June 1973. This means, to allow for the settling of the linseed oil in the tanks, that it must have been delivered by the end of the first week in May 1973. The terms of the purchase contract are that the brokers require payment by the end of the month of delivery, not on the date when the order is placed. The company can place an order nine months in advance, hence guaranteeing their supplies, yet will not have to pay for it for at least 9 months, since it should never be physically delivered any earlier.

The buyer has several alternative ways of obtaining this oil and fixing the price on 28 November 1972, if he wishes to make a purchase on that day. He could decide to buy the oil now for delivery in May at a price of £118.75 per ton. Alternatively he could decide to buy the oil now for delivery in an earlier month, say February. He would then pay for it at the end of February and incur the storage and interest carrying charges himself through March, April and May. His total price would be £115 plus storage charges etc. for 3 months. At the extreme he could buy it at £111.25 per ton for immediate delivery and incur 6 months of storage charges.

In effect he must determine whether it is cheaper to store linseed oil in his own storage tanks, by fixing an early month for delivery, or store it at Rotterdam, by buying the more distant month at the higher price. If the price margin between two successive months is higher than his own monthly stockholding cost, a purchase should be made for delivery in the earlier month; on the other hand if the margin is less than the monthly stockholding cost the purchase should be made for delivery in the later month. The margin between prices for delivery in 2 successive future months varies, not only between the particular pair of months as seen above, but also over time. Table 7.1 shows four examples of daily prices over the years 1969–72. It is seen there that the price margin is not always positive. On some occasions the price is cheaper for delivery at later times in the future. This arises because there is a 'temporary' shortage and the market expects the situation to improve as time goes on. In fact for over 2 years, from January 1969 to February 1971, the price margins were generally negative. Under these circumstances it is best to have the oil delivered as late as possible. The company pays a lower price and incurs lower stockholding costs at the same time. From March 1971 to the time of the study, early 1973, the monthly price margins remained positive.

During the 6 months from October 1971 to April 1972 the average price

TABLE 7.1 Prices for delivery in future months

Example	Delivery Month						
	Spot	June	July	Aug	Sept	Oct	Nov
May 1969	111.00	110.50	110.50	110.00	109.00	108.25	108.25
	Spot	July	Aug	Sept	Oct	Nov	Dec
June 1970	113.75	113.00	112.75	112.50	112.25	111.00	110.50
	Spot	Jan	Feb	Mar	Apr	May	June
December 1971	89.50	90.00	90.50	91.00	91.50	91.50	91.50
	Spot	May	June	Jul	Aug	Sept	Oct
April 1972	86.50	87.00	88.00	89.00	89.50	90.50	91.50

margin between months, as a percentage of the price, was 0.9%. In 90% of cases it was less than 1.1%. If the storage cost in the company tanks is 12% per annum then the monthly cost is 1%. In this case the price margins and monthly stockholding costs are almost in balance, so that the total costs do not depend on the month of delivery. The largest monthly price margin over the period was 1.25%. Hence if the storage cost exceeds 15% per annum (1.25% per month), the stockholding costs always exceed the price margins. In such circumstances it would be more advantageous to have the oil delivered as late as possible in order to lower the stockholding costs. For CFC an interest cost for capital tied up in stocks of 15% was generally used. Hence the stockholding costs will be at least 15% per annum. Thus over the whole 4 year period of price data available, the monthly cost of holding one ton of linseed was always larger than the saving made in purchase cost by taking delivery one month earlier.

This result leads to a great simplification in the formation of models to derive efficient purchasing policies. It states that it is never worthwhile purchasing material in advance of usage and storing it in the company's own tanks. Essentially it means that it is cheaper to store linseed oil at Rotterdam than at the company. Material should *always* be scheduled for delivery as late as possible. This result applies to any ton of linseed oil for which there is a choice of delivery dates. For example if there is a choice, in meeting the company's consumption requirements to ensure no shortages occur over a 2 monthly period say, of having *either* 600 tons delivered in the first month and 400 tons in the second month *or* 400 tons delivered in the first month and 600 tons in the second month, the market prices for the 2 monthly delivery positions are always such that the second alternative gives lower purchase and stockholding costs than the first. This is exactly the same as choosing the alternative which gives the lower amount of physical stocks to be carried. If there were no constraints on the delivery order sizes then the obvious policy is to have each month's usage delivered by the beginning of the previous month, to allow the oil to settle in the tanks. Because of the constraint 'excess' stocks have to be carried to avoid shortages. One way to have deliveries made as late as possible is to minimise the amount of physical stocks that are held by the company. This could also be expressed as minimising the average level of stocks held. However provided alternative policies are compared over the same time period this is the same as comparing the total physical stock held. In terms of modelling the situation this approach is somewhat easier than dealing directly in terms of the average stock level.

The overall problem of determining a schedule for the delivery of future purchase orders and deciding when and at what price to purchase them can thus be split into two distinct problems. Firstly a schedule of future purchase orders in terms of their size and future delivery date should be determined on the basis of minimising the physical stocks carried by the company. There is no need to attempt to forecast the different prices that will be paid for the alternative different delivery months. They are not required to determine the best schedule for future deliveries of linseed oil, since this depends only on the levels of the physical stocks held. Secondly, the amounts scheduled for each month's delivery should be purchased at the lowest possible cost over the period allowed by the chosen active buying period considerations, using an appropriate tactical day-to-day buying policy. This second problem is discussed later on in the book in Chapters 11 to 13. Note that possible purchases of each particular scheduled month's requirements can be considered separately from those of any other month, since as we have seen no interactions occur. It follows that buying one particular month's delivery on a particular day will not automatically lead to buying other months' deliveries on the same day.

This simplification, of separating the overall problem into two distinct sub-problems, the scheduling of future deliveries and the actual buying of each month's delivery quantity, is likely to apply to all U.K. importers of agricultural materials. Since harvesting occurs over a short period the producing countries must have large stockholding facilities available. If the material is transported in large ships from the producing countries to some central port in Western Europe, then transshipped in smaller vessels to neighbouring ports and/or other countries because larger ships give lower transport costs, the transshipment port must also have large storage units available. There are appreciable economies of scale in stockholding costs from larger stockholding units. Furthermore bank interest rates in the U.K. over the past years have generally been amongst the highest in the world.

This chapter now examines the specific problems in determining the schedule of future delivery orders that minimise the physical stocks carried. As mentioned in Chapter 3 there are two other constraints that must also be taken into account in this problem, one self imposed and one imposed by the suppliers. The self imposed constraint is that the company will not itself ever sell linseed oil that it has bought previously. Secondly the suppliers do not allow the company to cancel an order once it has been placed. However the supplier will allow the company to call delivery forward, provided sufficient notice is given, two months notice for advancing a delivery by one month. The purchasing director was sure that there would be no difficulty in advancing delivery on frequent occasions. He also felt that on occasion, particularly if his storage facilities were getting full, he could manage to persuade his supplier to delay delivery for one month. On occasions the company has placed an order originally for delivery in a month with an option to bring it forward a month or postpone it for a month. This tends to happen when CFC's buyer would like to have delivery either very early in a month or very late in a month. At some stage as the delivery time gets closer and closer the supplier required the buyer to finally decide whether he wishes to exercise the option of advancing or delaying delivery by one month.

7.2 COMPANY PERFORMANCE

The purchasing director and the buyer usually plan a schedule of linseed oil deliveries for the next 12 to 15 months ahead. Revisions to the schedule are usually made each quarter, after the revisions to the sales forecasts. Sometimes the only change is to cover the further 3 months that come into the schedule each quarter. Revisions to the schedule can also be made at other times if for example an unusual event leads to a significant change to the forecast future consumptions or if the actual opening stocks in a month are significantly different to those predicted at the time of the previous schedule. This latter event occurs if the actual consumption over the last 3 months was vastly different to that forecast earlier. Thus the scheduling of future deliveries by the company is a dynamic process always covering 12 months or so ahead of the time of each revision. Because of the constraint that orders once placed cannot be cancelled, only small changes can be made to those orders such as advancing or delaying delivery by a month. Extra orders can be easily added to the schedule. However if no order has yet been placed with the supplier then it can be freely rescheduled either in delivery time or size. Thus the freedom of action to make changes to an earlier schedule will vary over time in practice, depending on how far ahead of its delivery a purchase is made.

The planned schedule of deliveries and hence monthly stocks of linseed oil for the whole of 1972 and the first 2 months of 1973 are shown in Fig. 7.1. This was produced towards the end of 1971. At the start of January 1972 there were 1270 tons of linseed oil in stock at the company. Remember that payment occurs at the end of the month of delivery so that deliveries in a month are not counted into stocks until the following month. The average level of stocks anticipated to be carried with this initial schedule over the period January 1972 to April 1973 was 599 tons. If we assume a cost on the capital tied up in stocks of 15% per annum on the average price paid over the period, approximately £100 per ton, then the cost of stockholding would be over £12,000. In determining the average stock level it has been assumed that usage occurs at a regular daily rate over a month.

It is to be expected that the actual stocks carried by the company over the period January 1972 to April 1973 will be rather different to those predicted in the initial planned schedule. Actual monthly usages will vary from those originally forecast so several revisions may be made to the

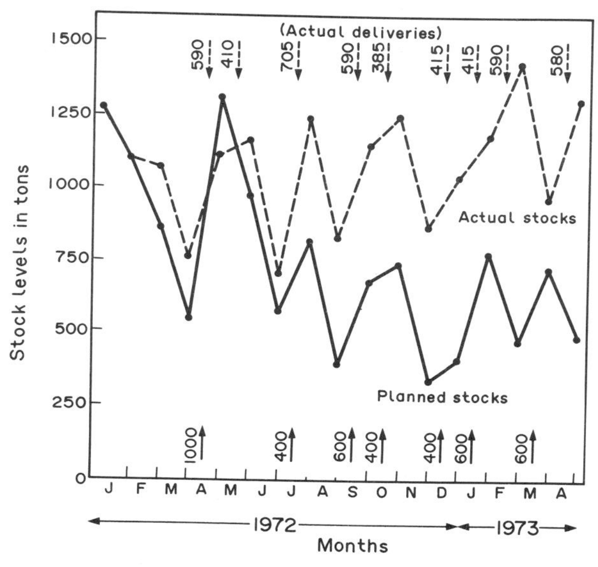

FIG. 7.1 The Company's original planned schedule of delivery orders and monthly opening stocks and their actual performance

schedule. The actual level of stocks was in fact vastly different to those anticipated in the original schedule produced towards the end of 1971. The actual average stock level was 925 tons, almost 55% higher than anticipated. On the same assumptions as before this gave an increase in stockholding costs of at least £6,500 to over £18,500. There were 800 tons more in stock at the beginning of May 1973 than was anticipated in the original 1971 schedule. The stock levels at the start of each month together with the actual deliveries that took place are also shown on Fig. 7.1. The actual amounts delivered are as shown in the figure. It can be seen that the amounts delivered by the supplier can be up to 20 tons below or 15 tons above the standard order quantities.

The large increase in the stocks actually carried above the levels originally anticipated arose from several unexpected or unusual events and inadequate rescheduling of deliveries. The first significant unexpected event was the coal miners' strike in February 1972. This resulted in a plant shutdown for much of February because of power shortages. The actual usage was only 10% of that originally forecast. Thus the actual stocks at the end of April were over 200 tons higher than predicted in the initial planned schedule. During April the forward monthly production plan was revised for the remaining months of 1972. Over the following five months it was felt that about half of the lost February production would be made up. The miners' strike had a nationwide effect on industry and individual consumers alike. Possibly some of the sales of linoleum made from the linseed oil were expected to be lost or perhaps it was impossible because of capacity constraints etc., to programme in even more production.

A revised schedule of deliveries was produced in April. The major change made was to schedule an extra 400 tons for delivery in November 1972. The original schedule contained a 400 and 600 ton order for delivery in January 1973. The first of these had the option to be delivered one month earlier. In the calculation of the stock levels for the original schedule it was found that the opening stocks for January 1973 were virtually zero and insufficient to meet January's forecast consumption. Accordingly it was assumed that the option was exercised and the 400 tons delivered in December 1972. With the extra 400 tons for November 1972, this 400 tons was put back to January and the other 600 tons delayed until February. The loss of the February 1972 consumption, even allowing for the resulting increases to the forecast consumptions for the remainder of 1972, together with the extra 400 tons scheduled for November 1972 delivery caused an increase in the stocks carried from month-to-month. Furthermore the lost production in February was not actually made up in later months. Although there was an increase in the early summer months there was a fall in the later summer and autumn months. In fact the actual consumption over May to November 1972 was virtually identical to the total forecast requirements as made towards the end of 1971 and about 200 tons less than the revised forecasts of April 1972 predicted. Thus substantial extra stocks were carried over this period.

A further unexpected fall in consumption of around 100 tons occurred in December 1972. An equivalent increase took place in March 1973. It is probably due to some special large order for the final product being delayed by the customer for three months. This caused a further small increase on the overall average level of stocks over the period considered.

A further major cause of the increase in the level of stocks carried was due to a most unusual purchase by the company of 300 tons of linseed oil in May 1972 for delivery in July, only 5 weeks later. This was unusual in two ways. It breaks the delivery constraint that orders must be of 400 or 600 tons and also the minimum cover constraint of not buying anything within 2 months of its scheduled delivery. Unfortunately no records had been kept concerning the details of this particular purchase. It appeared to be a 'special offer', significantly below the then prevailing prices by £7 per ton. It was probably part of some shipment to another consumer, who had decided it was surplus to his requirements. He was thus willing to sell it off at a bargain price.

It was an extremely good purchase apparently. Unfortunately the company did not revise its schedule of future deliveries after this purchase.

At the beginning of June 1972 no actual purchases had yet been made for the orders scheduled for delivery in November 1972 and January and March 1973. The schedule for deliveries after October 1972 could have been freely revised. The November 1972 delivery order should have been removed from the schedule as well as one of the January delivery orders. This was not done and the company bought according to the revised schedule produced in late April. Hence this low priced unusual purchase of 300 tons effectively raised stock by 300 tons over the next 9 months at least. The extra stockholding cost incurred was at least £12 per ton thus more than cancelling out the savings in purchasing cost of £7.

These examples all show the need for revising the future delivery schedule at regular intervals or whenever some unusual event has occurred. The company does not appear to have revised their schedules in a satisfactory manner. They appear to have taken the previous schedule and only added in extra orders or delayed or advanced existing orders by one month. It would appear that it would be better to produce a new revised schedule constrained only by those delivery orders for which purchases have already been made and ignoring those other forward deliveries scheduled earlier. This is better done with the help of more formal scheduling procedures. Accordingly in the next section more formal rules for producing a schedule that minimises the expected average future stock levels are derived.

7.3 THE FORMULATION OF A MODEL TO GIVE THE DELIVERY ORDER SCHEDULE

It has been shown that the schedule of future delivery orders should be derived on the basis of minimising the average future stocks of linseed oil that will be carried by the company. One obvious question is over what period of future time the level of stocks should be minimised. This period of time is usually known as the planning horizon.

The use of an infinite planning horizon implies the assumption that the production of the product using linseed oil will continue indefinitely. This would seem an unwise assumption. The final product has competitors and may be replaced in its end use by something else following technological changes. Alternatively the company could decide to get out of that market altogether because the product is no longer economically viable*. Inevit-

*At a later date after the study had been completed and this work written up and presented at an OR Society National Conference, the price of linseed oil became so high that it was completely uneconomic to carry on producing the product manufactured from it. The company changed to using a cheaper vegetable oil. Interestingly the company had said at the time of the study that such a change was completely impossible. As is often the case, 'necessity was the mother of invention'. This retrospectively provided ample justification for the choice made of a limited time planning horizon.

ably some limited length of planning horizon should be considered. Furthermore reasonable forecasts of the final product sales can only be made for a limited period ahead. Currently the company forecasts sales up to 3 years ahead. For most of the second and third years ahead it only estimates annual sales, rather than monthly sales. Thus the estimates of the monthly sales in the third year ahead will obviously contain much larger errors than the usual 10%.

If we work out successively the delivery order decisions for increasing lengths of the planning horizon e.g. 3, 6, 9, 12 months etc., then the amounts scheduled for delivery in each month will change. However at some stage the amounts scheduled for the initial months of the future period will eventually stabilise and remain the same no matter how much longer we make the planning horizon. This is because the effect of those initial decisions will only persist for a certain length of time. Precisely at what stage such stability in the initial decisions occurs varies from situation to situation. A rough rule of thumb is that the decisions for the first 6 months remain stable for horizons of 12 months or longer, for the first 12 months for horizons of 24 months or longer etc. This is of course assuming that the forecasts of the future sales are unbiased and within about ± 15% of the actual values. Hence if the limited planning horizon is adopted as a rolling planning horizon, then the decisions made will be quite close to those that would have been made with an infinite horizon, provided at the current point in time we only implement the decisions recommended for the first few months of the planning horizon. A rolling planning horizon means that each time the situation is reviewed the best decisions over the planning horizon forward from that point in time are made.

The company policy is that purchases can only be made to meet forecast usages up to nine months ahead of the current time. This means that if an order for 600 tons should be scheduled for delivery in 10 months time then company policy prevents the purchasing director and buyer from making such a purchase at the current time. Hence only the purchase order deliveries scheduled for the next 9 months forward in time can be implemented at any time. Taking this into account and all the points made earlier suggests that a planning horizon of 24 months is appropriate to the linseed oil situation. After discussions with the company this was adopted as the basis for scheduling future linseed oil deliveries.

With a limited planning horizon there will be some effects whose treatment needs further consideration with regard to evaluating alternative policies. The discussion leading to the limited planning horizon suggests that the expected stocks at the end of the horizon should be zero or as low as possible. This is because any surplus linseed oil at that time is assumed to be of no value to the company. The severe limitations on the size of delivery orders means that it is most unlikely that a schedule can be produced giving a zero stock at the end of the planning horizon. Furthermore planning to have a zero stock level at the end of the planning horizon and minimising the average stock held over the planning horizon are not usually compatible objectives. This can be seen in the three period example given below in Table 7.2. It is assumed in this example that linseed oil is used at a uniform rate over the period for the calculation of the average stock level.

Two alternative policies are given to meet the usage requirements stated. Policy 1 gives the lowest average stock held per period but leaves some surplus stock at the end of the three periods. Policy 2 leaves no surplus stock at the end of the third period but incurs a higher average stock level. The question arises whether the surplus stocks at the end of the third period are money invested that is completely lost. They can of course

TABLE 7.2

	Period	1	2	3	Stock at end of horizon	Average stock level per period
	Usage	300	100	200		
Policy 1	Deliveries	400	0	400	200	200
Policy 2	Deliveries	600	0	0	0	266.7

be resold. Possibly this might be done at a profit above their original purchase price or alternatively it might incur a large loss. Although it may be possible to make realistic estimates for a three period example, for a 2-year planning horizon it will be completely impossible to say which outcome will occur. However in view of the special purchase made by the company in May 1972 from another linseed oil user, discussed in the previous section, a loss below the prevailing prices appears most likely to occur. It seems more prudent then to assume that any surplus stocks at the end of the planning horizon have to be disposed of at a small loss, of around £7 per ton as in May 1972. In comparing alternative policies the stock levels should be converted into money on the basis of the average price and the policies compared relative to their total stockholding costs plus the loss in disposing of any surplus stocks expected at the end of the planning horizon.

However certain other practical aspects of the problem situation also need to be taken into account in evaluating alternative policies. As was mentioned earlier, actual usages of linseed oil vary about the forecast values and in addition the purchase order quantities when delivered are allowed to vary by up to 5% of the specified amount. Thus aiming to achieve a zero stock level at the end of the horizon runs the risk in actual practice of being short of material. Remember the prime objective of the purchasing function is to buy material to satisfy the planned needs of production and avoid any production losses due to shortages of material. From the point of view of security of supply, it is better to plan to finish with some safety stock at the end of the time horizon to cover the expected variations in usages and deliveries. Policy 1 in Table 7.2 gives a more realistic end of horizon stock value than Policy 2. The best procedure to follow is to determine a schedule of deliveries to minimise stock-holding and work out the expected stock at the end of the horizon. If this end stock appears a reasonable safety margin, neither too little nor too excessive, the schedule should be accepted. Otherwise a new solution can be determined by increasing the usage in the final period by whatever is thought a reasonable safety margin for end stock, to guard against large usage variations.

From this general discussion it follows that the underlying structure of the problem situation contains the following components which must be included in any model for decision-making.

(i) The planning horizon is divided into 24 successive equal monthly periods each with a deterministic demand given by the forecast usages. The differing numbers of days between some months are ignored in calculating the average stock level over a period.

(ii) Linseed oil arrives in the quantity requested in the month specified. Suppliers are in fact allowed to deliver up to 5% above or below the quantity requested in practice. Normally the deliveries are within 2% of the order size requested, so that it is a reasonable approximation to ignore such small variations.

(iii) Demand in each month must be met when it occurs and cannot be back ordered. If the linseed oil is not available for use at the beginning of a month, that month's production is lost and cannot be made up in later months. The company policy is not to allow this to happen.

(iv) The total stock at any time cannot exceed the maximum storage capacity. The deliveries cannot be arranged for an exact day within a month. Hence the only way to ensure this constraint is always satisfied is to insist that the opening stock plus the quantity delivered in each month does not exceed the maximum tank storage capacity.

(v) Oil delivered in one month is available, without loss, to be used to meet consumption from the beginning of the month following the month of delivery. The 1% loss in 'foots' is small relative to the possible variations in the delivered order size, see (ii), so assuming a 1% loss is a pointless accuracy.

(vi) Stockholding costs are not usually incurred by the oil whilst it is settling in the tanks ready for usage, since the normal purchase conditions are that payment takes place at the end of the month of delivery. Linseed oil delivered during a month is thus assumed to enter the active stocks only from the beginning of the month following the month of delivery. For calculating stock levels, the oil is assumed to be available from the first day of the month following delivery. Stock is assumed to be depleted uniformly over a month.

The scheduling of linseed oil deliveries is a sequential decision problem of making a decision month-by-month as we move forward through time over the 24 month planning horizon. The decisions that will be made each month depend not only on the future demands to be met but also the consequences of earlier decisions. There are 24 variables to be evaluated, which are the amounts to be scheduled for delivery each month. Each month there are, realistically, four different alternative sizes for the delivery order. These are 0, 400, 600 or 800 tons. Given the maximum monthly demand is about 500 tons there is unlikely to be a need for an order size of 1000 tons or more.

If we try to plan the orders forward through time then the alternative choices mount up at an exponential rate. The particular choice made in the first month gives the opening stock for the following month. Each of the four choices in the first month will lead to four different levels of the opening stocks at the beginning of the second month. In the second month there are four choices of the delivery orders for each of the four alternative opening stocks. So considering the 2 months together at the start of the first month we have $4 \times 4 = 16$ alternative choices to evaluate to give us the delivery orders for the first and second months. For the whole 24 month period we will have different choices amounting to four multiplied by itself twenty-three times to evaluate to determine the best delivery orders for each of the 24 months. This amounts to 280,000,000,000 choices. It is thus clearly impracticable to evaluate all of such a large number of choices.

We could of course invent various simpler 'heuristic' decision rules to derive the delivery orders for each month. A heuristic decision rule is essentially an enlightened guess as to what is a good procedure to use. Several alternative heuristic rules could be compared to see which gives the lowest stock levels. Unfortunately this heuristic approach does not enable us to know whether we could do much better or not. We have no standard with which to compare the alternative 'heuristic' rules.

Fortunately there is an operational research technique, called dynamic programming, which can give us the very best possible solution to our problem. It is based on the 'Principle of Optimality' put forward by Bellman in 1957. This states:

'An optimal policy has the property that, whatever the initial state and the initial decision are, the remaining decisions (over the successive future stages—months in our problem) must constitute an optimal policy with respect to the state resulting from the first decision.'

The states in our problem are the values of the stocks of linseed oil available for use in the linoleum manufacturing process at the beginning of a month. Let the opening stock at the beginning of a month be denoted by S, the amount of linseed oil used that month by d, and the delivery order made that month to be X. Because the delivery order must settle in the tanks before usage and because payment for that order is made at the end of the month of delivery it is not counted into the stocks until the start of the following month. Hence the opening stock at the beginning of the following month will be $(S + X - d)$. The average stock level over the month will be half the sum of the opening and closing stocks for the month, i.e. $\frac{1}{2}S + \frac{1}{2}(S - d) = S - \frac{1}{2}d$. Let $F(S)$ be the total stock carried measured in ton month units for this current month and over the remaining months until the end of the planning horizon. The objective as defined earlier is to find the delivery quantities to minimise this function $F(S)$. The purchasing manager has to make the best decision on the value for X, this month's delivery quantity, and then the best decisions for next month, assuming that the opening stock is $S + X - d$, relative to minimising the total stock carried from next month to the end of the planning horizon. This is the operational implication of Bellman's Principle of Optimality for the linseed oil problem.

It is computationally convenient to jump to the final month of the planning horizon and then work backwards month-by-month to the initial month of the planning horizon. A subscript is added to the variables to denote that they refer to the month which is n months from the end of the planning horizon. Hence for a 24 month planning horizon, the initial month has subscript 24, the second month a subscript of 23, the third month a subscript of 22 etc. A model of the situation can then be written mathematically as

$$F_n(S) = \min_{X_n} \{S - \tfrac{1}{2}d_n + F_{n-1}(S - d_n + X_n)\}$$

where the minimisation is subject to the conditions

(i) $X_n = 0, 400, 600, 800, 1000$, etc. (multiples of 400 and 600 from the ship size constraint on orders.)

(ii) $X_n + S \leqslant 2500$ – Maximum storage capacity.

(iii) $S \geqslant d_n$ – demand must be met when it occurs and cannot be back ordered.

(iv) $S + X_n \geqslant d_n + d_{n-1}$ – The present stock plus this month's delivery must cover demand over this month and the next, since next month's delivery must settle in the tanks for one month prior to use. where initially

$$F_1(S) = S - \tfrac{1}{2}d_1 \text{ provided } S \geqslant d_1.$$

$$= \text{'}\infty\text{'} \; S < d_1$$

The symbol '∞' denotes a very large number. It shows that a very large penalty 'cost' will be incurred if all of the demand is not met in time. This ensures that demand is satisfied at the end of the planning horizon.

7.4 THE MODEL SOLUTION—THE DETERMINATION OF THE DELIVERY SCHEDULE

The model has to be solved numerically for each given set of forecast demands for the 24 months of the planning horizon. The numerical values of $F_n(S)$ for all values of n and S are calculated and the optimal value of X_n for each combination of n and S recorded. Note that X_n is the delivery order to be made in the nth month before the end of the planning horizon. This depends only on n, the particular month, and the opening stock S at the start of that month.

The delivery schedule for linseed oil for the 2 years 1972 and 1973 is derived by substituting the forecast demands for the 24 months of these 2 years in the model for $d_{24}, d_{23}, \ldots, d_2, d_1$. The values used are the forecast demands as made by the company towards the end of 1971. The solution values for X_n for each month as a function of the opening stocks of that month are given in Fig. 7.2. The opening stock is measured along the vertical axis. Each column shows the best value for X_n, the delivery order size, for particular values for the opening stock. Consider August 1972, with an opening stock of 490 tons. Draw a line horizontally across the figure for a stock of 490 tons on the vertical axis, to see where it intersects the column for August 1972. It is seen that at this opening stock level the best value for X_n is 600 tons. So if the opening stock in August 1972 is 490 tons then the lowest stock levels are obtained by having 600 tons of linseed oil delivered then. The full policy to be followed for August 1972 can be summarised from the figure as

have 400 tons delivered if the opening stock lies between 410 and 469 tons

have 600 tons delivered if the opening stock lies between 470 and 669 tons

have 400 tons delivered if the opening stock lies between 670 and 739 tons

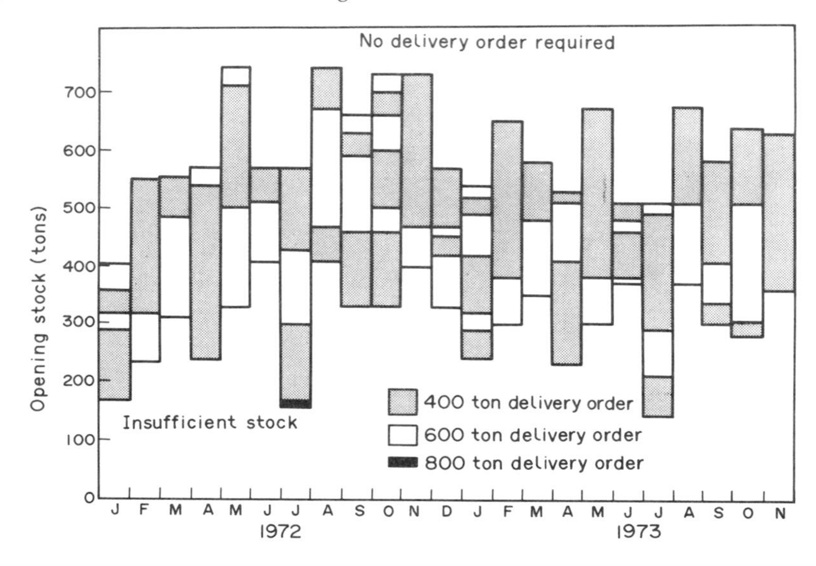

FIG. 7.2. The policy to give the linseed oil delivery order schedule over 1972 and 1973 as a function of the monthly opening stocks. (The solution to the dynamic programming model)

have nothing delivered if the opening stock is 740 tons or more.

As can be seen from the figure in some months the policy is much more complex than for August 1972. If the opening stock for any month lies above the column boxes in the upper region of Fig. 7.2, then nothing should be delivered in that month. The opening stocks can never be allowed to fall below the bottom of the column boxes into the lower unshaded region of the figure, since this means there is insufficient stock to satisfy the current month's demand. A shortage would arise contrary to the company's policy. The complexity and variety in the policies for determining the delivery orders each month arise from the large variations in the demands from month-to-month.

Figure 7.2 gives the solution to the model, the policy relating the orders to be delivered in a month to that month's opening stocks. However orders have to be placed well in advance of their delivery. The schedule for future order deliveries that will minimise the future level of stocks to be carried is obtained by applying the policy of Fig. 7.2 successively month-by-month over the 2-year period starting with the opening stocks anticipated for January 1972. These opening stocks were 1270 tons. The column for January 1972 in the figure shows that nothing should be delivered in January with this level of opening stocks. The January usage is forecast to be 170 tons, so that with nothing being delivered the opening stocks for

February will be 1100 tons. The column for February in the figure, the second column, shows that nothing should be delivered in February if the opening stocks are 1100 tons. The February usage is forecast at 235 tons so that the opening stocks for March 1972 will be 865 tons. The figure shows that nothing should be delivered in March. The March usage is forecast at 315 tons so the April opening stocks will be 550 tons. The column for April, the fourth, shows that 600 tons should be delivered in April for an opening stock of 550 tons. Since the forecast April usage is 240 tons the opening stocks for May will be 550 − 240 + 600 = 910 tons. The same process can be repeated for May and all other months of 1972 and 1973. This repeated procedure gave a 400 ton order for delivery in both July and August 1972. This means that the opening stocks for August 1972 were exactly equal to the forecast consumption. It seems more realistic to have at least some spare stock so both orders have been scheduled for delivery in July 1972. A minimum of 10 tons spare stock per month has been assumed to be necessary since the purchasing director expected opening stocks to fall to this level in his schedule.

The resulting delivery orders in each month together with the monthly opening stocks expected are shown in Fig. 7.3 for the period January 1972 to April 1973. Note that this is the time period covered in their initial schedule by the company, see Fig. 7.1. The average level of stocks over the whole period anticipated with the planned schedule of Fig. 7.3 is 486.5 tons. This is about 112.5 tons less than the level expected in the company's

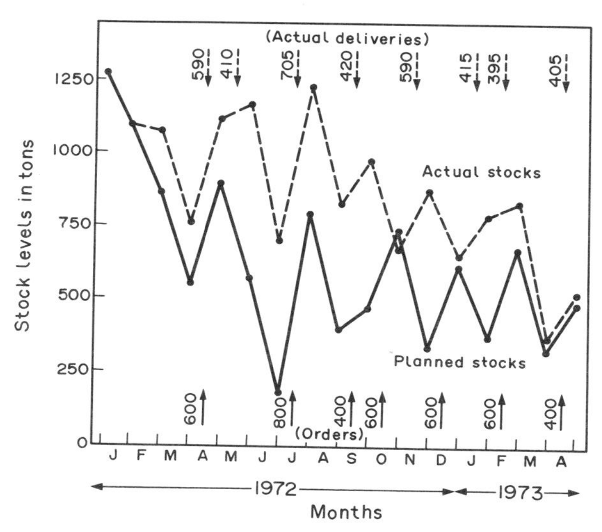

FIG. 7.3 The DP model's initial planned schedule of delivery orders and monthly opening stocks and its actual performance

initial planned schedule. Assuming an average purchase price of around £100 per ton, together with a 15% interest charge on capital tied up in stocks, the dynamic programming model policy gives a saving of £2250, or 19%, on the stockholding costs of the original company schedule. Note that both the model schedule and the initial company schedule plan exactly the same opening stocks of 490 tons for May 1973, see Figs. 7.1 and 7.3. Thus the results are not biased by one system buying more linseed oil over the 16 months considered than the other.

The solution to the dynamic programming model was derived using a microcomputer, an Apple machine. One afternoon was required to write the computer programme correctly. The solution to the problem in the form summarised in Fig. 7.2 took 45 minutes of running time on the Apple computer. It could of course have been achieved much quicker on an expensive large main frame computer. However the Apple computer system used to obtain the solution only costs between £1500 and £2000, 1980 prices. This purchase cost is less than the potential annual savings in stock holding costs etc. that the company could have achieved. Over the 16 month period illustrated the DP model would have had to be solved no more than four times, with different forecast future consumptions, so a total computer running time of only about 3 hours would have been needed. If time was hired from a computer bureau the cost would have been extremely small.

In 1973 during the actual study the model had to be solved on a large main frame computer, since there were no appropriate microcomputers available then. The running time to obtain the model was much longer than on equivalent large computers today. Since dynamic programming is essentially a logical procedure for evaluating all possibilities, any dynamic programming model requires extensive computational effort. This explains why dynamic programming models are not a good approach for practical operations if hundreds of different items are held in stock and have to be purchased. However it is surely a worthwhile approach for a small number of items, particularly with the advent of cheap good microcomputers. Even when many items are being stocked the Pareto curve or 80/20 rule seems to apply. Only a small number of the items, say 20%, make up the vast proportion of the total annual purchasing expenditure, say 80%. If faced with a similar situation to the linseed oil ordering problem it will surely be worth using a similar dynamic programming model for up to about 20 significant items.

7.5 UPDATING AND REVISING THE DP MODEL SCHEDULES

The delivery schedule just described as the solution to the DP model also needs to be revised in practice just as the purchasing director revised his schedule over time. There are two different situations that arise which must be treated in different ways. The decision policy given in Fig. 7.2 will give the optimal decisions at the revision times over the next 6 months provided the forecasts of future consumptions are unchanged. So if the only change is that the actual stocks at the beginning of March 1972 are different to those anticipated with the initial schedule and the future consumption forecasts remain unchanged, then the revised schedule can be calculated in exactly the same way as described in section 7.4 using Fig. 7.2, but starting with the actual stocks in March 1972. If the forecasts of future consumption are changed then the DP model must be rerun with these new forecasts to give a revised Fig. 7.2 for the policies for the 24 months forward from the current revision time. The procedure described in section 7.4 is then applied to the revised Fig. 7.2.

We can thus apply the policies updated by these means to the period January 1972 to April 1973 to see how the buyer could have reacted better to what actually happened and what level of stocks would have been carried. In order to give a fair comparison with the actual performance it has been assumed that purchases for delivery in different months ahead are bought in the same months that the company purchased those delivery months. If it is a month when the company did not have a delivery an appropriate purchase time has been assumed based on the actual purchase times for the immediately preceding months. At each revision time no orders for delivery over the next three months can be modified in size or delivery date. Delivery orders already purchased for 4 months or more ahead are allowed to be advanced or delayed a month. Such a change can only be made once and no further change to such an order has been allowed in later revisions to the schedule. If a delivery order, scheduled at an earlier time, has not yet been purchased then this order is ignored and does not constrain in anyway revisions to the schedule.

Over the period considered in the initial analysis there were in fact apparently only two major revisions to the forecast future consumptions by the company. The first was in April 1972, as mentioned earlier, where after the effect of the coal miners' strike in February 1972 the consumption forecasts were increased for the remainder of 1972. The other revision to the forecasts was towards the end of 1972 when the consumptions for 1973 were increased. So revised versions of Fig. 7.2 were produced in April 1972, to give the policy for the 2 years to March 1974 and in November 1972 to give the policy for the 2 years to October 1974.

It is assumed that the initial schedule derived from the DP model in the previous section was produced during October 1971. By this time the company had already purchased linseed oil for delivery in April and July of 1972. A fairer comparison with the actual performance achieved by the company is to derive an initial schedule, which will be modified subsequently, accepting these orders that have already been purchased. The April order was for a 400 ton and a 600 ton order to be delivered. The initial DP schedule, see Fig. 7.3, required 600 tons to be delivered in April but nothing more until July. It has, therefore, been assumed that the second 400 ton order for April delivery could be delayed for one month until May. In fact this is precisely what the purchasing director himself did in practice. The 400 ton order already purchased for delivery in July 1972 is accepted since such an order is in the initial DP schedule. The only difference that these orders already purchased earlier in the year make to the initial schedule derived from the DP model, see Fig. 7.3, is to bring forward by 2 months a 400 ton order planned for July delivery to May 1972. However this adds 50 tons to the average stock levels over the period January 1972 to April 1973 implied by the initial DP schedule, i.e. an extra stockholding cost of £1000. The large effect of what appears a minor change to the schedule shows the importance of planning the schedule properly.

A new policy was produced for April 1972, i.e. a revised version of Fig. 7.2. By this time the 400 and 600 tons scheduled for delivery in September and October 1972 had been purchased. So these are a constraint on revising the schedule. The policy can be used to evaluate whether the special purchase of 300 tons in May 1972 for delivery in July 1972 of about £7 per ton below the prevailing prices is a good or bad decision. The first step is to work out a revised schedule for the next 12 months accepting this

TABLE 7.3 The revised schedule in May 1972 accepting the special July order. (All numbers in the table are in tons)

Year	Month	Opening stock	Demand	Closing stock	Order
1972	May	1110	360	750	400
	June	1150	455	695	0
	July	695	180	515	700
	Aug	1215	450	765	0
	Sept	765	335	430	400
	Oct	830	335	495	0
	Nov	495	415	80	600
	Dec	680	305	375	0
1973	Jan	375	275	100	400
	Feb	500	330	170	400
	Mar	570	415	155	400
	Apr	555	230	325	0
	Averages	745.0	340.4	404.6	

Average stock level = ½ (745.0 + 404.6) = 574.8

TABLE 7.4 The revised schedule in May 1972 refusing the special July order. (All numbers in the table are in tons)

Year	Month	Opening stock	Demand	Closing stock	Order
1972	May	1110	360	750	400
	June	1150	455	695	0
	July	695	180	515	400
	Aug	915	450	465	0
	Sep	465	335	130	400
	Oct	530	335	195	600
	Nov	795	415	380	0
	Dec	380	305	75	600
1973	Jan	675	275	400	0
	Feb	400	330	70	600
	Mar	670	415	255	0
	Apr	255	230	25	400
	Averages	670.0	340.4	329.6	

Average stock level = ½ (670.0 + 329.6) = 499.8

300 ton order and secondly to work out a revised schedule excluding this order. These are shown in Tables 7.3 and 7.4. Accepting the 300 ton special offer leads to an increase in the average stock levels of about 75 tons. The extra stockholding costs thus incurred, at £1125, are equivalent to about a £4 per ton premium on the price paid for the special order. Since the savings on the prevailing prices for the order is £7 per ton it should be accepted. In working out the revised schedule it has been assumed that 600 ton order for delivery in October, purchased at the end of December 1971, could be delayed for one month until November. The 600 ton order scheduled for December 1972 and February 1973 plus the 400 tons for April 1973, see Fig. 7.3, are replaced by orders for 400 tons to be delivered in January, February and March 1973.

The only change to the schedule at the revision in November 1972 was to delay the 400 ton order scheduled for March 1973 delivery by one month.

The actual monthly opening stocks that would have occurred with the actual consumptions are shown in Fig. 7.3. These can be directly compared with the original planned schedule based on the forecast consumption values. The actual average stock level over January 1972 to April 1973 was 753.1 tons compared to the value of 536.5 tons anticipated with the initial schedule of Table 7.3. This is an increase of 40%. This increase is a result of the forecasting errors, the unexpected losses in consumption in February and December 1972 plus the special purchase of 300 tons in May 1972 for delivery only 7 weeks later.

The actual average stocks carried by the company using their current policy over January 1972 to April 1973 were 925 tons. Thus the company in practice carried 172 tons in stock over the whole period more than they need have done, i.e. an unnecessary increase of 23% in stocks. On the same assumptions as discussed earlier this amounts to an extra stockholding cost of at least £3440 over the 16 month period illustrated. In addition the company had over 700 tons more linseed oil in stock at the beginning of May 1973 than necessary. This would further increase the stockholding costs in the succeeding months.

Comparing the relevant parts of Figs. 7.1 and 7.3 shows that most of these extra stockholding costs were incurred after September 1972. The earlier period is dominated to a considerable extent by the constraints imposed on the previous schedule by the company's early purchases for 1972 usages. This would suggest that in practice the DP model will give higher savings in stockholding costs per year that the value achieved over 1972 and early 1973.

7.6 ALTERNATIVE HEURISTIC POLICIES

Since the dynamic programming model gives the best possible schedule for future delivery orders to minimise the level of stocks carried, it can be used as a standard with which to compare alternative simpler policies. The criterion is to try to find a policy giving similar savings in costs but which has a simpler procedure to calculate the order schedule. One obvious policy is to have a constant order size of 400 tons, the smaller delivery quantity, whenever a delivery is required in a month. The policy to determine whether a 400 ton order should be delivered in a month or not must satisfy all of the constraints discussed in previous sections and imposed on the dynamic programming model. Such a policy takes the following form:

'If the opening stocks of a month are less than that month's demand plus the following month's demand then an order of 400 tons should be scheduled for delivery in that month, otherwise nothing should be ordered for delivery that month.'

The schedule that would have been produced by this policy in late 1971 to meet the forecast usages for 1972 and early 1973, given in Fig. 7.4, is not very different to that given at that time by the DP model, see Fig. 7.3. They are also quite similar over the other months of 1973. The two planned schedules have the same opening stocks of 490 tons for May 1973 and also the same closing stocks at the end of December 1973 of 105 tons. The average stock level over the 24 month planning horizon of 1972 and 1973 was 433 tons for the DP model and 453 tons for this simple policy. The difference in stockholding cost is only £600. Thus the simple policy

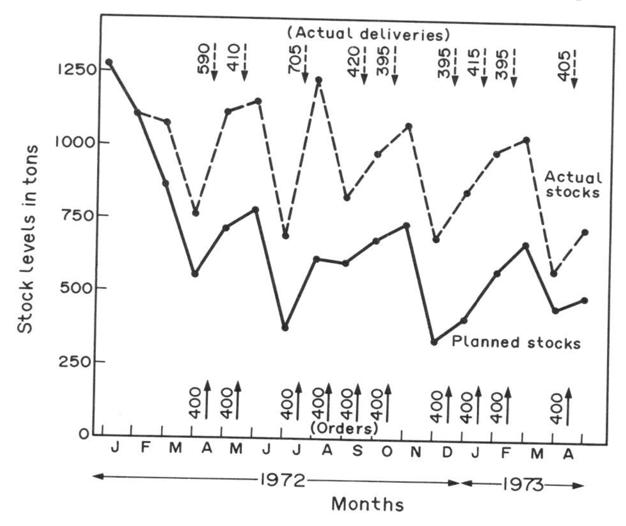

FIG. 7.4 The initial plan and actual delivery orders and opening stocks for a constant 400 ton order policy

performs well at the initial planning stage and may be preferred because it does not need a computer to calculate the policy.

In practice this simple policy does not work quite so well. Because of its fixed delivery order size of 400 tons it does not adapt efficiently to the changing situation in terms of the revised forecasts and the actual stock levels. This simple policy was applied to the actual situation over the period January 1972 to April 1973 and revised under the same starting conditions and constraints as in section 7.5 for the DP policy. The orders delivered and actual opening stocks are shown in Fig. 7.4 also. The average stock level over January 1972 to April 1973 was 815.5 tons compared to the 753.1 tons of the DP policy. In addition the simpler policy had 200 tons more in stock at the beginning of May 1973 than the DP policy. The extra stockholding cost is over £1240. A further point to bear in mind is that the stocks over the first 9 months of the period were identical for both the simpler policy and the DP policy since the deliveries up to October 1972 are conditioned by the buyer's early purchases in 1971 for deliveries over 1972. Thus this £1240 extra cost in stockholding is incurred over a period of only 7 or 8 months. Hence over a full year it would most likely be larger.

Thus because the simpler policy does not apparently adapt well to the changing practical circumstances the dynamic programming model policy is recommended as best. The computational effort to work out the schedules, once Fig. 7.2 has been derived for the DP policy, is virtually identical for the two policies. The extra effort is a computer run on a minicomputer of about 45 minutes, four times a year. This is surely worthwhile because of the cost savings in practice that will be achieved and better adaptability of the DP model to the changing situation.

7.7 SAFETY STOCKS TO COVER ERRORS IN FORECASTING FUTURE USAGES

The DP model gives as its solution the best way to schedule future orders to meet the forecast demands at the lowest possible levels of excess stockholding. The forecast demands are treated as if they will occur with certainty. As mentioned earlier even without unexpected events, the actual usages will naturally vary stochastically above and below their forecast values. Thus if the planned opening stock for some month is only slightly above the forecast usage and if the actual demand is above the forecast, because of such natural stochastic variations, a shortage of linseed oil may occur. An earlier attempt to guard against this was to plan to have some surplus stock at the end of the planning horizon. However further consideration shows that this is likely to be inadequate since this surplus stock will probably only be bought for delivery towards the end of the planning horizon. It does not really provide any protection against shortages in the earlier months of the planning horizon. Revising the schedule at frequent regular intervals helps to lessen the probability of such shortages occurring. However this is more concerned with adapting to the consequences of unexpected events or significant revisions to the forecasts. Furthermore nothing can be done to modify the delivery orders for the first 3 months of the planning horizon at each revision time. It is possible by chance that the actual usages in all 3 months could be above their forecast values, giving a high probability that a shortage will occur. It would therefore seem more prudent to deliberately plan to have some safety stock in reserve at the beginning of each month to guard against such variations.

The size such a safety stock should take depends upon the forecasting errors that would normally be expected to occur. The large errors that occur due to unexpected events, such as the coal miners' strike and the apparent delay of a large order from December 1972 to March 1973, should be excluded from any analysis of forecasting errors. Such unexpected large changes or significant increases in the general forecast level of future usages will be covered by the continuous revision of the delivery order schedule. Within the period of the next 3 months the large errors are anyway likely to be overforecasts. Since little can be done in the next 3 months to increase linseed oil supplies without paying premium prices most likely, a large underforecast means that a special extra order for the final product has been received. At such short notice obviously the price will surely be adjusted to include the extra purchase costs. Because any large errors are thus likely to be overforecasting errors this in itself builds some surplus safety stock into the system. This effect as mentioned earlier is one cause why the actual stock levels were so much higher than expected in the planned schedules.

Unfortunately no records had been kept of the forecasts of linseed oil usages prior to December 1971. The only data available at the start of this research project was thus for the fourteen months January 1972 to April 1973. Some of these month's suffered from unexpected events as already mentioned, so they should be excluded from the analysis. The remaining data sample is thus very small so that no real statistical analysis can be carried out. The largest underforecasting error was just under 15%. In these circumstances one simple policy would be to plan the delivery order schedule such that the anticipated opening stocks in any month are at least 15% higher than the forecast consumption for that month. This covers the worst situation that occurred over the period for which data was available.

To derive a schedule that achieves this objective it is necessary to modify the original dynamic programming model of section 7.3. The third constraint in the model must be modified so that the opening stock is at least 15% above the forecast usage. Hence it should now become

(iii) $S \geqslant 1.15\, d_n$.

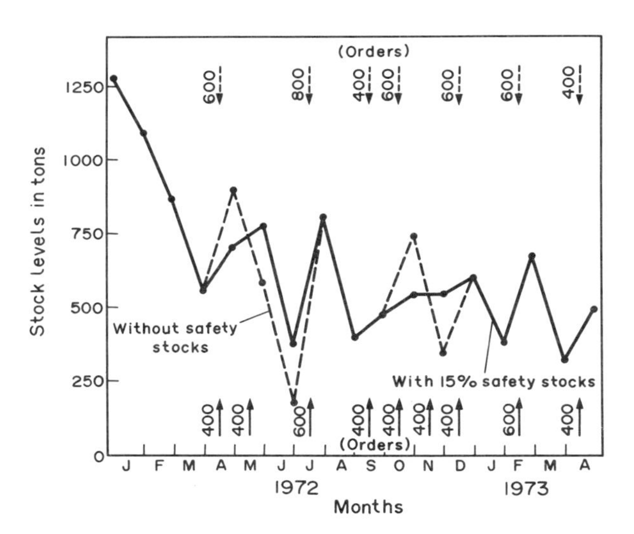

FIG. 7.5 The initial planned DP schedule modified to give a 15% safety margin at the start of each month

The modified dynamic programming model now needs to be resolved to produce a new policy exactly in the same form as Fig. 7.2. The best delivery order schedule to meet the forecast usages over the 24 month planning horizon of 1972 and 1973, starting with an opening January 1972 stock of 1270 tons is derived by exactly the same procedure as described in section 7.4, using the new version of Fig. 7.2.

The resulting planned initial schedule of orders and the expected monthly opening stocks are shown in Fig. 7.5 by the continuous line. For comparison the initial DP schedule of Fig. 7.3 is shown by the broken lines. As can be seen there are few differences between the two schedules. In only 5 months are the opening stocks different, although there are more differences between the delivery orders. The average expected stock level over the period January 1972 to April 1973 is 499 tons, only 12.5 tons more than the original DP schedule. Thus ensuring an adequate safety reserve to guard against moderate forecasting errors only increases the average stock levels by a very small amount. It is thus a better policy to follow. Note that it gives lower average stock levels than the purchasing director's own initial schedule but better protection each month against demand forecasting errors.

8

Medium-Term Price Forecasting

8.1 TYPE OF MODEL NEEDED

As discussed in Chapter 4 an efficient company purchasing policy requires good forecasts of the general trends of future price movements so that the stock cover or active buying period can be adjusted as market circumstances change. The forecasts must be able to predict the periods when reversals in the trend are likely to occur, i.e. when minimum or maximum price levels will be reached. Most statistical forecasting systems are essentially linear in nature, although they may include cyclical or seasonal components. However, as illustrated for cocoa in Figs. 4.5 and 4.6, a consistent single seasonal price pattern rarely exists for a commodity. The linearity means that the current trend estimate is extrapolated continuously into the future for ever. Hence, although such statistical methods may be appropriate for short-term price forecasts over 1 or 2 months for use within a tactical day-to-day buying policy, as is illustrated later in Chapters 12 and 13, they are clearly not suitable for strategic purposes. Descriptions of the application of statistical methods to longer term price forecasts are rare. Jarrett (1968) has discussed their application to average monthly values for forecasting the future prices of wool, for use by the Australian Wool Board in selling wool. His results suggest that good forecasts are not likely to result from this approach in general. Labys and Granger (1970) in their book, *Speculation, Hedging and Price Forecasting in Commodity Markets*, have a substantial section cataloguing a long series of failures in attempting to determine purely statistical forecasting systems for many different commodities.

Forecasts which enable price reversals or changes in trend to be anticipated well in advance can only be generated from models which try to explain how the particular commodity market behaves with regard to price determination. The market prices are the results of individual buying and selling decisions made by buyers and sellers all over the world. These decisions are determined by their own particular circumstances, their need to buy or sell materials and their assessment of whether the market will become better for them in the nearby future. The commodity market is merely a mechanism to enable buyers and sellers to agree a price at some particular time. The question to be answered is why a particular price is found acceptable to buyers and sellers at one time whilst completely different prices become acceptable at other times.

Although economists have traditionally regarded commodity markets as an example of 'a pure economic system' little successful applied work has been carried out. Three published studies should be mentioned however as exceptions to this. Researchers in this field would find it worthwhile reading up these studies in detail. These are a simple analysis of maize prices by Working (1958), an econometric analysis of the tin market by Desai (1966), and a detailed study of the cocoa market by Weymar (1968). Working, for the crop years 1921/2 to 1938/9, demonstrated a correlation coefficient of about −0.8 between the annual changes in the logarithm of crop year price and the logarithm of crop size. Desai regressed the London Metal Exchange Tin price, over 1948–61, against the ratio of stocks to demand. By including a dummy variable for the years of the Korean War the regression equation explained 76% of the price variance.

Weymar in his book *The Dynamics of the World Cocoa Market* argues on theoretical grounds for the 'Supply of Storage' theory. Through a formidable economic analysis he develops a model where the average price of cocoa over a crop year is dependent on a single variable, the ratio of the carryover stocks (the surplus of supply over demand) to the grindings (the actual consumption). Over the years 1952/3 to 1963/4 a substantial correlation is proved. Weymar recognises the importance of studying the current market expectations of economic variables in addition to the historic values established at a later date. The current expectations are calculated in a very tortuous manner. When a weighted average of market expectations and historic values of the carryover stock to grindings ratio is regressed against price, there is an improvement in fit over the pure historic values analysis. The validity of the model was assessed by goodness of fit tests for the period over which the model parameters were derived. However Tewles *et al.* (1974), through a private correspondence with Weymar, were able to tabulate and calculate the pertinent variables up to the crop year 1971/72, so extending Weymar's analysis by 8 years. The results showed that the model was an excellent fit to describe cocoa price behaviour. However, as shown in Chapter 2, commodity prices changed radically in 1972/73. Substantial modifications to the model would be required for it to explain price behaviour since 1973.

Economists have usually been concerned mainly with long-term market behaviour and all too often with attempting to explain commodity markets in terms of preconceived economic theories. The problems facing decision-makers, either buyers or sellers or merchants, in the short- and medium-term arising from the large and uncertain price fluctuations have been ignored. The general approach and the models that economists generally use, require knowledge of the real underlying values for supply and demand. These values are never known at the current time but only many months or even years later. Economists hence spend a great deal of effort in attempting to find or estimate the real underlying values in order to make real time forecasts. This is a very difficult and frequently impossible task. Hence economists often spend a great deal of time complaining of the problems of inadequate and inaccurate data. All too often they give up at this stage assuming that their theories are right and that the problem of practical price forecasting is insoluble because of the data problems. For a producer or consumer it is of little use complaining of the inaccuracy and unreliability of the data on production, consumption, stocks, etc. Buying and selling decisions must be made on the basis of whatever information is currently available.

An alternative approach used by this author in several studies is to turn the argument about the unreliability and inaccuracy of the data on its head. It is suggested that it is precisely this inaccuracy and everchanging view and knowledge of the state of the market at any time that is a major cause of the sometimes excessive price fluctuations that occur. Primarily it is the market's general beliefs or 'expectations' as to the present and future state of supply and demand that determine prices currently and over the near to moderate term future periods. The actual values of supply and demand factors are irrelevant to price determination since the people in the market are continually making decisions on their 'expectations', their present unreliable inaccurate knowledge. They will only know the real actual values at some later stage when this information is too late to affect their decisions. This hypothesis implies that, if in fact supplies of any commodity are abundant relative to demand but the market expectations are for an exteme shortage, prices will increase, notwithstanding the actual situation of a real surplus. It is only when the market expectations or beliefs change to accept the real situation that prices will decrease because of the abundance of supply. The subsequent fall in price will often then be quite rapid compared to what it would have been if the nature of the real situation had been appreciated earlier. Rapid price movements then have a tendency to overdo the reaction so prices fall even lower than perhaps is really implied by the relative supply/demand situation. The large fluctuations in price arise to a major extent from the delays and distortions between reality and reality as perceived by the market at the time.

8.2 A CONCEPTUAL MODEL OF A COMMODITY MARKET

A general conceptual model of the world market economy for a raw material would take a form similar to Fig. 8.1. Price is generated as the result of the interplay of supply and demand. It will itself then lead to

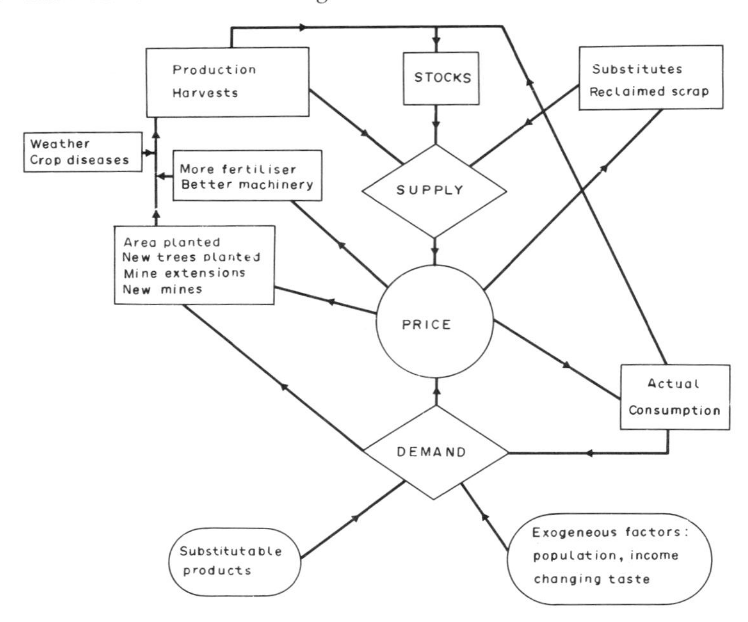

FIG. 8.1 A conceptual model of the world market economy for a raw material

modifications to many of the future values of the parameters in the system. Thus the model has many feedback loops within it. The precise nature of these feedbacks and their response times will constrain the type of forecasting model required.

The major source of supply of a commodity is the production of new material from the new crops each year or the mine outputs. A secondary source is the stocks of material carried over from previous years. At times they have been as much as 40% of the new year's production. In the nonferrous metals markets reclaimed scrap is a third, further, highly significant source, of more importance than carryover stocks. In a few cases direct substitutes exist such as coberine as a replacement for cocoa butter.

A farmer can choose the amount of land he will plant and the amount of seed he will sow on the land. Thereafter he is almost entirely at the mercy of the weather and crop diseases. His direct control of the amount that will grow and ripen for use is very small. At harvest time the farmer, having invested time and money, virtually always feels it is worthwhile harvesting and selling the whole crop, whatever the prices in the market. It is extremely rare for a farmer to leave his crop to rot in the fields because of low market prices. Governments in their agricultural policies may complicate the situation by setting up agencies to buy up surplus crops at guaranteed minimum prices. However farmers invariably complain that these minimum prices are below the cost of production. So the general point still stands that production at harvesting time depends on the crop size not the level of the prevailing prices for the material. Similarly mine operators normally attempt to run their mines and smelters at a regular output rate. If prices are low or demand is low they tend to sell the output at the low price or stock the material hoping for better things next year. Thus within the crop year for agricultural materials or calendar year for metals, production either cannot be, or by deliberate decision is not made adaptive to fluctuations in prices over that period. The one exception to this is reclaimed scrap metal. As prices increase it becomes profitable to reprocess general scrap to sort out the particular material required. Scrap dealers become more energetic in seeking out fresh supplies. In this case the supply does respond relatively quickly, within several weeks, to high or low prices.

What low or high prices do is to affect future production. For annual crops, for example maize, soyabeans or beet sugar, high prices around planting time will motivate farmers to increase the sown acreage. This increase may come from land left fallow the year before or may be land switched from other crops. Low prices have the opposite effect. For commodities such as cocoa the cycle is longer. High prices can motivate farmers to plant new cocoa trees. However it will be at least 5 years before those trees produce any new cocoa. Increasing mine output significantly requires a large capital investment and 2 or 3 years to come to fruition. Developing a new ore deposit requires much longer, nearer to the 5 years for the new cocoa trees.

The area planted is only one component determining the size of the crop. The yield achieved per acre planted, or per tree cultivated, is just as important. Late spring frosts, heavy rainfall or unexpected droughts whilst the crop is growing, damage due to insect pests or plant diseases can lead to enormous variations in the yields from one year to the next. Estimates of the size of the crop made over the period from planting to the final gathering-in of the harvest, months later, can thus vary widely. Sometimes man-made barriers such as wars and strikes retard the movement of supplies from producers to consumers. When trying to forecast the size of the crop a year or more ahead, prior to planting, these stochastic effects impose a large uncertainty on the likely crop size that generally obscures any systematic feedback due to the earlier prices.

On the other side of the equation the distinction between demand and consumption is an important step to understanding market behaviour. Consumption in any year is what user companies or individuals actually consumed. Demand is the amount of the material that, based on their own economic and financial situation generally over the year, they would have liked to consume. If there are ample supplies of material and prices remain at reasonable levels then consumption will reach the total demand level. On the other hand, shortages of material may mean that all such demands cannot physically be met. Prices will increase and act as a rationing device which temporarily removes certain elements of demand. Either an alternative material or product becomes cheaper and is thus preferred, or the final consumers of the manufactured products made from that material just cannot afford to buy them, or at least only in lower quantities than they would have preferred.

Although demand, at least temporarily, is not affected high prices will reduce the rate of actual consumption. The response time for this to occur can vary enormously, for example from 2 months or less for soyabeans to over nine months for cocoa. It depends on how far ahead of usage the initial consumer buys his raw material and how long are his manufacturing and distribution lead times. Manufacturers prefer to maintain some stability in their final product pricing. So rising or falling prices do not lead immediately to increases or decreases in the final product prices. Initially at least consumption is not affected. The decision by a manufacturer to change his raw material or to stop manufacturing a certain product, necessarily incurs various extra disruption costs. He will not wish to be making such decisions at frequent intervals. A sustained period of high prices and an expectation that this will continue in the future is usually required before such changes are made. For all of these reasons actual consumption does not always react quickly to high or low prices.

The effect on demand rather than consumption will be even further delayed. The time lag for price to significantly affect demand is rarely less than a year. Thus high prices in the market in a crop year will not affect demand in that crop year, although it may affect actual consumption. Demand in subsequent years may be affected. A second year of high prices will certainly lead to a cut back in demand.

In some countries farmers are obliged to sell all of their crop to a government marketing company at a price fixed by the government. These prices often are not readily adjusted to reflect changes in the world market prices. Similar policies may also apply to consumption. Such policies either eliminate or delay further the feedback of price on supply.

Governments may try to control the acreage planted directly by financial incentives or taxes. More usually governments intervene after the crop is harvested. They may set up a new extra demand, a government stockpile, if prices fall below some minimum level. This then becomes a further source of supply in future years, to be released if prices rise above a pre-set level. These policies are usually more motivated to support farmer's incomes than to moderate price fluctuations in the market. A further factor behind them is to increase or maintain a country's self sufficiency in home produced food.

International commodity agreements have similar provisions to finance stockpiles or control the exports of producer countries. In all these cases attempts have been made to impose more rapid feedbacks of price on the supply/demand factors. However all they hope to do is maintain market prices within very wide minimum and maximum price limits. Such agreements have not had much success in the past. Maintaining prices above a minimum level by controlling supply, i.e. stockpiling part of it, is perfectly possible, particularly on a national basis for rich countries, e.g. the U.S.A. and the EEC. Keeping prices down in a period of shortage is likely to be impossible in practice. Immense stocks of the commodity would have to be carried at a huge stockholding cost. It is doubtful on current production and consumption trends that such large stocks could actually be generated. Massive investment in agriculture would be required.

8.3 AN HYPOTHESIS FOR PRICE DETERMINATION IN A COMMODITY MARKET

The market is an amalgam of buyers and sellers. It is a place where decisions are continuously made to reach an acceptable price at the current moment. Essentially a commodity market acts as a 'price determination mechanism'. The assertion is that the buyers, and presumably sellers, make rational decisions based on their current purchase and stock cover situations and their beliefs as to the future state of the market. The assumption of the conceptual model is that prices are generated by the interaction of the supply offered by the sellers as a group and the demands made by the buyers as a group. As illustrated and discussed in Chapter 3 buyers rarely make decisions to buy for consumption needs more than a few months in advance of usage. This time span is much much shorter than the response time for the market prices to affect future supplies and demands. Thus the feedback effects of the prices are irrelevant to the decisions they make. Past prices may have influenced current production and demands but as we have seen these effects are overwhelmed by the various stochastic factors that exist. As far as they are aware when they make their decisions on a price, supply and demand are exogenous variables. Their values will be determined by many factors but these are all independent of current prices.

For decision-making to be rational there must exist some way of assessing or measuring the relative state of the market and then a common mechanism to translate this into a price. Vast amounts of information circulate in the market which are presented to each buyer and seller individually. If they are going to agree a price they must agree on the general state of the market and the price appropriate to that state. At any time several buyers and sellers over the world are agreeing to transactions at that same price or at levels very close to it. As this is happening all of the time the market must form a 'collective expectation', or a majority 'expectation', of the relative state of supply and demand. There must also exist a common mechanism to convert that 'expectation' to a price acceptable to the buyers and sellers. If this were not so then market prices on a day, or within a week, should be fluctuating as much as market prices do over a quarter or a year. There would thus be no agreement even on what today's price level should be.

The market has explicitly or implicitly determined some way of measuring the current relative state of the market supply and demand factors. It can then compare this state with the states that have occurred in the past. If this state is the same as has occurred several times in the past then the common price determination mechanism will generate the same 'price' as occurred at those times in the past. This hypothesis implies that the market has through its experience over time formed a (1–1) relationship between each particular state of supply and demand and 'price'. This relationship must persist for long periods of time. Alternatively, in this era of floating currency exchange rates and general price inflation the 'price' factor may itself be changed over time in a systematic fashion. Such systematic changes are however independent of the supply/demand changes in that commodity's situation.

A useful analogy is that we must search for some instrument that will measure the state of the market, rather like a thermometer measures temperature. A thermometer is a rather precise measure. More appropriate is a barometer. The air pressure and changes in pressure give an indication of the state of the weather we can expect in general terms, such as set fair, unsettled, stormy etc. Such descriptions have affinities with the way commodity markets are often discussed. So to forecast prices we need to determine a 'market barometer' that 'measures' the current state of the market and which will then tell us the general level of future prices we can expect. The general hypothesis is that a single such barometer exists that explains the greater part of the variation in price. Secondary 'measures' may assist in the fine tuning of the price movements.

8.4 MARKET EXPECTATIONS

If a collective expectation exists then it must be based on information that is generally available to all the members of the market. Hence it must be published information that circulates freely in the market. It cannot be based on the actual values of the crop size or the crop year consumption since these are not known until much later. The expectations must be formed on the basis of the various estimates that are made, particularly for production, the reported consumption over the last few months and discussions and comment in the market trade press. It is this author's view that quantitative estimates of the market expectations can be derived from these sources, either directly or via simple models that buyers and sellers might intuitively use.

In most markets the merchants and dealers attempt to provide up-to-date advice on expected production, consumption, stocks etc. In addition government agencies, international organisations such as FAO and private forecasting companies provide various forecasts. In the case of agricultural commodities such as cocoa and sugar, production estimates are first provided part way through the harvest, then continually revised at intervals throughout the rest of the crop year and into succeeding crop years. For many of the grains and oilseeds such as maize and soyabeans, where the U.S.A. is a dominant producer and consumer, crop estimates are first made soon after planting has occurred. They are then continually revised at approximately monthly intervals over the following year. These estimates can change unexpectedly by very large amounts because of bad weather or the outbreak of crop diseases, as the following example illustrates. At the beginning of July 1974 the forthcoming U.S. maize crop for the 1974/75 crop year was predicted to be 160 million tons. By mid August this estimate had been reduced to 124.4 million tons, as a result of dry weather. Over 6 weeks an abundant supply situation had been transformed to one of an extreme shortage. A similar situation over a similar number of weeks occurred over February to March 1978, when the Brazilian soyabean crop was unexpectedly reduced by 35%.

In the case of consumption, estimates are much sparser and tend to be made later in the crop year than production estimates. Values of the actual consumption in many important countries are issued on a quarterly basis. Thus revisions to forecasts of yearly consumption tend to be made at the end of each quarter. For many commodities U.S. consumption figures are produced more regularly on a monthly basis. The time lag to publication usually varies between 1 and 3 months. For mined materials monthly values for the important producing and consuming countries are issued regularly, usually some 2 to 3 months in arrears. The availability of such data varies enormously from commodity to commodity. Generally such data is published frequently for the western industrialised nations. It is available, but at much later intervals for the less developed countries and, until recently, rarely officially published for the Soviet Union or China.

In some markets a single source may have become established as the most accurate estimator of the present state and expected future state of production, demand, stocks, etc., e.g. Gill and Duffus for cocoa, the

United States Department of Agriculture for grains and soyabeans world-wide. In that particular market its estimates become the market view of supply and demand. In other instances it may be necessary to determine how estimates from different sources should be combined to give a single figure which appears to represent best the present view of the market, e.g. Licht and Czarnikow for sugar. Alternatively, it may be necessary to model the way the market forms its expectations of various market factors, particularly demand. The approach used by many merchants for example is to average past consumption figures to assess the underlying rate of growth. The average annual growth is added on to the latest value to estimate next year's consumption. If monthly or quarterly data on consumption is published then the growth on the same month or quarter of last year can be extrapolated to the whole crop year to give an estimate of the underlying rate of demand. Such models must be based on the way buyers and merchants discuss the issue in conversations or in discussion articles in the trade press. The model must be formally stated and specified in terms of the information known to exist at the times the model is used to make forecasts.

The assertion, that market expectations can be quantitatively measured on the basis of published numerical estimates or simple formal averaging models, means that we can properly evaluate alternative forms for the price determination models over past periods, without biasing the expectations from hindsight knowledge. The author's approach means that a great deal of time must be spent reading through the past daily issues of the trade press newspapers such as *The Public Ledger*, the commodity section of the *Financial Times* or even more specialist commodity journals with some care, to find the estimates and figures circulating at the time. Weymar's cocoa model was partially based on expectations but these were generated by very complex mathematical models.

In some rare instances changes in expectation must be based on an interpretation of qualitative views rather than numerical estimates. An example of this which had a significant impact on prices occurred for copper in the late sixties. In late 1968 and early 1969 there was a strike of the United States dock workers. The price of copper immediately increased on the world markets, both the London Metal Exchange and the New York Exchange. According to published market comment at the time it was thought that the United States was a net exporter of copper and hence there would be a copper shortage due to reduced supplies. Prices continued to increase over the period of the strike, since obviously the longer the strike lasted the larger the shortage would be. However, when the dock strike ended the price of copper again went up in world markets. It was now thought that the United States was a net importer of copper, hence with the ending of the dock strike there would be an increase in consumer demand from the United States. Thus over the period of the strike the market had reversed its view as to whether the United States was a net importer or a net exporter of copper. At no time were statistics produced to support either view of the situation.

8.5 MARKET EXPECTATIONS AND PRICE MOVEMENTS

The price for each state of supply and demand is in a sense an equilibrium price, in that it is the price which the buyers and sellers collectively find acceptable. The acceptance of new information by the market explicitly changes market expectations and gives rise to a new 'equilibrium price level'. Because new information takes time to be absorbed by the totality of people in the market and its effects evaluated, and because of the natural inertia of a complex system such as a commodity market, prices do not jump immediately to the new price level consistent with the change. It is more that a change in market expectations gives a new impetus and direction to price movements. Prices after a lesser or greater immediate reaction begin to move towards this new price level with several reversals and rallies on the way. There may be an appreciable time lag of up to 6 months before prices finally reach the level consistent with the new expected state of supply and demand factors.

In most cases, the points in time at which significant changes in expectations occur are known with reasonable certainty, coinciding with critical periods at the sowing, growing and harvesting stages of the crop and when government statistics on consumption, imports or exports are published. Often before a price movement is fully completed new information may become available which leads to revised market expectations and hence a completely different pattern of price movement. Rumours, which are given great prominence in daily and weekly market reports, will only have a short-term effect of increasing the fluctuations about the underlying basic market price trend unless they are substantiated and hence lead to a change in market expectations.

As we have seen the expectations of supply and demand factors can change significantly over the crop year as new information becomes available. Through the market behavioural mechanism these changes will lead to changing and fluctuating prices over the crop year. Prices can thus be expected to be more volatile in years when changes to estimates of supply and demand are frequent and large and relatively stable when these estimates change little over time. Thus the dynamic situation with regard to price is a result of the dynamic situation in the supply and demand situation, not the other way round.

The time scale over which prices in a market adjust to a new situation varies from commodity to commodity. The better organised the market and the more frequently the data is published the greater the proportion of the eventual price movement that occurs soon after the change. Market reaction times have in general shortened in recent years compared to the sixties, say. The reaction time depends also on the nature of the change. Prices react quicker if the situation worsens to an even more extreme shortage than to an improvement brought about by increased supplies. To some extent the markets react to a 'fear' of a shortage but require the 'hard physical evidence' that there is an increased surplus before falling to significantly lower price levels.

8.6 A MARKET BAROMETER

The appropriate form for the barometer may differ from commodity to commodity depending on its particular production characteristics and distribution patterns. Different forms may be appropriate at different times of the year, since information becomes available on a piecemeal basis. The form in which market comment is presented and the way buyers and sellers discuss the state of the market will give valuable insight into possible precise forms for the price determination model. An important market characteristic for consideration in any model is that market factors are generally discussed in relative terms. The forecast crop is high relative to anticipated consumption, stocks are high or low not in terms of tonnage but in terms of the number of future weeks' consumption they will cover. This implies that the significance of an increase in production of 100,000 tons, say, will depend on the existing estimate of the relative abundance or shortage of supply over demand. If the market generally views the world in relative, i.e. ratio, terms then it would seem only sensible in developing market behavioural models to specify the price determination mechanism in relative terms. Thus specifying price as a linear function of production, consumption and stock would seem inappropriate in general, notwithstanding the advantages of the simpler statistical methods required.

The following quotation, taken from a report prepared some years ago by the economists of the Commodities Division of UNCTAD on the production, consumption and price prospects of a particular commodity, is surely a classical example of how not to derive market models. 'However, because of the difficulty of using ratios (i.e. nonlinear forms of variables) [this having been indicated as the best approach earlier] the level of stocks and consumption are used as separate explanatory variables instead of their being combined in a ratio.' A simple econometric model giving price as a linear function of production and consumption as separate variables is often derived. Forecasts of future prices are obtained by forecasting production and consumption separately and then inserting the forecast values into the estimated relationship to give price forecasts. If production and consumption are estimated as linear functions of time then the price forecasts will merely be linear extrapolations of the recent past into the future. This process will be effectively the same as extrapolating price into the future via the statistical forecasting methods discussed earlier. No explanation of market behaviour has in fact been achieved.

There are three alternative supply/demand functions that are likely to figure prominently as possible candidates in a study to derive 'market barometers'. The first is the actual supply of the commodity that is physically available at the moment to meet demand. This will be essentially the stocks carried over from previous years. Later in the crop year it will include the new harvest less any consumption that has already occurred. A second approach is to extend the initial model to cover the forthcoming crops, not yet harvested, that will shortly be available to meet demand. It should include all crops that will be harvested within the crop year, including countries from both the Northern and Southern hemispheres. The barometer is thus the expected crop year surplus of production and carryover stocks above the likely expected demand during the crop year. The final approach is to concentrate attention primarily on that part of the supply that is traded between different countries. An implicit assumption here is that by one means or another a producing country will ensure that its home demands are satisfied. The demands of those countries needing to import the commodity relative to the surpluses in the exporting countries become the prime motivator of the upward or downward price movements in the international commodity markets. As has been observed it is the marginal supplies and demands that appear often to determine prices. For some commodities world production may be synonymous with world exports etc., so the second and third models will be virtually identical. Obviously alternative ways of calculating supply and demand factors for each approach exist. Further choices arise as to which countries should be included in the models. The choice is dependent on whether a 'world' market or different 'regional' markets exist, also whether various countries isolate themselves by consuming only whatever they can produce themselves. Perhaps a sufficiently good model can be derived by only including a small subset of the producers and consumers of the material. A substantial number of alternative models may therefore need to be examined to find which of the above possibilities is most appropriate for each material.

The first stage in any study is to attempt to determine possible market barometers on the basis of historic data. The second stage is examine the price movements over the year and the timing of new estimates for supply and demand factors. In this way a set of critical points of the year is established. Either significant changes to expectations are likely to occur there or each point initiates a particular pattern of price movement. Then several different models are derived relating the prices at future critical time periods to the expectations value for the market barometer at each critical time over the crop year. The forecast of price movements over the different months of the crop year is based on interpolation between the prices forecast for each critical time period of the year. It is often necessary to develop a detailed understanding of the general background of the commodity of interest, either at the very start of the study or during the modelling process.

Two case studies of the application of the theory and approach developed in the chapter to producing price forecasting models are presented in the following chapters. These two particular examples have been selected because they arose in the studies of two of the case situations described in Chapter 3. The first example covers linseed oil and was developed in the study of the buying problems of CFC. The main emphasis of the chapter is on the development of an appropriate 'market barometer', including the modification of the 'price factor' to take account of changing relative currency values. The second example covers the world maize market (or corn as it is called in North America). This study is described in rather more depth than that for linseed oil. The various steps necessary to develop models of the price formation process in the market and then to develop a medium-term future price forecasting model are presented in some detail. In world terms maize is one of the more important commodities, much more so than linseed oil. Hence, in a sense the chapter provides a 'cookbook recipe' on how to carry out such studies, at least for other agricultural materials.

The maize chapter concludes with a detailed description of how the models developed in the study were used for forecasting prices over a whole crop year. The crop year chosen is some years after the time at which the models were first developed. No changes were made to the original model in the intervening period. This concluding section shows how the price forecasts need to be adjusted to cope with unexpected and unusual changes to the market situation. It illustrates clearly that forecasting in practice is an adaptive process. The market must be continuously monitored with care. It demonstrates that in this area it is more efficient that the analyst and the decision-maker work hand-in-hand on a continuous basis to implement the results of the study.

9

A Market Barometer for World Linseed Prices

9.1 THE LINSEED MARKET SITUATION

This model was developed during a study to derive methods to improve linseed oil buying policies. The problem situation was described in Chapter 3. Part of the overall study has already been described in Chapter 7. It should be noted that this study was carried out in the first half of 1973.

Linseed oil is a mixture of the glycerides of various unsaturated fatty acids. It is a golden yellow-brown, viscous liquid. It is known as a drying oil, since on exposure to air the oil thickens and hardens, forming a crust owing to the absorption of oxygen. It is classified as an industrial oil, and not regarded as an edible oil, with the sole exception of India. Its most important uses are in the manufacture of protective coatings such as paints, linoleum, printing inks, varnish and oilcloth. However the success of vinyl paints and vinyl floor coverings has considerably reduced the market for linseed oil in recent years. The other naturally occurring drying oils, tung and perilla, compete with linseed oil, but they have special uses for which linseed oil is not suitable and are only produced in very small quantities. Thus in practical terms their impact on linseed oil is extremely small. Linseed oil is produced by the crushing of linseed or flax seed. The extract is heated and pressed into cakes and is used as a supplement to the coarse feed grains, for example maize and barley, in animal feed mixtures. The cake (or meal) is rich in protein and minerals. It is thus a substitute and a minor competitor to other materials such as groundnuts, soya bean meal, rape seed meal, etc.

The plant linseed, otherwise knows as flax seed, is a variety of common flax. Apart from the oil and meal which are obtained from the seed, the fibre in the stem of the plant can be used to make linen and twine. Linseed is grown on almost every continent. The principal producer countries are the U.S., Canada, Argentina, the U.S.S.R. and India, although only the first three are exporters. Different varieties of linseed can be grown depending on whether a high seed or a high fibre yield is required. Linseed grown for its seed output in the three exporting countries gives between

34% and 36% of its weight as linseed oil. It is usually sown in early spring and harvested in early autumn. Thus the harvest takes place during the period July to September in Canada and the northern U.S., North and South Dakota and Minnesota. In the Southern hemisphere, in Argentina and Uruguay, linseed is usually sown in August and September, (their spring) and harvested during the period December–January. In terms of production the major crop competing for the farmer's acreage at the time of planting is spring wheat or rapeseed. Changes in the acreage devoted to linseed plantings are often attributed to variations in the relative prices of linseed to these competitors.

The linseed market is less well supplied with on-going supply and demand data than many other commodity markets. Various sources of historical data exist, the Food and Agricultural Organisation of the United Nations, the Commonwealth Economic Committee and Frank Fehr and Co. (a vegetable oilseeds broker). The only source of data continuously over the crop year is a weekly publication by Ista Mielke & Co. entitled *Oil World Weekly*. This provides analysis and reports for the three major linseed producer and exporter countries, Argentina, Canada and the U.S., on the level of carryover stocks and the expected crop sizes from both official government and unofficial sources. Estimates are first made when the crop is sown, and continue through the growing and harvesting period. The final determination of crop size is made only several months after harvesting is completed. The timing of the publication of revised estimates at critical stages of the crop year is well known in advance and remarkably constant from year-to-year. Of interest is that discussions of crop prospects, etc., have occurred much earlier in time and more regularly in recent years than in the past, reflecting the greater importance attached by the market to quantitative estimates.

On the demand side the situation is somewhat different. Reasonable statistics are provided on the home consumption of linseed by the three major exporters, covering industrial consumption and seed kept back for sowing for next year's crop. In recent years the *Oil World* has provided quantitative estimates of this home consumption early in the crop year. For other countries, including the importing countries, no data on actual consumption or stocks is available from any source. The only data on demand are the annual imports of the importing countries or the annual exports of the exporting countries. Numerically world imports and exports in any year are not the same. In the data published by the FAO for example they can differ by up to 15% in any year. Even over long periods of time the differences between imports and exports do not balance out. Thus two different methods of measuring overseas demand for linseed exist. It was found that for the final form of the model to explain market behaviour, the net import demand for the major importing countries,

Western and Eastern Europe and Japan, was the most appropriate measure for demand.

9.2 A MARKET BAROMETER FOR LINSEED

No on-going estimates of net import demand are provided in any regular data source. It was thus necessary to model this factor on the basis of the methods used by merchants in other markets and in the occasional discussion articles published on the prospects for linseed. The approach used generally is to average past values to assess the underlying rate of growth. The average annual growth is then added on to the latest known values to estimate next year's demand. Markets in general tend to give more weight to the more recent rather than earlier data and only include the past 3 to 5 years in their averaging processes. This approach is thus akin to using exponential smoothing methods with a smoothing constant of about 0.5. An essential point to appreciate here is that we are trying to measure the market's expectations of the 'demand for imports'. This is not necessarily the imports that will occur, since all demand for imports may not be satisfied. In times of scarcity this is certainly so. Demand for imports may exceed supply. Hence the demand to supply ratio may take a value greater than one. It is asserted that values greater than one are meaningful and that a ratio value of 1.3, say, will have a different discernible effect on the prices that will be quoted in the market than a ratio of 1.1.

On the supply side Canada only exports linseed, mainly to Europe and Japan, whilst Argentina exports only linseed oil, mainly to Europe. Neither consumes appreciable quantities at home, merely keeping back seed for next year's plantings. The U.S. has a considerable home consumption of seed and oil but because of her very large crop can export large quantities of both seed and oil. Consumers in the importing countries hence have a choice of buying either imported linseed oil or buying oil produced at home from imported seed. In order to properly assess trade in linseed it is thus necessary to combine oil and seed on some standard scale. This is achieved by converting seed into its oil equivalent, normally, 35%, and adding this on to trade in linseed oil direct.

The essential logic of the model was assumed to be that prices were primarily determined by the ratio of demand to supply. Demand was measured as the expected net imports of Europe and Japan as mentioned earlier. Supply was measured as the expected export surplus of the three major producer countries. This was the amount remaining after their own home consumption had been deducted from their combined crops and stocks accumulated in previous years. A ratio approach is used since a change of 100,000 tons in supply can only be assessed in terms of its effect relative to demand. It is more logical to say that the surplus of supply over demand covers 20 weeks import consumption rather than the surplus is 100,000 tons say. A difference rather than a ratio implies that a surplus of 100,000 tons will have the same effect now as some years ago when demand was only half its present value. The model is a simple price determination mechanism with supply and demand determined in an exogenous fashion as discussed earlier.

Initially it was hoped to determine direct price relationships between linseed oil prices and a market barometer. However, it was impossible to find any direct relationship which would form the basis of a good predictor. All had too much scatter. It was noticed then that the outlying points of such relationships occurred when the linseed oil price relative to the linseed price was at one or other of its extreme values of 1.4 or 1.8, well away from the average of 1.6. The approach was then modified to a two stage procedure, firstly to develop a model to explain the linseed price behaviour and secondly a model to convert the linseed price to a linseed oil price. The average quarterly prices for linseed, linseed oil and linseed meal are given in Fig. 9.1 for the period 1960 to the end of 1972.

The linseed crop year used by the market is the period July to June. This is a compromise between the harvesting dates for the crops in the Northern and Southern hemispheres. In April a first estimate of the acreage sown with linseed in Canada and the U.S. is published. At the same time the final estimate of the current year's Argentinian crop is published. Since

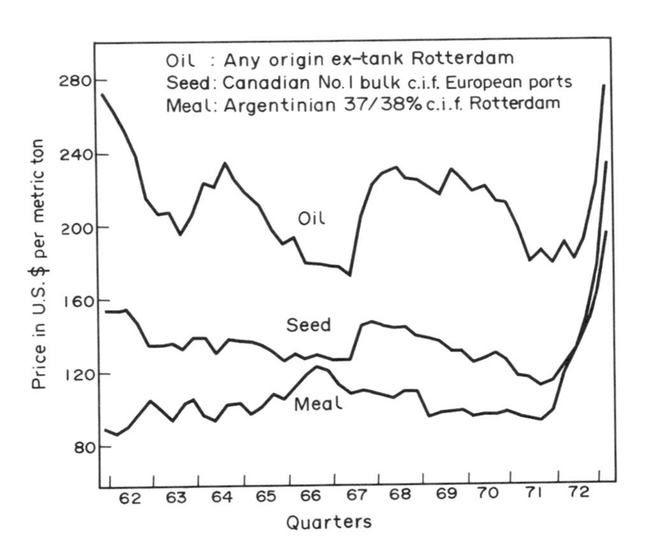

FIG. 9.1 Average quarterly prices for linseed, linseed oil and linseed meal over 1960 to 1972 inclusive

1969 the *Oil World* has published a discussion on the likely prospects for the forthcoming crop year in late April or early May. So, on an assumption of an average yield plus little change in Argentina's next crop, a preliminary idea of whether the next crop year supply/demand situation will be better or worse can be generated. However the first reliable estimates of the crops in Canada and the U.S. are available in September as harvesting begins. In late August and early September the acreage sown in Argentina (in the Southern hemisphere) is announced. Hence September is the first time in the crop year when reliable estimates of the supply of linseed for the crop year can be made. Hence it is thus the first time forecasts based on reliable crop information can be made.

The values of the market barometer, the demand to supply ratio, as calculated in September of the crop year are plotted against the average crop year price in Figs. 9.2 and 9.3 for the years 1960/61 to 1971/72. The demand to supply ratio is measured along the horizontal axis in both graphs. In Fig. 9.2 the average price is plotted along the vertical axis whilst in Fig. 9.3 it is the reciprocal of the average price. Each point on the graph represents a different crop year. If all the points on the graph lie near to a straight line then this indicates there is strong correlation between the price measure and the September market barometer value. Since all the points, with the exception of the point for the 1962/63 crop year, lie virtually on a straight line this shows there is very strong correlation between the market barometer and prices in both graphs.

Note that the three solid circles on the figures are for years when no ongoing estimates of the crop size over the various months of the year

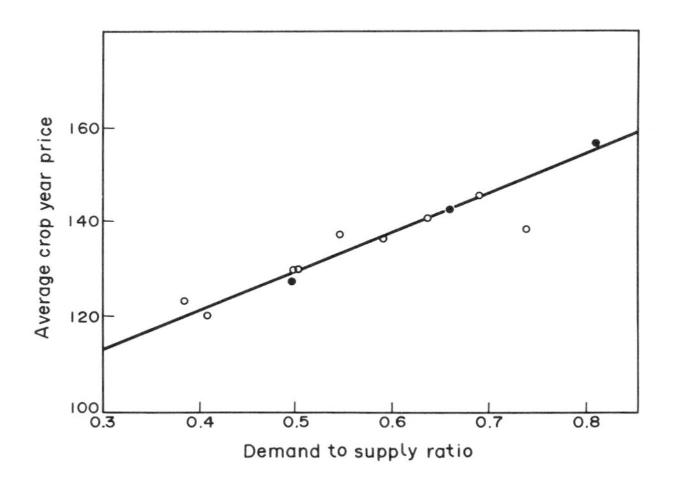

FIG. 9.2 Average crop year linseed prices versus market barometer based on September supply/demand expectations

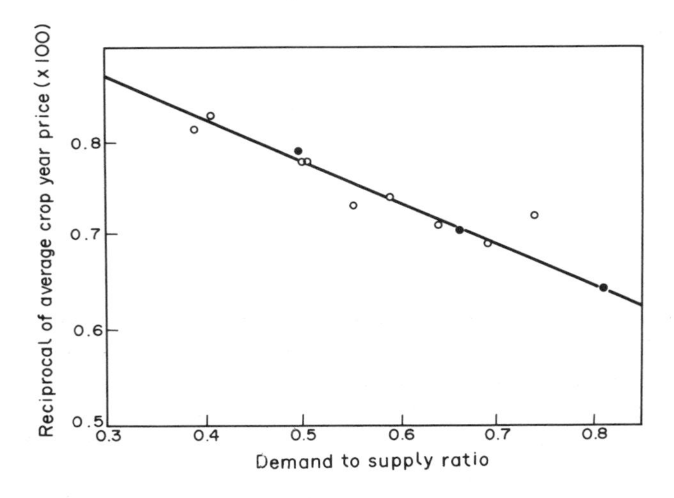

FIG. 9.3 Reciprocal of average crop year linseed price versus market barometer based on September supply/demand expectations

could be obtained. These years are 1960/61, 1961/62 and 1966/67. In these cases the historic crop estimate had to be used for the crop estimate in September. The other components of the market barometer were calculated as usual. These years were included in the analysis to increase the data sample.

There is virtually no difference between using price or its reciprocal as the price measure in terms of the goodness of fit for the data over the period covered. However, if the value of the market barometer exceeds 1.0 then significant differences will result. *A priori* one would expect a nonlinear relationship between price and the market barometer, since the market is far more volatile to small changes in times of scarcity than in times of plenty. Thus one would choose Fig. 9.3 as the best model of market behaviour and the basis for future prices forecasting, even though over the time period covered no statistical evidence exists to confirm this choice. Statistical analysis confirmed that the relationship derived from Figs. 9.2 and 9.3 explained 96% of the price variation, excluding the 1962/63 crop year.

If X represents the ratio of the expected imports to the expected surpluses in the producer countries then the regression equations for the two figures showing the numerical relationship between price and the market barometer are:

linear model

$$\text{average crop year linseed price} = 88 - 82.5X$$

nonlinear model

$$\text{average crop year linseed price} = 100/(1.00 - 0.45X)$$

In neither of these two models were there any significant serial correlations in the residual errors of the price predictions from year-to-year.

The 1961/62 crop year had the highest price and highest market barometer value over the period considered. This single point may have had an unduly large effect on the regression analysis. However including and excluding this point from the regression analysis resulted in only an insignificant change to the parameters in the relationship obtained. A similar result was obtained if the 3 years 1960/61, 1961/62 and 1966/67 were excluded from the regression analysis.

Interestingly when the same regression analysis of the two models was carried out using the actual historic values for both the demand and the supply factors, the amount of the price variation explained by the regression equation decreased to 92% compared to the 96% of the September expectations model. A further important point was that the residual errors of the models using historic data were significantly serially correlated. These two results thus provide confirmation that the expectations approach explains price movements better than the use of historic values. In the expectations model the net imports over the previous 2 or 3 years influence this year's price, since the demand factor in the model is estimated as some weighted average of the net imports of past years. If the expectations model is correct, then since the model based on historic data uses the net imports in the years of interest only, the only way previous years' net imports can influence the historic model is by serial correlation in the price prediction errors.

A further important development was that if the average price over the third quarter of the crop year, January–March, was used as the price measure, the same pattern of strong correlation existed. This is an important result since it means that we can now forecast a profile for the way prices will move from September to the end of the crop year rather than an average price over a period. This type of forecast then provides the basis for adjusting the stock cover and buying period as discussed in Chapter 4. If prices are forecast to increase from September over the rest of the crop year then the stock cover should be increased to its maximum value and the active buying period for all the months to be covered reduced, requiring those monthly deliveries affected to be purchased over

September and October say. A 2 month buying period is preferable to a single month to give more opportunities for the tactical buying policy to take advantage of the price fluctuations about the overall underlying trend. However, if detailed analysis of past price behaviour shows that prices over September are always lower than October or vice versa then a one month active buying period should be used. In addition the forecast gives an upper limit to the prices we are prepared to pay. If any daily price offer exceeds the forecast average price over the third quarter of the crop year no purchase should be made. As we move forward in time the stock cover should be maintained at its maximum value, until in late April or early May we can make our first estimate of the average price over the following crop year to see if prices are expected to continue increasing or will begin to fall.

To carry out thè procedures just described a whole series of forecasting models should be derived, relating the market barometer values calculated at different points in time over the crop year to the prices at later months or quarters of the crop year and the first quarter of the following crop year. In general each quarterly price model will have different parameter values. These will be a result of the existence of consistent seasonal price patterns. For linseed these effects were small. As might be expected the nearer to the quarter the forecast is made the better the prediction. Table 9.1 shows the predictions of the average quarterly price over January–March as would have been made in late September and mid December of each crop year. All prices are in U.S. dollars per tonne and rounded to the nearest 0.5 dollar. The forecasts are given for the linear model. The nonlinear model gave virtually identical results. The monthly crop estimates for the 1966/67 crop year could not be discovered in any published source in the U.K.

TABLE 9.1 Forecasting the January to March Quarter's Price. (The prices shown are in U.S. dollars per metric ton)

Crop year	September surplus estimate	December surplus estimate	Actual January/March average price
62/63	149.0	144.5	138.0
63/64	136.0	137.5	137.0
64/65	133.0	136.5	137.0
65/66	129.0	130.0	130.0
66/67	—	—	126.0
67/68	144.5	144.0	145.0
68/69	140.5	136.0	138.5
69/70	129.0	128.0	125.0
70/71	119.0	117.0	118.0
71/72	121.0	123.0	123.0

during the course of the study. Thus this year had to be left out of a comparison of the fitted values (i.e. forecasts) with the actual prices.

In all cases except 1967/68 the December forecast error is less than the September forecast error. Interestingly if the price forecasts are reduced or increased from September to December then the actual prices are generally a continuation of this trend. As can be inferred from the price predictions there were usually only small changes in the estimates of production, stocks and consumption estimates over the first half of the crop year. Little new information was published between September and December to enable modifications to the forecasts of the importing countries' demands to be made.

Another method of checking whether the model explains price well is to compare it with the random walk hypothesis. The random walk hypothesis states that the best forecast for future prices is the current price. Hence for the estimate made in September it would be the average price over September. The sum of the squared deviations between the Jan/Mar. average price and the September price is 471.25. The sum of the squared deviations between the Jan/Mar. average price and the forecast of this price as made in September is 164.25. This sum of squared errors is only just over one third of those of the random walk hypothesis. This again confirms that the expectations model does explain well the price movements that do occur in the linseed market.

9.3 LINSEED, LINSEED OIL AND LINSEED MEAL PRICE RELATIONSHIPS

Virtually all linseed produced is crushed to provide the linseed oil and meal. All three products are traded on the world markets. The linseed crushers appear to follow a policy of earning their revenue from the crushing process and not from speculation on price differentials. Thus as can be seen in Fig. 9.1 there is a close relation between the prices of the three products. As the price of meal rises relative to the price of linseed, the price of oil relative to linseed falls and vice versa. Examining Fig. 9.1 suggests that the meal to linseed price margin changes before the oil to linseed margin. This is to be expected given the policy followed by the linseed crushing mills, since linseed meal competes with other oil seeds as a minor source of proteins and minerals for animal feeds. This is confirmed by comparing the time series of the difference between the average quarterly prices of oil and linseed and the difference between linseed and meal as shown in Fig. 9.4. A close relation exists between these margins but of great importance is that the oil to seed margin lags two quarters behind the seed to meal margin. Statistical analysis over the crop years 1966/67 to 1971/72 shows a strong correlation. Interestingly the constant in

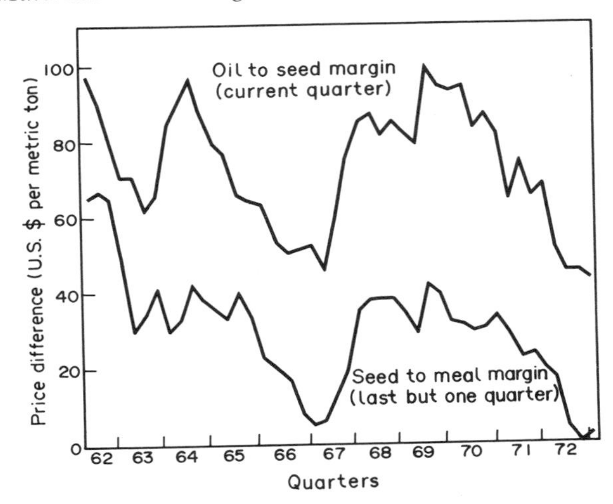

FIG. 9.4 The margin between the linseed oil and linseed prices compared to the margin between linseed and linseed meal prices

the statistical relationship obtained, 40 U.S. dollars, is the same as the prevailing crushing margins charged by the oil crushing mills (as at the time of the study in early 1973).

This relationship at, for example, the end of September, predicts from the then known information the value of the oil to seed margin over the period January to March of the following year. We can then use this value as our prediction for the oil to seed margin over the whole of the crop year. Hence in September we can convert our linseed price forecast into a linseed oil price forecast. The first part of Table 9.2 shows these values compared to the actual margins for those years used in fitting the relationship. The second part of the table shows the forecasts for the crop years 1963/64, 1964/65, 1965/66 and 1972/73, periods not used in deriving the statistical relationships. The accuracy of these forecasts to the actual values is good. Note that the four forecast years cover the full range of the September seed to meal margins that occurred for the fitted years.

The effect of the changing oil to seed price margin is to alter the magnitude of the changes in linseed oil prices over time and not the direction compared to the changes in linseed prices. Thus as regards the operation of the stock cover policy, the best values for the stock cover and buying period at any time can be derived from the linseed prices just as well as the linseed oil prices. The second stage of converting linseed forecasts to linseed oil price forecasts is unnecessary for determining the

TABLE 9.2 Forecasting the oil to seed price margin

Crop year	Average over 1st quarter of crop year seed to meal	Forecast of oil to seed margin	Actual oil to seed margin	
			Crop year	3rd quarter of crop year
Fitted estimates				
66/67	5	48	50	52
67/68	35	86	76	84
68/69	34	85	82	82
69/70	32	83	95	93
70/71	33	84	79	81
71/72	19	66	65	68
Forecasts				
63/64	30	80	75	84
64/65	35	86	80	79
65/66	23	71	62	63
72/73	2	44.5	50	43

optimal stock cover, although it may be useful to have a forecast of the oil price one might expect to pay.

The proof of whether any model explains market behaviour lies not in the goodness of fit to the period used in the analysis but in its ability to forecast over the future for circumstances not included in the original analysis. Rather than having to wait for the future a possibility is to use only part of the past for the model analysis. The remainder then becomes the 'future' over which the estimated model can be tested. Interestingly, excluding the last 4 years of historic data from the analysis to derive Fig. 9.2 and 9.3 does not significantly change the relationships obtained. Thus good forecasts of the prices for these 4 years would have been obtained using the model.

9.4 EFFECT OF THE DEVALUATION OF THE U.S. DOLLAR

On the basis of information available in early September 1972 the nonlinear model of Fig. 9.3 gave a prediction of 159 U.S. dollars per tonne for the average price over the 1972/73 crop year. One underlying assumption that had been used in the historic analysis was that the U.S. dollar parity has remained fixed. Note that the dollar had been a basic international monetary standard for many years and the pricing basis for many commodities. However in August 1971 the U.S. imposed a 10% surcharge on all imports which effectively led to a formal 10% devaluation of the dollar by early 1972. Thus the actual price predictions for the 1972/73 year should be adjusted to give an average value for the year of 175 U.S. dollars per metric ton (1972 values). The linseed price over July and August of 1972 was just over 140 U.S. dollars. Thus a substantial price increase could be expected over the next few months. The stock cover should be increased to its maximum value and all the scheduled order deliveries within it bought over a buying period of 2 months, September and October. The average price over the second quarter of the crop year was 177 dollars, very close to the forecast yearly price.

At the beginning of 1973 a further serious world monetary crisis occurred. The dollar was again devalued by 10% and further deteriorated following the 'floating' of currency exchange rates. The general floating of all currencies against each other has meant that there is no longer a generally accepted stable money unit for world trade. Commodity prices can change because of changes in currency values as well as changes in the supply/demand circumstances of the commodity itself. The average crop year linseed price for 1970/71 per metric ton was 123 U.S. dollars, 450 Deutsch marks, 683 French francs or 445 Dutch guilders. Using the August 1973 exchange rates the average price over the second quarter of 1973, the final quarter of 1972/73 crop year, was 270 U.S. dollars, 643 Deutsch marks, 1123 French francs and 705 Dutch guilders. The U.S. dollar price has increased by 120%, the Deutsch mark price by 43% the French franc and Dutch guilder prices by about 60%. The question is which of these different increases really reflect the changes in the linseed market situation between the 1970/71 and 1972/73 crop years. Taking the changes in the Dutch and French prices as the standard then if currencies had kept the same relative values as in 1970/71, the price in U.S. dollars in the second quarter of 1973 would have been about 200 dollars. Thus for the world pricing of linseed the U.S. dollar in 1973 would be at least 35% higher than its value over the 1970/71 crop year.

The U.S. dollar has been chosen as the standard pricing medium for international trade in linseed. Although the U.S. as a major exporter of linseed might accept lower earnings for its linseed because of economic problems at home, the other two major exporters Argentina and Canada surely would not. They would seek to obtain the same return in foreign currency terms after the U.S. devaluations as before. They would thus increase their asking prices in dollar terms to offset the decline in the value of the dollar. On the other side, presumably importers are willing to continue paying similar prices in their own currency after the dollar devaluations to those which they paid immediately before, because in general the importers consume the products manufactured from linseed and its oil and meal within their own countries. The devaluation of the

U.S. dollar has not affected the consumption circumstances within the importer's country. The importers would thus be prepared to pay on the international markets the same own currency price converted into the new dollars. Thus both from the viewpoint of rival exporters and importers the dollar price for linseed should increase by whatever devaluation of the U.S. dollar has occurred.

The problem is that in an era of floating currency exchange rates the rates are determined by the interplay of relative supply and demands on the commercial exchanges rather than by government edict. The dollar will decline by different amounts against different currencies as we have seen, and perhaps even gain against some. In general the devaluation of a currency is determined as a trade weighted index basis, covering all of the countries with whom the U.S. trades, who are members of the International Monetary Fund. On this basis the linseed dollar price for mid 1973 should be multiplied by 1.20 to give the 1973 dollar value. However, the international trade in linseed is not split amongst importing countries in the same proportions as the total trade of the U.S. The vast bulk of the trade in linseed is with Western Europe and Japan. A trade weighted index based on linseed imports is thus probably more appropriate. Alternatively a small subset of the linseed importing countries can be chosen as most representative and the devaluation of the dollar against their currencies used as the standard. During this period the dollar suffered worst against the West German deutschmark but much less against weaker economies such as Italy and the U.K. Holland, France and Japan all had an approximately equal 'middle' amount of appreciation in their currencies relative to the dollar by mid 1973. The dollar devaluation against these three currencies was used as a standard for standardising the linseed dollar price in the study. This was preferred to the averaging approach on the basis of the relative ease of calculation. Thus for mid 1973 the linseed price forecast by the models of Fig. 9.3 must be multiplied by 1.35 to give the current linseed prices. Up until the financial crisis of February 1973 a value of 1.10, reflecting the earlier dollar devaluation should be used. The predicted corrected price for the first half of the 1972/73 crop year thus becomes 175 dollars and for the second half year becomes 215 dollars. The average prices over the four quarters of the 1972/73 crop year, on a July/June basis, were 148, 177, 220 and 270 dollars. The linseed price rose from 220 dollars to 310 dollars by the end of the fourth quarter due to very low sowings of linseed, which would give the crop for 1973/74, in all three exporting countries, particularly the U.S.

In late April and early May of 1973 initial crop estimates based on the acreage sown were discussed in the *Oil World*. The average price over May was 280 dollars. Using the *Oil World* estimates as the basis of expectations, the price predictions using the behavioural market model of Fig. 9.3 give a range of 350–450 dollars for the average price over the crop year 1973/74. This is much higher than the current prices so again the stock cover should be increased to its maximum level and all scheduled deliveries bought within the 2 month buying period, May and June.

The expectations following the first crop estimates for 1973/74 for Canada and the U.S. and the sowings for Argentina were of an even worse scarcity of linseed supplies. This gave a revised price prediction of around 470 dollars for the average crop year price and the third quarter average price. The average price over September was 388 dollars, down on the August value of 485. The model predicts increasing prices over the next 9 months so again the stock cover should be increased to its maximum value etc. Over the 1973/74 crop year the successive average quarterly prices were 440, 427, 510 and 450 U.S. dollars per metric ton giving a crop year average price of 460. Thus the model prediction of around 470, at a time when prices were just under 390, appears a good one.

At the beginning of May 1974 the *Oil World* published a discussion of prospects for the 1974/75 crop year on both harvests and import demands. Using these figures as a basis for expectations gives a forecast for the 1974/75 crop year price of between 320 and 420 dollars. The average price over May was 460 dollars. Thus prices are predicted to fall over the next 12 months. Hence with falling prices expected the stock cover should be reduced to its minimum value and scheduled deliveries bought as late as possible. This policy should be continued until September when the first harvest estimates for the U.S. and Canada and the first sowings estimate for Argentina become available. A revision or reinforcement of this policy will then result. (Note that the forecasts discussed in this paragraph and the previous one were made by the author after the study had been completed and use the model as described and developed in early 1973. No changes were made to the parameters of the original model.)

The price forecasting model developed for linseed from the behaviour over the sixties has performed well for the seventies. However because of the collapse of a stable international currency as a basis of the world's financial system the price given by the model must itself be dynamically adjusted. It should be stressed that this adjustment is independent of the events in the linseed market. However the problem of forecasting future linseed prices now requires forecasts of future relative dollar values. In a sense we need to determine a standard 'linseed dollar' and determine its relation with the commercial U.S. dollar. For linseed it apparently depends on relative currency values. For other commodities it may need to take account of general price inflation in some subset of the world's economies, inflation in wage rates, prices of other competitor materials, oil prices etc.

10

Price Determination and Price Forecasting in the World Maize Market

10.1 THE MAIZE MARKET WORLD BACKGROUND SITUATION

Maize, or corn as it is called in the U.S., is the most important of the coarse grains grown. It is grown in many countries of the world and is more widely distributed than any other cereal. In many countries of Africa, South America and Asia maize is grown or imported for direct human use. In North America, Europe, Russia, Japan and Brazil it is mainly used for animal feed. Since 1963 the use of all grains for animal feeding purposes has grown faster than the use for direct human consumption. Because of this world maize production has grown enormously over recent years, see Table 10.1. From about 200 million tonnes in 1965 it grew at an annual average rate of 5.5% over the next 10 years to reach 310 million tonnes in 1973/74. By 1978 it had risen to 360 million tonnes. Over this period, as shown in Table 10.1, the U.S. has generally provided about half the world's production of maize. Its crop was 100 million tonnes in 1963/64, 144 million tonnes in 1973/74 and 175 million tonnes in 1978/79. The next largest producer China, at around 25 million tonnes, has only produced about one sixth of the U.S. output. Table 10.1 illustrates vividly how the U.S. has dominated the world's production of maize.

Farmers continually expanded their crops from year-to-year because they knew consumption was growing. This was a direct result of the world's growing appetite for meat and dairy products. This was originally most apparent in the U.S., but has since spread to other countries. During the sixties the main increase for U.S. maize farmers was for home consumption. U.S. usage of maize for animal feed rose from 70 million tonnes over the late fifties to 80 million over the early sixties, to nearly 90 million tonnes over the late sixties, to reach a peak of 107.5 million in 1972/73. Between 1963/64 and 1972/73 animal numbers in the U.S. increased by 9%, but the amount of maize fed per animal was up by 23%. More maize produced more meat at a faster rate.

TABLE 10.1 World production of corn or maize in millions of metric tons***

Crop year	U.S.	Argentina	Brazil	Mexico	South Africa	France	China	India	Italy	Bulgaria	Hungary	Yugoslavia	Rumania	Indonesia	U.S.S.R.	World total
1961	3,598	210	373	219	220	*1963-*	*1960-*	168	155	45	107	179	215	91	500	7,465
1962	3,606	180	400	215	240	*Avg.*	*64*	180	128	61	128	208	194	126	385	7,470
1963	4,019	211	370	253	168	*(3.6*	*Avg.*	179	136	68	140	212	237	105	335	8,060
1964	3,484	202	465	295	177	*M.T.)*	*(425)*	179	156	81	138	274	263	161	360	7,835
1965–6***	104.2	7.0	11.4	8.0	5.1	3.4	21.4	4.8	3.3	1.2	3.6	5.9	5.9	2.4	6.4	218.9
1966–7	105.9	8.0	12.8	8.2	9.6	4.3	21.7	4.9	3.5	2.2	3.9	8.0	8.0	3.7	6.8	239.4
1967–8	123.5	6.6	12.8	8.0	5.2	4.2	25.5	6.3	3.9	2.0	3.5	7.2	6.9	2.4	8.0	255.5
1968–9	113.0	6.9	12.7	8.5	5.0	5.4	23.5	5.7	4.0	1.8	3.8	6.8	7.1	3.1	7.1	243.3
1969–70	119.1	9.4	14.2	6.5	6.2	5.7	24.3	5.7	4.5	2.4	4.8	7.8	7.7	2.3	10.1	260.0
1970–1	105.5	9.9	13.5	8.7	8.6	7.6	26.4	7.5	4.8	2.4	4.0	6.9	6.5	2.9	7.8	255.1
1971–2	143.3	5.9	12.9	9.0	9.6	8.8	25.3	5.0	4.5	2.8	4.7	7.4	7.9	2.6	7.1	292.0
1972–3*	141.1	9.2	14.3	8.1	4.2	8.2	22.0	6.2	4.8	2.9	5.5	7.9	9.5	2.0	9.8	286.8
1973–4**	144.2	9.2	14.6	9.5	8.0	10.1	25.0	6.8	5.1	2.9	5.9	8.1	9.6	2.6	13.0	308.5

*Preliminary. **Estimated. ***Data prior to 1965–6 are in millions of bushels.
Source: Foreign Agricultural Service, U.S.D.A.

A further market for U.S. maize farmers was the startling growth in the U.S. exports. During the late fifties U.S. annual exports averaged only 6 million tonnes. They rose to an average 13 million tonnes over the early sixties and remained relatively stable over the late sixties at around 14 million tonnes. The seventies saw a period of rapid growth. Exports rose to 19.4 million tonnes in 1971/72 and 32.5 million tonnes in 1972/73 and 1973/74. Further large increases have taken them to 49.5 million tonnes by 1977/78. Over the decade to 1973/74 U.S. exports increased from 13.6 to 32.5 million tonnes, an increase of 130%. The other major exporting countries, Argentina, France and South Africa, only increased their exports from 7 to 11 million tonnes, an increase of about 50%. Thus the U.S. dominates the world export situation even more than the world production situation. Only France amongst the rival exporters appears to have increased production consistently, helped by the Common Market's Agricultural Policy.

Western European maize imports have grown regularly over the sixties and early seventies but the more startling growth has occurred for Japan and the U.S.S.R. Both these latter countries have increased their livestock herds etc., at fast rates in recent years. Russian grain crop failures over the seventies have caused large fluctuations in U.S. maize exports from year-to-year. Russian imports from the U.S. for example were 3.0 million tonnes in 1973/74, 1.0 million tonnes in 1974/75 and 11.8 million tonnes in 1975/76. A further factor over the seventies has been the increase in maize imports by the middle rich countries of the Third World such as Mexico and Taiwan.

In a similar way the U.S. has dominated the world carryover stock situation as well. Over the early sixties the carryover stocks from one year to the next were on average 39 million tonnes for the U.S. compared to 200,000 for Argentina, 850,000 for South Africa and 425,000 for France. Over the second half of the sixties the comparative values were 24 million tonnes for the U.S., 1.3 million for Argentina, 1.2 million for South Africa and 700,000 for France. By 1973/74 the U.S. carryover stocks had fallen back to only 12.5 million tonnes.

The relative values of the stocks in the various exporting countries suggests that the other exporters by their marketing policies are able to export all the maize they wish. Importers then turn to the U.S. for their supplies. The U.S. is the main but in a sense the residual supplier of maize to the world.

To give a similar situation with regard to the proportion of total consumption covered by the carryover stocks as in 1963/64, the 1973/74 U.S. carryover stocks would have had to reach a level of 65 million tonnes. To create such a stockpile in a short period is clearly impossible. These relative carryover stock figures illustrate the radical change in the world maize situation that has occurred over the decade between 1963/64 and 1973/74. From a period of an ample surplus of maize and low prices in the range $1.10–$1.30 per bushel the situation has been transformed into one of high maize shortages where prices could range between $3 and $4 a bushel. This change in the maize situation from surplus to shortage has been in part deliberate, but mostly an unavoidable direct result of the world's increasing consumption of meat, two poor U.S. maize crops and Russian grain crop failures.

As the production, export and carryover stock figures indicate, essentially the world maize situation *is* the U.S. maize situation. The major U.S. maize market at Chicago is the centre for determining world maize prices. The dependence of the outside world on the U.S. for its maize import supplies has increased significantly over the recent past. Only the U.S. appears able to increase its production to keep pace with world demands. By the same token exports have become a more dominant factor in the U.S. maize market. From a level of 7% of total usage in the late fifties, exports rose to 12% over the sixties. By 1971/72 this had increased to 15.5% and continued increasing to 21% in 1972/73 and 31.5% by 1977/78. To develop models of the world maize market we must therefore analyse the U.S. maize situation in some detail.

10.2 THE U.S. GOVERNMENT FEEDGRAIN SUBSIDY AND SUPPORT PROGRAMMES

The feedgrain programmes operated by the U.S. government to subsidise U.S. agriculture have played a major role in the operations of the maize market for over 25 years. On occasions they have effectively determined the prices on the 'free' market in Chicago. The general aim of these programmes has been to attempt to keep production and consumption generally in line and provide a reasonable standard of living for U.S. farmers. In the fifties acreage was limited by the Corn Allotment programme and also by encouraging acreage to be retired from production under the Soil Bank programme. These programmes failed to achieve their objectives. At the beginning of October 1961 the U.S. carryover stock had risen to 51.2 million tonnes, enough to cover six and a half months at the then prevailing consumption level. Prices had fallen to under $1.00 per bushel on Chicago. All but 3 million tonnes of these stocks were owned by the U.S. government itself.

By 1961 the financial burden on the U.S. government to subsidise these stocks had become too great. A new programme was announced which required farmers to divert acreage out of maize to qualify for loan support to finance their stocks. The farmers were paid directly for the acres diverted to other crops. Each farmer was allocated a nominal base acreage, being the average of that planted in 1959 and 1960. To join the programme he would have to divert at least 20% of this base to other crops. Only then would he be eligible to sell all his maize to the government's Commodity Credit Corporation (CCC) via the loan support system if he wished. Under the price support system, a 'farm loan rate' was set each year. A farmer may place his corn under 'loan' immediately it is harvested. At some later stage, he can redeem it if he wishes by repaying the loan plus various interest and redemption charges. Alternatively, he can keep the 'loan' and forfeit the maize to the CCC. Because prior to 1961 there were few conditions that farmers had to meet to sell their surplus maize to the CCC and because production had outstripped consumption over the fifties, the government via the CCC had become very large owners of maize stocks.

The CCC was allowed to sell maize in years when there were shortfalls in supply. The prices at which the CCC could sell its maize have changed over time according to changes in government policy. Over the late fifties and early sixties, with the very large carryover stocks of maize, the operations of the CCC virtually fixed the market price. Prices were depressed if it sold heavily, as it did in 1961/62. If it sold only small amounts, the market price rose to the level required to make it worthwhile for farmers to redeem their maize from 'loan' to the CCC and resell it in the market. Later the CCC was restricted to only selling maize if the price rose above the loan plus reasonable carrying charges. The different policies in selling maize of the CCC from year-to-year effectively determined the general market price. This continued over the first half of the sixties whilst the CCC had large stocks to sell.

During the early sixties the reduced U.S. production remained below consumption for most years. The CCC was thus able to sell off most of the accumulated stocks. The amount of maize the farmers needed to place under loan support also declined.

At the beginning of the 1961/62 crop year, about one third of the total supply had been owned or subsidised by the U.S. government but this had fallen to only one thirtieth at the beginning of 1973/74. In October 1973 76% of the carryover stocks were in commercial hands compared to only 6% in October 1960.

The disappearance of the government owned stocks was due to the success of the 1961 Feed Grain Programme in controlling maize production via acreage limitations, the high level of exports in 1965/66 and the U.S. crop failure of 1966/67. By the beginning of 1967/68, the carryover stock was down to 20.9 million tonnes compared to 51.2 million tonnes at the beginning of 1961/62. For the first time, at the beginning of 1967/68 over half the carryover stock was in commercial hands. Over the sixties and early seventies, usually between 65 and 66 million acres were planted with maize and about 20 million acres diverted under the Feed Grain Programme. The large increase in production in the U.S. over this period has mainly arisen from improved yields due to expansion in the use of nitrogen fertilisers, better seed and improvements in planting and harvesting machinery.

In the future the U.S. government policies will always provide a floor to the market. When there is a large surplus the policy will keep prices near to this floor for most of the year. This indeed happened again in the 1978/79 crop year. Figure 10.1 shows upper and lower forecast limits for the price of maize on Chicago over the whole crop year. These forecasts were made in August 1978 on the basis of the prevailing loan rate alone. No predictions of the supply/demand situation for 1978/79 were used in the forecasts. As can be seen prices moved exactly as forecast right up until June. News then of poor Russian crop conditions leading to a large Russian maize import need for 1979/80 led to a substantial increase in prices because of a likely world shortage.

As we are concerned with predicting price for maize into the future, our analysis of past market behaviour is best confined to those years when a commercial market was in operation. A model for price determination in the world maize market will thus mainly be based on past behaviour for the years from 1967/68 onwards.

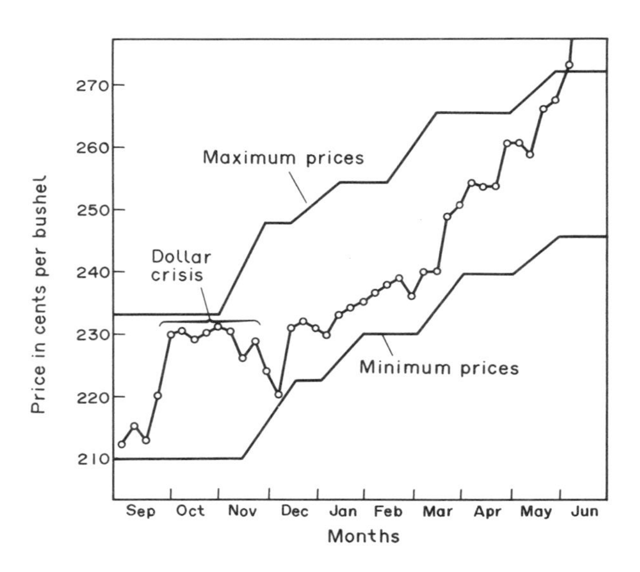

FIG. 10.1 Price forecasts for the 1978/79 crop year

10.3 A MARKET BAROMETER FOR MAIZE

The U.S. maize situation is illustrated diagrammatically in Fig. 10.2, where the hindsight values for the total supply, U.S. domestic use and exports are shown. This shows clearly the difference between the 2 years of high prices 1966/67 and 1970/71 and the 3 years of high prices from 1972/73 to 1974/75. High prices result from a shortage of maize. The figure shows that in both 1966/67 and 1970/71, the shortage arose from a large drop in the total supply compared to the previous years, of 10 and 13 million tons respectively. This was caused by poor crops in both years. In 1972/73 the total supply rose by 10 million tons compared to 1971/72. Domestic use rose as it has whenever total supply has increased. The shortage which resulted this time, was caused by an unusually large, unprecedented increase in exports. In 1973/74, the total supply fell slightly by 5 million tons, as did domestic consumption. Exports however stayed at the high level of 1972/73. Again the shortage was caused by large consumption demands. In 1974/75, there was another poor crop and the total supply fell by 30 million tons to the same level as 1970/71. Thus the shortage this time was initially caused by a large drop on the supply side. Domestic use dropped drastically by 25 million tons to well below the level of 1970/71. Exports, on the other hand, stayed at the high level of the previous two years. The total consumption was almost 10 million tons above the 1970/71 level, thus bringing about a far more serious shortage than in 1970/71.

The above discussion shows that any 'barometer' attempting to measure the state of the maize market must take account of changes in both the

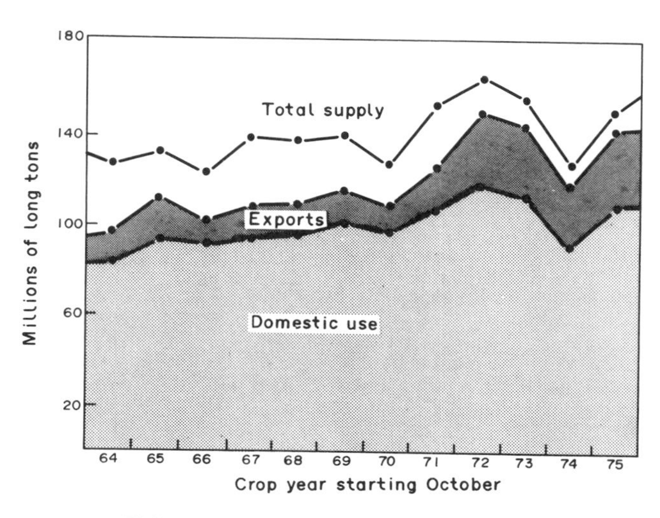

FIG. 10.2 The general U.S. maize situation

supply of maize and the demand for maize. A simple measure that does this is the market surplus, being the simple difference

$$\text{Total Supply} - \text{U.S. Domestic Use} - \text{Exports}$$

In Fig. 10.2, this is the unshaded area between the total supply line and the exports line. However it is more meaningful to measure the surplus in terms of the time it will last rather than its absolute value. If the use has increased by 25% then the same absolute market surplus will indicate a larger shortage. Thus a better market measure is the estimated market surplus as a percentage of the prevailing crop year usage, given by the expression

$$100 \times \frac{(\text{Total Supply} - \text{U.S. Domestic Use} - \text{Exports})}{(\text{U.S. Domestic Use} - \text{Exports})}$$

As a proportion rather than a percentage this is the amount of time (in years) the surplus will last at the present rate of usage, before new supplies of maize are required. It is plotted in Fig. 10.3, together with the average crop year Chicago prices. (Note it is the hindsight values known after the end of the relevant crop years.)

Comparing the market surplus with the crop year average shows a good connection between the two series. As the market surplus falls, prices rise and vice versa. For example, the market surplus increases from 1970/71 to 1971/72 and the price falls. The market surplus falls to its lowest level in 1972/73 and stays virtually unchanged in 1973/74 and 1974/75. Similarly

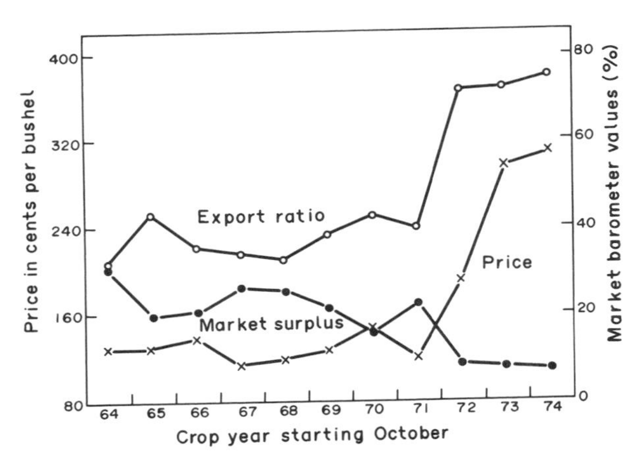

FIG. 10.3 Market barometers and annual prices 1964/65 to 1974/75

prices reach their highest levels over 1973–1975. Certainly over the period 1965/66 to 1972/73, a very close relationship exists between the market surplus and the prices. Initially in the study, a great deal of effort was devoted to analysing this market barometer. Using its changing values over the crop year due to the changes in the supply and demand estimates showed that it provided a good explanation of market prices up to the beginning of 1973. However, even allowing for the devaluation of the U.S. dollar, it did not prove to provide good predictions for the remaining years 1973–1975 inclusive.

It should be noted that this barometer treats U.S. domestic use and exports on the same basis as interchangeable demands on the U.S. total supply of maize. As has been discussed earlier, in recent years exports have become a dominant factor in the market. An alternative approach is to treat exports and U.S. domestic consumption separately. Firstly the U.S. supply is used to meet its domestic consumption. This then leaves an 'export potential supply' available to meet the demand for U.S. exports as

$$\text{Export Potential Supply} = \text{Total Supply} - \text{U.S. Domestic Use}$$

A 'barometer' could then be the percentage of this 'Export Potential Supply' that the actual exports or export demand takes up, i.e. the 'market barometer' is the Export Ratio defined as

$$\text{Export ratio} = \frac{100 \times \text{Exports}}{\text{Export potential supply}}$$

The values of this 'barometer' have been determined on a hindsight basis, and are also plotted on Fig. 10.3. Excluding 1965/66, they also show a good correlation with the average crop prices. In this case, prices increase as the Export Ratio increases and vice versa. The relation is directly with price, rather than inversely, as with the market surplus. The correlations with price appear from Fig. 10.3 to be as good up to 1972/73 as for the market surplus.

The Export Ratio values indicate a more extreme shortage situation over 1972/73, 1973/74 and 1974/75 than do the market surplus values. They are thus more likely to explain the extreme high prices in those years than the market surplus values. This suggests strongly that the Export Ratio is the better market barometer, although the two appear much the same up to 1971/72.

The analysis so far has been based on the actual historic values, which are only known exactly some years after the event. The estimates of supply and demand factors in the market change radically over the year and hence so do the values of the market barometers. For example the hindsight

Export Ratio for 1972/73 was 71.8 indicating an extreme shortage. In August, November, February and May of the crop year the values of the Export Ratio, based on the prevailing market estimates, were 42.0, 49.4, 54.0 and 64.0 respectively, which are all very different to the hindsight value. Prices over 1972/73 moved from $1.53 per bushel in November to $2.40 per bushel in the following September. As Fig. 10.3 shows the hindsight values of the Export Ratio for 1972/73 and 1973/74 were identical. The prices however were very different. The Export Ratios for 1973/74 based on market estimates were 66.3, 73.2, 73.8 and 75.6 from August to May. These are very different to the corresponding values in 1972/73 and much closer to the hindsight value. The current market estimates thus 'explain' the price differences beteen 1972/73 and 1973/74. The years 1967/68 and 1971/72 had the same Export Ratio of 30.5 in August, prior to the start of the crop year, and for much of the crop year. Yet the hindsight values were very different at 34.0 and 40.5 respectively. It is thus not surprising that the prices in the 2 years were very similar. These examples illustrate the need to examine expectations and the flow of information to the maize market.

10.4 MARKET EXPECTATIONS

The supply of U.S. maize comes from two sources, the new crop and the carryover stock from earlier years. Maize is planted in April and early May in the U.S. Harvesting begins in September and is generally completed by the end of December. Surveys of farmers intentions are made prior to planting, followed in June by a later estimate of the acreage that was actually planted. These can be converted to a crop estimate assuming average yields. Unless a large reduction in acreage planted from that of the year before occurs, these surveys have virtually no impact on the market. Actual crop estimates are first made in July and are revised month by month through to January. No further revisions are made until November of the following crop year. July and August are the critical parts of the growing period, when lack of rain can reduce yields. Diseases tend to occur as large outbreaks during August. The July estimate is thus very tentative. August is usually regarded as the first reliable crop estimate, although large changes can and do occur between August and January.

The major source of crop estimates is the United States Department of Agriculture (USDA). It usually publishes its estimate on the fourteenth day of the month based on surveys made at the beginning of the month. Great care is taken to ensure that these estimates are kept secret prior to publication. They are made available to everyone in the market at the same time. Via telex etc. they are circulated throughout the world within minutes of being announced by USDA. A private forecaster, Conrad

TABLE 10.2 USDA crop estimates in millions of long tons

Crop year	Jul	Aug	Sep	Oct	Nov	Dec	Jan
64/65	97.7	97.1	91.0	89.1	88.5	88.7	88.7
65/66	97.8	102.4	103.6	104.5	103.2	103.2	104.3
66/67	105.5	99.5	102.3	102.4	103.3	103.3	102.6
67/68	112.7	116.3	117.3	117.9	117.9	117.9	118.0
68/69	111.0	114.0	115.9	114.7	111.0	111.0	109.4
69/70	107.2	107.7	107.8	108.8	111.1	111.1	114.5
70/71	120.5	117.3	110.1	104.7	102.6	102.6	102.7
71/72		133.6	131.6	135.0	138.8	138.4	
72/73	126.0	123.7	128.1	129.1	135.0		136.8
73/74	147.0	141.5	144.2	144.1	142.0		141.1
74/75	145.0	124.1	124.9	117.9	115.5		115.5

The gaps in the table are months when USDA did not give an estimate.

Leslie, has established himself as an alternative source of good crop estimates to which the market pays great attention. In recent years the National Corn Grower's Association (NCGA) has begun making forecasts, although, as a representative of the maize farmers, its forecasts tend to be somewhat more conservative. The Leslie and NCGA forecasts appear a few days before those of USDA. Their major impact is in giving a guide to the likely change in the USDA estimates. The current USDA crop estimate becomes the current market expectation of the crop size. The monthly USDA crop estimates are given in Table 10.2 for the years 1964/65 to 1974/75, together with the hindsight values in the final column.

Carryover stock estimates are provided by USDA when they issue forecasts of consumption and exports. Over the growing and harvesting period, estimates have always been made in July/August and November. The November estimate is based on a survey of all the stocks of maize that were held in the U.S. on 1 October. The July one is based on forecast consumption and export values. In some years stock estimates are made in other months. The changes in carryover stock estimates can be very large from July to November, sometimes larger than the change in the crop estimates.

Prior to 1973/74 no forecasts of the total U.S. crop year domestic use or exports were made until November. By then the previous crop year's domestic use and exports were known reasonably well and could presumably form the basis for the forecasts. These forecasts were usually updated in the following February and May or June. These values are shown in Table 10.3. Note that USDA only started making these forecasts at the end of 1966. Thus during the period July to November whilst the new crop was

TABLE 10.3 USDA forecasts of maize utilisation during the crop year in millions of long tons

Crop	November		February		May/June		Next November	
year	domestic	export	domestic	export	domestic	export	domestic	export
64/65							84.2	13.75
65/66							95.2	16.7
66/67	94.1	20.0	98.0	14.4			91.6	11.5
67/68	98.3	13.75	97.7	16.0			94.7	15.9
68/69	96.0	16.3	96.3	15.5	102.2	13.1	98.0	13.4
69/70	99.2	15.0	100.4	14.4	103.2	13.75	102.0	15.4
70/71	98.9	13.1		11–12	99.0	12.5	98.1	12.9
71/72	105.0	15.0	107.5	15.0	107.6	16.25	107.2	19.0
72/73	112.4	25.0	117.5	25.0	120.0	25.0	118.4	31.5
73/74	117.2	28.1	115.5	28.1	117.5	30.0	115.7	31.1
74/75	98.2	22.5	96.5	24.4	92.5	26.9	90.8	28.6

growing and being harvested forecasts of the consumption demands for the crop were not available. All the market had to work on were forecasts for the old crop year consumptions as made in May or June. Since 1973/74 USDA has issued a forecast in August, at the end of the old crop year, of the domestic use and exports for the new crop year.

The crop year for maize worldwide is taken as the U.S. marketing year, 1 October to 30 September. The U.S. government, by survey methods, collects data on the stocks of maize everywhere in the U.S. at the end of each quarter of the year. From the estimates of the total supply of maize available for the crop year and these quarterly stock values, the total amount of maize consumed at home and exported in each quarter can be calculated. Monthly statistics on U.S. maize exports are published regularly by USDA about a month in arrears. Hence the total maize used within the U.S. as well as the total exports over a quarter can be calculated. The first quarter usages can be estimated at the beginning of February, the first half year usages at the beginning of May, the first three-quarters usages at the beginning of August. At each of these times actual usage data is available to enable the forecasts of crop year total usage to be revised.

One approach to determining expectations is to use simple extrapolation methods. At the end of January 1968 the published statistics showed the first quarter U.S. domestic usage of maize for 1967/68 was 28.9 million tonnes. The first quarter usage for the previous year, 1966/67, was 28.7 million tonnes. These figures suggest total crop year usage will only be up 0.7% in 1967/68 to a total of 92.6 million tonnes. The first quarter U.S. exports were 4.42 and 3.51 million tons in 1967/68 and 1966/67 respectively. By extrapolation the crop year exports for 1967/68 will be

$$\frac{\text{1st. quarter exports for 1967/68}}{\text{1st. quarter exports for 1966/67}} \times \text{(Crop year exports for 1966/67)}$$

i.e. $\frac{4.42}{3.51} \times 12.2 = 15.4.$

Looking at Table 10.3 we see that both the crop year domestic use and export forecasts are different from those made by USDA at the same time. Thus we have two alternative ways of depriving 'market expectations' of the demand levels. Other alternatives would be to take account of the past 2 or 3 years rather than just one. Furthermore since exports are published, at least recently, on a monthly basis, forecasts of crop year total exports could be revised monthly. Obviously both approaches would need examination as bases for market behavioural models. In addition market comment as made at the time needs to be taken into account. Other relevant information is published by USDA such as the numbers of animals of different types on feed at various times of the year etc.

A further source of information on likely exports became available after August 1973. From that date all export sales, even if delivery would not take place for many months in the future, had to be registered with the U.S. government as soon as they were made. Not all these sales in the event necessarily became exports. Some were resold back to U.S. consumers or cancelled. These 'export commitments' are published weekly by the U.S. government. This is a further source of information on which to estimate crop year export demand. This was a reaction to the Russian 'grain raid' of May/June 1973. The Russians had a disastrous crop which they kept secret from the rest of the world. Within a few days a 'discreet' trading mission quietly bought up many millions of tonnes of wheat and maize to be imported into Russia over the following few months. A moderate shortage had been transformed into an extreme shortage virtually overnight. When the purchases were announced the market prices soared. U.S. farmers complained they had been cheated, in that they sold grain at much lower prices than they would have done if they had known the size of the Russian demand. These purchases reduced the U.S. maize supplies to very low levels over the final quarter of the 1972/73 crop year, barely sufficient to cover U.S. domestic consumption. Export controls were imposed by the U.S. government to ensure the situation would not be repeated in the future and to improve the flow of information to the U.S. maize market. Later a more detailed agreement was signed with the U.S.S.R. to control even more closely Russian purchases of U.S. grain in any year.

10.5 PRICE MOVEMENTS

Examining the monthly Chicago prices for the eleven crop years from October 1961 to September 1972 shows clearly that there are definite seasonal movements in prices. For example, the lowest price over a crop year is usually in October or November. Seasonal price indices can be calculated for all months as follows:

$$\text{Percentage price change} = 100 \times \frac{(\text{This month's price} - \text{last month's price})}{\text{Last month's price}}$$

The values can then be averaged over a number of years to derive a 'general pattern' for the seasonal movements of prices over the crop year. The results for the eleven crop years are shown below.

Av. Percentage Monthly Price Change

Oct.	Nov.	Dec.	Jan.	Feb.	Mar.	Apr.	May	June	July	Aug.	Sep.
−2.8	−2.7	4.8	2.2	−0.7	0.8	1.2	2.1	0.1	−0.6	−1.6	0.4

Note that negative values indicate a fall in price from the previous month, whilst positive values indicate an increase. Thus in general over the crop year October to September, prices fall heavily through October and November and then increase through December and January, even more than they fell earlier. Prices then rise slowly over February to May, followed by a slow decline to September. However extending the series to cover two further crop years, up to September 1974, rather different results are obtained, as shown below.

Av. Percentage Monthly Price Change

Oct.	Nov.	Dec.	Jan.	Feb.	Mar.	Apr.	May	June	July	Aug.	Sep.
−3.1	−1.5	6.4	2.3	−0.2	0.4	0.4	3.6	1.3	1.7	1.5	−1.7

The addition of the two further years changes significantly many of the values. This suggests that the last 2 years 1972/73 and 1973/74 behaved very differently in their seasonal price movements to the earlier 11 years. The year 1972/73 and 1973/74 are years not only of higher prices but of larger variation from month to month.

Figure 10.4 is a graph of the monthly prices for May to November for the crop years from 1966/67 to 1975/76. As can be seen, taking an average over a large period obscures what is really happening. Distinctly different patterns of behaviour can occur, falling into three groups.

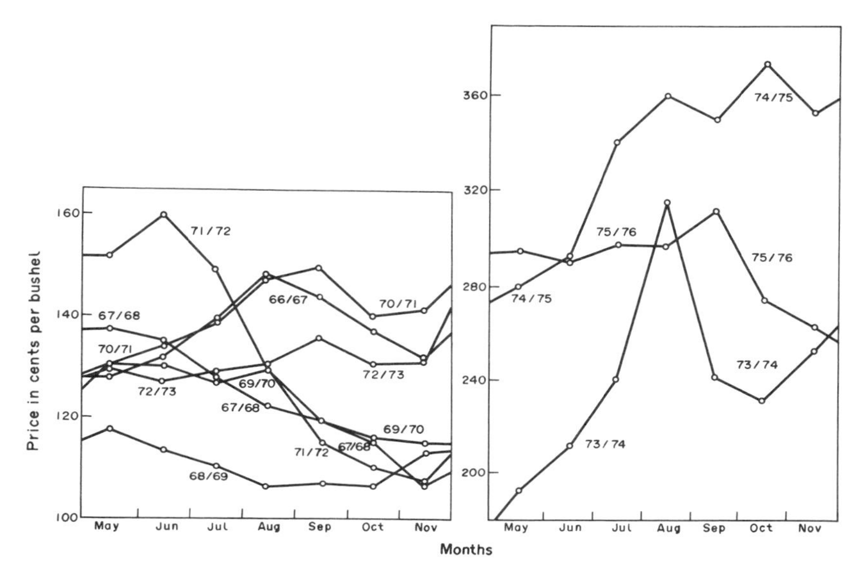

FIG. 10.4 Seasonal price movements over May to November (the old/new crop transition)

(i) The first group includes 1967/68, 1968/69, 1969/70 and 1971/72. Prices generally fall continuously from May/June to reach a low in October/November, around 105 to 115 cents per bushel.

(ii) The second group include 1966/67, 1970/71 and 1972/73. Prices rise continuously from May to reach a peak in August/September and then fall to a low in October/November. However this low is at a higher level than the starting point in May.

Prices are below 160 cents per bushel and above 100 cents per bushel for both these groups.

(iii) The final group includes 1973/74, 1974/75 and 1975/76. These years have very high prices, compared to the other two groups. The prices in 1974/75 increased from May through August, like group (ii) years, but continued increasing thereafter to reach a peak in October/November. Note the abnormal price spike in August 1973 which was probably due to the fears of export controls being imposed. If this August price is excluded the year is similar in behaviour to those of group (ii). The 1975/76 year rises to a peak in August like the group (ii) years, but then falls to a minimum through October and November at a lower level than the starting point in May, more like the group (i) years.

An interesting point is that the crop year 1969/70 in group (i) was also rather similar to 1972/73 in group (ii) in its behaviour from May to August; prices remained level and then increased from July to August. However circumstances presumably changed and thereafter prices fell continuously through to November.

Figure 10.5 shows the monthly price movements over the period November to June for each crop year. Again the price movements demonstrate different patterns of seasonal behaviour.

(i) The crop years 1967/68, 1968/69, 1969/70 and 1971/72 are again remarkably similar in behaviour. Generally a small rise in price of about 10 cents per bushel occurs from November to January. Prices then remain stable until April, with a further small surge in price through May and June. In 1967/68, the surge in May and June did not really occur. In 1971/72 there was an unusually large increase in price from November to December, which was probably due to the 10% dollar devaluation of December 1971.

(ii) The second group of years, 1966/67, 1970/71 and 1972/73 are again remarkably similar in seasonal behaviour for most of the period. Prices increase significantly from November through January and then decline slightly over the months to June. The prices in 1970/71, after a temporary surge of 8 cents in June, fell back by 10 cents in July. The 1972/73 crop year prices, after behaving exactly the same as 1970/71, diverged in April to increase consistently from 158 cents per bushel in March to reach 243 in June.

(iii) The crop year 1973/74 and 1974/75 are completely different to the others in being very much higher in price. Their seasonal movement is also different. Prices in 1973/74 increased from 253 in November to peak at 308 in February. They then dropped back sharply to 266 in April but rallied to 282 in June. In 1974/75 prices rose slightly from 353 in November to 361 in December and then continuously fell to 294 in March. They then stayed on a plateau around 295 for the remaining months up to June. Interestingly enough, although starting at very different prices in November, they were at much the same level in March and again in June.

Thus the two figures show that, although the price levels and the changes from month-to-month may be different, there are typical patterns of seasonal price movements. However the fact that the group (i) and group (ii) years have virtually opposite patterns of behaviour from June to November means that assuming a single pattern of seasonal price movement, and always choosing particular months as the time to buy, is not a good basis for price forecasting and the making of purchasing decisions.

Examining Figs. 10.4 and 10.5, which show the seasonal price movements over the crop year, demonstrates that the critical points in time are early July, August/September, November, early February and the end of April. Knowing the prices at these times specifies which type of yearly

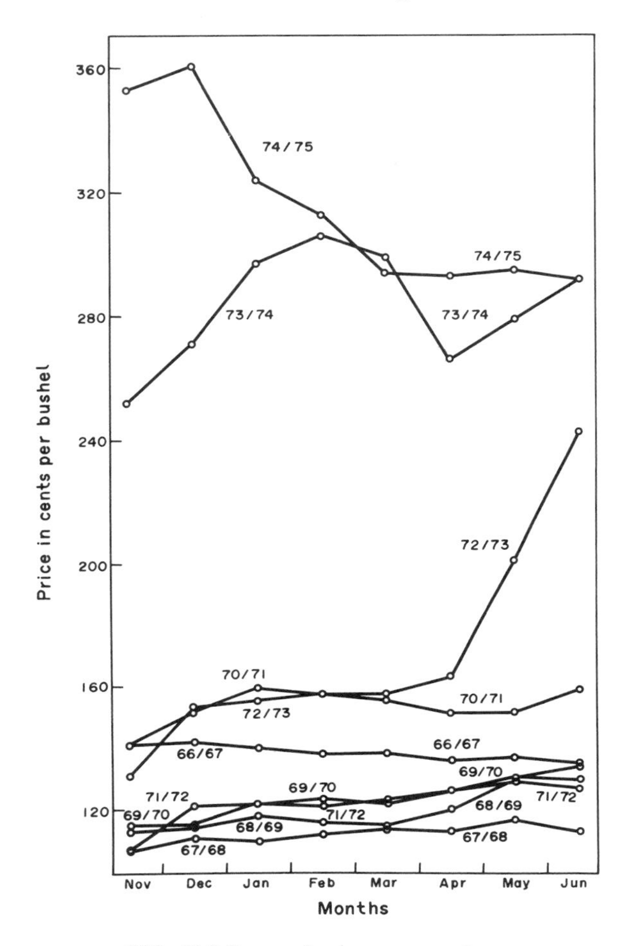

FIG. 10.5 Seasonal price movements over November to June (the major part of the crop year)

price movement will occur. Prices at other times can be obtained by simple interpolation.

These are also critical periods of the year with regard to the flow of information to the maize market, as discussed earlier. Early July is the time when the new crop acreage planted is known, enabling tentative estimates of the new crop to be made. It is effectively the end of the old crop year. August and September are the months of the first reliable crop estimates. In November the actual carryover stocks from the previous crop year are known and hence reliable estimates of the usages are available. November has traditionally been the month when the first USDA forecasts for the new crop domestic usages and exports are made. The end of January or early February is when statistics on the U.S. domestic use and exports over the first quarter of the crop year are published. This enables revision to forecasts of the crop year totals to be made. Similarly the end of April is when the first half year usages are known.

10.6 PRICE DETERMINATION AND PRICE FORECASTING MODELS

Models for price determination and hence procedures for price forecasting in the maize market must therefore correspond to these critical time periods. They show that forecasting will be a continuing process. As we move forward over time from one critical time period to the next we must make forecasts for the remaining future critical periods of the crop year. Thus several different forecasting formulae will need to be calculated. In August and September, for example, we need to have formulae for forecasting prices in November, January, April and June/July, the end of the old crop year. In early July we can form a tentative view whether prices are likely to fall or rise over the old crop/new crop transition, rather than a precise forecast. This will assist in deciding whether to accelerate or delay purchasing decisions.

By the middle of August each year, when the first reliable new crop estimate is made, statistics on the domestic use and exports during the first 9 months of the old crop year are available. Until 1971 USDA did not issue a revised forecast of the old crop year usages and hence the old crop carryover in August. Their most recent forecast was that given in the previous May or June, which only made use of the first half-year usages. Using the simple extrapolation methods described earlier more up-to-date forecasts of the old crop year usages can be made based on the first 9 months usages. These are given in Table 10.4. The expected carryover stocks can be generated by subtracting these forecast usages from the estimated total U.S. maize supply for the year. It would seem likely that the market must use these old crop year usages to evaluate the new crop situation, since no other information is available. Two exceptions to this are the years 1967/68 and 1971/72 which both follow years of crop shortages. With a large increase forecast for the new crop, market comment suggested that a return to the higher usages of the year prior to the shortage was more likely than a repeat of the low usages of the shortage year. Taking these points into account, together with the August crop estimates, gives the market expectations for the various supply and demand factors shown in Table 10.5 for the years up to and including 1971/72. The second entry for 1970/71 includes the September crop estimate. Maize blight struck the U.S. crop in late August and led to a

TABLE 10.4 August estimates of crop year consumption and new crop total supply

		Old Crop			New Crop	
Crop year	Source	Domestic use	Export	carryover	Crop estimate	Total supply
66/67	EXTRAP'N	91.0	11.25	21.7	116.3	138.0
67/68	EXTRAP'N	94.4	15.3	28.9	113.9	142.8
68/69	U.S.D.A.	99.7	12.5	26.3	107.7	134.0
69/70	EXTRAP'N	103.0	15.3	24.7	117.3*	142.0
70/71	EXTRAP'N	98.5	12.0	17.5	133.6	151.1
71/72	U.S.D.A.	107.6	18.75	28.75	123.7	152.5
72/73	U.S.D.A.	115.7	31.1	19.4	141.5	160.9
73/74	U.S.D.A.	115.0	30.6	10.7	124.2	134.9

*Maize blight over large areas announced on August 20, causing talk of falls of 10% or more.

TABLE 10.5 August consumption forecasts and export ratio

Crop year	Total supply	Domestic use	Export potential	Exports	Export ratio
67/68	138.0	93.0	45.0	13.75	30.5
68/69	142.8	94.4	48.4	15.3	31.2
69/70	134.0	99.7	34.3	12.5	36.3
70/71	142.0	103.0	39.0	15.3	39.5
	134.8	103.0	31.8	15.3	48.0
71/72	151.1	102.0	49.1	15.0	30.5
72/73	152.5	107.6	44.9	18.75	42.0
73/74	160.9	115.7	45.2	30.0	66.3
74/75	134.9	105.5	29.4	23.8	81.0

large reduction in the crop size. In August 1972 USDA gave explicit forecasts of the new crop usages for the first time. Such explicit forecasts from so respected a source would undoubtedly overshadow the effects of the old crop year usages. For the last 3 years in the table the USDA forecasts are used, with one exception. The export registrations scheme had been started following the Russian grain raid of May 1973. In August 1974 the export registrations for 1974/75 were 23.8 million tons, substantially higher than the current USDA forecast. All the published market comment suggested that the market found this higher figure a more plausible value than USDA's estimate.

Figure 10.6 shows a graph of the lower of the average monthly prices over October and November on Chicago against the export ratio calculated

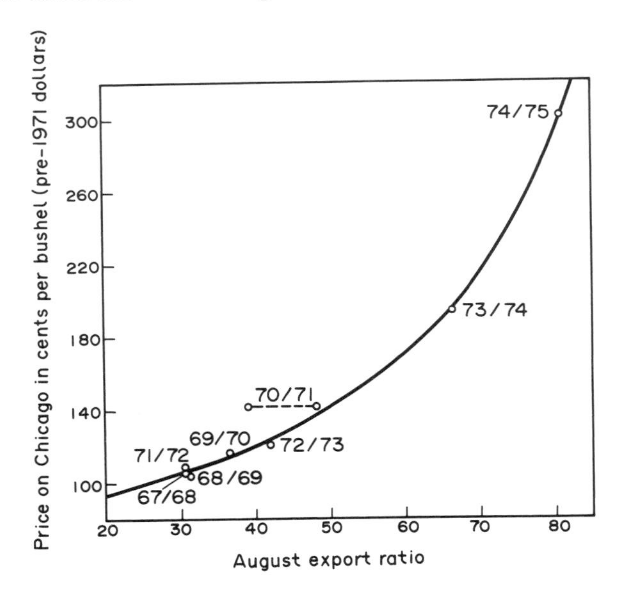

FIG. 10.6 October November low prices against August export ratio

in August. The figure shows a pattern of very strong correlation. The likely relationship is indicated by the continuous line in the figure. This line has an equation

$$\text{Price} = 8260/(108.5 - \text{Export Ratio})$$

In this case the actual monthly prices on the Chicago market have been adjusted for the devaluation of the dollar. Since U.S. maize is exported to many countries the adjustment* used is the official devaluation factor of the dollar relative to the SDR, the artificial money unit of the International Monetary Fund. The relationship shown in Fig. 10.6 becomes a basis for forecasting prices. For example in August 1966 simple extrapolation methods suggested that the 1966/67 U.S. domestic consumption would be around 90 million tons and that exports would exceed 17 million tons. The August crop estimate for 1966/67 gives an expected Export Ratio of about 44.0 whilst the lower September crop estimate gives a value of about 48.0. The relationship of Fig. 10.6 gives price forecasts of 128 cents per bushel**

*A U.S. maize export trade weighted average of the dollar devaluation against the most important maize importers (excluding Russia) gave virtually identical results.

**A bushel is a measure commonly used in the American grain trade. 40 bushels are equal to one imperial or long ton, whilst 39.37 bushels are equal to one metric ton (tonne).

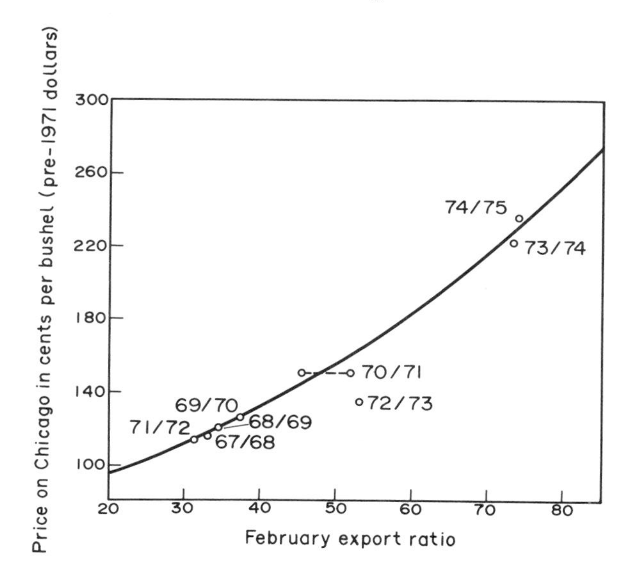

FIG. 10.7 April price against February export ratio

and 136.5 cents respectively for the Chicago market. The average price on Chicago in November 1966 was 131.0 cents per bushel, compared to 148 cents over August. Thus the model certainly fits the behaviour for the 1966/67 crop year, which was not included in the analysis to derive Fig. 10.6.

Another example of the same approach is given in Fig. 10.7. This time the average price over April, again adjusted for the decline in the dollar against other currencies, is plotted against the Export Ratio as estimated at the beginning of February. The forecasts of domestic use and exports have been derived using simple extrapolation models based on the actual first quarter consumptions modified on occasions by the published market comment at the time.

Two alternative points are marked on the figure for the 1970/71 crop year. In that year there was a U.S. crop failure similar to the situation in 1966/67. In 1966/67 the U.S. domestic use fell back strongly over the second and third quarters of the crop year. In the intervening years U.S. domestic use remained relatively stable over the first three-quarters of the crop year. The higher value of the Export Ratio is derived from comparing the first quarter U.S. domestic use with the usage over 1969/70 and 1968/69, whilst the lower value is from a comparison with the domestic usages over the 1966/67 crop year. The higher value implies, from Fig.

10.7, a price of 160 cents, whilst the lower value gives a price of 145 cents. Prices were around 160 cents over January and February, as would be expected assuming a repeat of the domestic usage patterns of the previous year, but fell back to 150 cents over April and May, when the second quarter usages showed that the pattern of 1966/67 was actually being repeated.

Again the figure shows a good relationship except for the 1972/73 crop year. The fitted relationship indicated by the continuous line on the graph has an equation

$$\text{Price} = \exp(4.225 + 0.0165 \text{ Export Ratio})$$

The results are less good if USDA's February usage forecasts are used. This suggests the market pays more attention to current actual usage rates than USDA's forecasts.

The high Export Ratio value for 1972/73 is due to a surge in exports, rather than a crop failure as in earlier years. A further world financial crisis had occurred in January and early February leading to another large dollar devaluation. In addition to the Chicago prices there exists another important maize price series. This is a price on an f.o.b. Gulf Ports basis, e.g. for maize loaded on to a ship ready for export. This latter price is essentially what importing countries are willing to pay. Usually the margin between the Gulf Ports and Chicago had been very close to 10 cents per bushel, representing extra transport costs to the Gulf Ports. Over the 1972/73 crop year the unadjusted real mid month prices were, in cents per bushel:

	Oct.	Nov.	Dec.	Jan.	Feb.	Mar.	Apr.	May	June	July	Aug.
Chicago	133	134	164	158	161	158	163	192	212	239	316
Gulf Ports	144	154	176	206	204	199	186	214	245	267	328

The margin between the prices was around 10 cents in October and then again in the following August. In between it rose to the high value of 48 cents in January. Chicago prices remained stable from December to April at around 160 cents per bushel. Gulf prices increased markedly to around 200 cents per bushel. Chicago prices then increased at a greater rate than the Gulf Port prices from May onwards. As mentioned the August peak was due to fears of export controls. It appears that initially Chicago prices did not react to the February dollar devaluation. A gap developed between home and export prices. The Chicago April price adjusted for the January rather than the April devaluation factor is 150 cents per bushel. This makes it much closer to the value suggested by the continuous line relationship of Fig. 10.7. The Chicago May price adjusted by the May devaluation factor is

159 cents, just above the value predicted by Fig. 10.7. The home and export markets have thus begun to move closer together to achieve their normal margin finally in August. The events of the 1972/73 crop year thus confirm the general hypothesis that world maize prices must be adjusted for the decline in the value of the dollar relative to other currencies as well changes in expectations of the maize supply/demand factors.

10.7 THE SUBSEQUENT MODEL PERFORMANCE

The previous section has described in detail the derivation of two specific price determination models, those relating the October/November prices to the Export Ratio based on the market expectations of supply and demand factors in August and the April prices to the market expectations at the start of February. A whole series of such models using the same approach was derived relating prices at the critical times of the crop year, as given in section 10.5, to the export ratio values based on the market expectations at the earlier critical times of the crop year. These models are then the basis for forecasting future prices. In some years they need to be modified using the U.S. agricultural support price model of section 10.2 which provides a 'floor' to the market prices when there is a very abundant supply of maize.

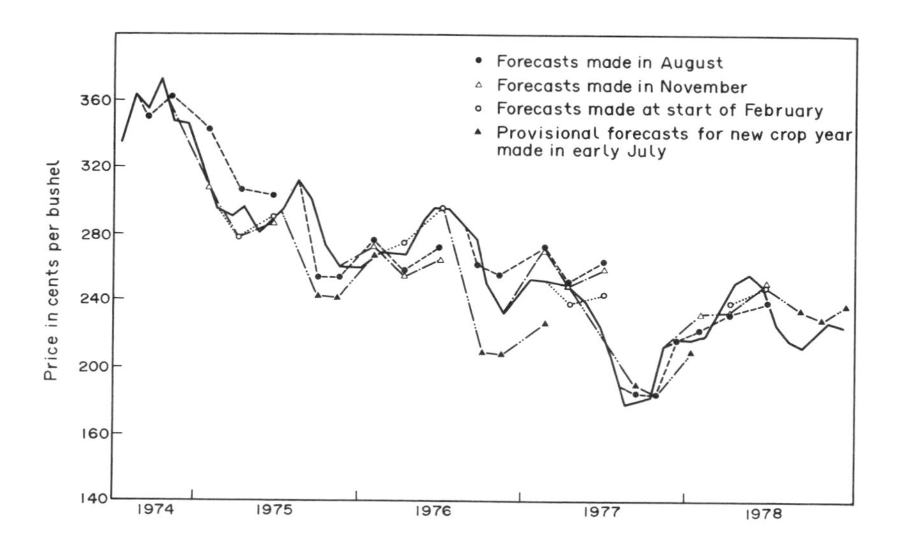

FIG. 10.8 The price forecasting performance for the crop years 1975/76, 1976/77 and 1977/78

The research described in the previous section has shown that the models explain past maize price behaviour extremely well. The acid test for their usefulness and validity is of course how they perform in the future. The forecasts made using the models, calculated in exactly the same manner as described in the last section, for the subsequent three crop years 1975/76, 1976/77 and 1977/78 are shown in Fig. 10.8. The figure also shows the fitted values for the last crop year of the period used in the analysis to derive the models, 1974/75, to give a comparison of the forecasting errors with the errors obtained in the fitted relationships. The very good forecasts made for a further subsequent crop year 1978/79 have already been discussed earlier. Figure 10.8 contains the forecasts of price movements for the whole of each crop year up to June/July, as made in August, November and the start of February. The majority of these forecasts were made in real time and all of them used only data as known at the time the forecasts were made. The price forecasts use the maize dollar adjustment factor calculated at the time the forecasts were made, not in the future months whose prices are being forecast. Thus for example, the forecasts made in August for October/November, January etc., use a maize dollar based on the value of the U.S. dollar relative to other international currencies as in August. They are not modified by what the maize dollar value actually was in October/November and January etc.

The forecasts shown in Fig. 10.8 are remarkably good at predicting in advance the general level of prices and the varying price movements over the crop year. One factor that should be stressed, is that the models for price determination and price forecasting do not contain the past prices as part of the forecasting formula, unlike very many of the forecasting models one sees in the economics literature. The worst forecasts are those made at the early times of the crop year for the June/July period, underestimating prices in 1975/76 and overestimating them in 1976/77. The underlying market situation changes over time so that one would expect the forecasts to be more in error the further ahead they are made. The 1975/76 crop year is a good example of this in that domestic usage and export were higher than expected earlier in the year. The February 1976 market expectations gave a higher value for the export ratio than in the previous August and November, so that as can be seen the price forecasts for June/July were higher than at the earlier times. The June/July prices are also somewhat different to those at other times of the year as they can be markedly affected in some years by the new crop situation. In 1977 for example, the acreage planted with maize, for the 1977/78 crop year, was higher than expected. An early drought as in 1974 or 1980 will give much higher July prices than the old crop year situation warrants.

Since the market situation is often changing, particularly over the old crop/new crop transition period and since information comes to the market piecemeal, the forecasts should be continuously monitored and updated if necessary. The forecasts given in Fig. 10.8 are based on the approximately quarterly critical times of the crop year identified in section 10.5. In practice the forecasts would be reviewed at least monthly and updated whenever necessary. An example of forecasting in practice is given in the final section of this chapter showing how the models described were used to make forecasts over the 1977/78 crop year, month by month. The section contains extracts taken from a set of regular reports produced by the author at the time for an outside organisation. These extracts illustrate the analysis and discussion necessary to generate the expectations of production, consumption, stocks, import requirements, etc., that are needed to calculate the export ratio at any time for use in the price forecasting models.

10.8 FORECASTING PRICES BEYOND THE END OF THE CROP YEAR

Reliable price forecasts can apparently be made from August onwards, when the first real indication of the state of the forthcoming maize crop can be made. Predicting future prices before August becomes a completely different kind of forecasting process however. A major factor in determining the new crop prices is the size of the crop. The crop size is the product of the area harvested and the yield per acre achieved. The yield depends on the weather over the period mid July to the end of August particularly, as well as the lack or outbreak of crop diseases. These factors by their very nature are completely unpredictable in advance. Their effects, if they happen, may be reasonably well predictable but forecasting that they will occur for this present year but not for the next year, say, is impossible, certainly at the current time. The most recent U.S. drought in the summer of 1980 was not anticipated until it happened, neither its start nor its severity. After the event some have said the cause was the eruption of the Mt Washington volcano earlier in 1980, but no-one in the weeks after the eruption first occurred predicted a serious U.S. drought. Curiously, there were U.S. maize crop disasters in 1966, 1970 and in 1974 but in no intervening years. However this apparent 4 year cycle was broken by a good crop in 1978.

Prior to late July forecasts of the expected crop size can only be made on the basis of some assumed average yield. In June USDA publishes a good estimate of the area that farmers have actually planted with maize. A remarkably constant proportion of the area planted is actually harvested for grain. The remainder is for grazing or direct feeding. The estimated area harvested multiplied by the average yield of the last few years, or better the average August and September yield estimates of the last few years, will give a prediction of the forthcoming new crop size. This together with estimates of the market expectations of the future demands can be used to calculate the export ratio and hence to make price forecasts. This was done for the crop years 1975/76 to 1977/78. These forecasts for the period September to January as made in June are also shown in Fig. 10.8. The forecasting performance is somewhat mixed as can be seen. In 1976/77 the future prices are badly underestimated; in 1975/76 the prices for September to November are again underestimated but by a lower amount whilst the forecast for January is very good; in 1977/78 the forecasts for all months are very good indeed. The June forecast for 1977/78 was for an abundant supply situation so that the prices were determined by the U.S. agricultural support price model of section 10.2. Even though they gave large forecasting errors, in an operational sense for purchasing the 1975/76 and 1976/77 forecasts are reasonable and useful, since they state that the new crop prices over September to November will be lower than currently in June and hence that it is better to wait 2 months or so, if possible, to buy requirements for the major portion of the new crop year.

The underestimates of prices for early 1975/76 and 1976/77 arose since the actual maize yields were lower than the average over the past few years. In these circumstances, when critical future events are by their nature unpredictable, price forecasting will have to be based on a series of alternative feasible scenarios. Estimates of the crop size should be made on the basis of various reasonable assumptions such as a crop disaster, a poor yield, an 'average' yield, a good yield and an excellent yield. Price forecasts should then be made for each of the alternative five scenarios. It is likely that the number of alternative price movement scenarios will be less than the crop size scenarios, since some of the latter will lead to very similar levels of future prices. Alternative strategic purchasing policies, for example buying heavily in June, or waiting until October to buy heavily, or buying at a uniform rate over June to September until the situation clarifies, should be evaluated for each crop scenario so that the potential gains and losses arising from the predicted price change from June to November are estimated. The company can then examine all of the various outcomes to decide which course of action best meets its general objectives. The choice will depend on the current state of prices as well as the new crop situation possibilities. For example if it is at the end of a year of shortage then the future prices will be lower in the nearby future months unless a real crop disaster occurs, whilst if it is at the end of an abundant year then prices are unlikely to be very much lower even if a very good crop occurs.

A similar approach will have to be taken in months earlier than June also. In these months an estimate of the area planted with maize will have to be based on surveys of farmers' planting intentions or models predicting likely plantings. However the major point is that prior to August price forecasts in the normal sense cannot be made except on the assumption of 'average' conditions. Only the effects of alternative assumptions of crop yields on future prices can be predicted. No one forecast can be made for the purchasing decision. This must come from an evaluation and discussion of the potential consequences of alternative policies in the various scenarios that can be envisaged. Then from August onwards good forecasts of the price movements over the next 9 months can generally be made. Because of the problems involved in this evaluation process it is probably unwise to anticipate buying the requirements for the new crop period, October onwards, before July of the old crop year.

10.9 USE OF THE MODELS FOR FORECASTING MAIZE PRICES FOR THE 1977/78 CROP YEAR

August

In August of 1977 USDA, the United States Department of Agriculture, predicted a U.S. crop of 154.75 million tonnes. Adding on their estimate for the carryover stocks from the previous year gave a total U.S. supply for the 1977/78 crop year of 177.65 million tonnes. USDA also forecast that domestic use would rise to 108.8 million tonnes, whilst exports would fall back to 36.8 million tonnes. U.S. exports to Western Europe in 1976/77 had been unusually high, because the hot dry summer of 1976 caused a partial failure to the French maize crop. On these forecasts the estimated export ratio value was 53.4. Published market comment at the time broadly agreed with USDAs export estimates but felt domestic use would be higher. Over May June and July the U.S. dollar had been relatively stable in value against other currencies. Its traded weighted value was about 86% of the pre 1971 value. Allowing for this, Fig. 10.6 gives a prediction of 174 cents per bushel for the price of maize over October and November. Other models gave a forecast price level for September of 184 cents and that prices would rise over December and January to reach around 210 cents per bushel in early February. Note at this point in time the price on Chicago for September delivery was 181.5 cents and 191 cents for December delivery.

However before making explicit forecasts we need to consider the effects of the U.S. government's feed grain subsidy programme, see section 10.2. The loan rate for maize for 1977/78 had been set earlier in the year at 175 cents per bushel. From the same historical analysis used to derive Fig. 10.1, this loan rate implies that prices should not fall much below 185 cents over September to November, rise to around 195 cents over December and continue increasing to around 206 cents in January/February.

The forecasts from Fig. 10.6 and the other similar models predict prices over September to November that are lower than the loan support rate, hence the loan support rate will hold up prices over this period. However note that the January forecast price is higher than that implied by the loan support rate. Combining this information together and allowing for a ± 5% range on the mean forecasts for forecasting errors and the day-to-day fluctuations, gives the following forecasts:

September/October/November	180–195 cents per bushel
December	185–205 cents per bushel
January/February	196–220 cents per bushel

September

USDA increased their crop estimate to 158.2 million tonnes. No revisions were made to the forecasts of domestic usages and exports. The revised price forecasts for the new value of the export ratio are all below the levels implied by the loan support rate. Hence no change should be made to the forecasts.

At the beginning of September, as a result of pressure from farmers over many months, the loan support rate was increased by 25 cents to 200 cents per bushel. The new price levels implied by this loan support rate are thus:

September/October/November	201–222 cents per bushel
December	213–235 cents per bushel
January/February	218–241 cents per bushel
March/April	228–252 cents per bushel
June/July	235–260 cents per bushel

Because of the large surplus prices are likely to remain mostly in the lower half of the above forecast ranges.

Prices reacted immediately to the proposed new support loan rate. For example, on the 29 August the price for September delivery was 182 cents per bushel. A fortnight later the September delivery price was 196 cents. The market had some problems in deciding how to react to the proposed new loan rate, because it was unclear when it would become law and from what date it would operate. So the market took some weeks to rise above the new $2.00 per bushel loan rate. The actual price movements over the period mid July to mid September are shown in Fig. 10.9.

October

The crop estimate was further increased to 160 million tonnes. News was confirmed during October that Thailand would have 1.5 million tonnes less

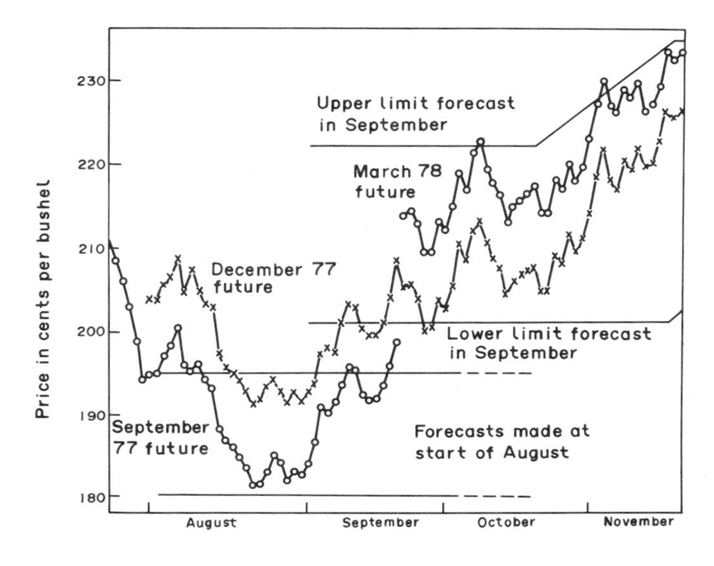

FIG. 10.9 Forecasts and daily prices on Chicago over July to November 1977

and Argentina 1 million tonnes less to export this year due to poor crops. (These are two important maize exporting countries.) Talks were held between the U.S. and U.S.S.R. governments to try to persuade the U.S.S.R. to import more U.S. maize. Rumours circulated in the market that the Russian crop was not as good as expected and that the U.S.S.R. would need to import more maize this year. In view of these events USDA raised its forecast of 1977/78 exports to 39.4 million tonnes.

The net result of these changes was to give an export ratio value of 53.5, identical to the value estimated in mid August. The price forecasts from the models are still all below the prices implied by the loan support rate system. Thus the September forecasts are endorsed.

November

The U.S. crop estimate was further increased to 161.7 million tonnes, an all-time record high level. The government of the U.S.S.R. confirmed the earlier rumours that a grain crop failure had occurred. The total Russian grain crop would be only 194 million tonnes, 19 million tonnes down on the target level. USDA estimated that the U.S.S.R. would be importing 20–25 million tonnes of grain this coming year and that she would now take up the U.S. offer of 15 million tonnes of grain, rejected the previous month. Other countries of Eastern Europe, particularly Poland, would have to increase imports from the U.S. Because of these changes USDA raised its forecast of crop year exports to 43.2 million tonnes.

The calculated export ratio value for November is thus 57.8, much higher than the earlier values. Since the summer the dollar had declined in value, particularly against the Japanese yen. Its value had fallen to 85% by October and continued declining through November. The mean forecast levels for future prices given by the models are 215 cents per bushel, 209 cents and 235 cents for January/February, April and June respectively. Note these are still below the loan support levels so the previous September forecasts are unchanged.

In their November Feed Situation Report USDA discussed their forecast of the domestic use of maize for 1977/78 in more detail. They predicted that the number of animals on feed would rise by 3.5% on the 1976/77 numbers to 78.8 million GCAUs (Grain Consuming Animal Units). During the previous 3 years on average 1.51 tonnes of feed grain had been fed to animals. At the beginning of the seventies, 1971/72 and 1972/73, the amount fed was higher at 1.74 tonnes. The difference is explicable by the prices received for the livestock products by the farmers compared to maize prices, e.g. a beef steer to maize price ratio of 26.8 for the early seventies compared to the 15.8 average over the last 3 years. Last year maize made up 80.5% of all feed grains fed to animals. Thus USDA's forecast of maize for animal feed use is:

$$(78.8 \times 1.51 \times 0.805) \text{ million tonnes} = 95.8 \text{ million tonnes.}$$

However USDA predicted favourable feeding margins for cattle, hogs and poultry over most of the year. With lower corn prices the beef steer to corn ratio will be much higher than 1976/77 and could reach 21.0. If such an improvement occurs simple interpolation gives a higher feed rate per animal, say 1.59 tonnes. This would give a crop year maize usage for feed of 100.8 million tonnes.

On this higher feed usage the export ratio value becomes 62%. The mean forecasts for January/February, April and June now become 234 cents per bushel, 224 cents and 251 cents respectively. Except for the April forecast, which is lower these are very close to the mid point of the forecast price ranges made in September on the basis of the new loan support rate. Thus even if a higher feeding rate per animal takes place it is unlikely to lead to a change to our September forecasts. Although prices are likely to be closer to the mid point, rather than the lower half, of the forecast price range.

December

No real change to the situation occurred, except USDA noted that prices received by farmers for their livestock had not risen as much as they had earlier predicted.

February

The January 1st stocks report showed that the first quarter feed use had been 32.3 million tonnes whilst the use for human consumption was 3.3 million tonnes. From a comparison with the first quarter usages for the last 2 years, this gives 101 million tonnes for feed over the crop year plus 13.5 for other uses, a total of 114.5 million tonnes. This is consistent with the higher usage suggested in the November report.

Various individual reports on cattle on feed and hogs suggested that animal numbers would be nearer 80 million CGAUs than USDAs November estimate of 78.8 million. As indicated in December livestock prices have not increased as much as expected so the feed rate per animal is unlikely to be as high as the 1.59 tonnes assumed in November. A value of 1.55 tonnes may be more appropriate. This gives a feed maize usage of 99 million tonnes, 2.0 million tonnes less than simple extrapolation. USDAs forecast of total domestic use was maintained at 109.4 million tonnes.

The first quarter exports were only 10.0 million tonnes compared with 12.7 million in 1976/77. However there are substantial differences between 1977/78 and 1976/77 which indicate a simple extrapolation of the first quarter exports over the whole year is inappropriate. Exports to Western Europe were forecast to be 5 million tonnes down on those of the drought year of 1976/77. Exports to Western Europe from the U.S. are usually higher in the first half of the crop year. An increase in exports to Eastern Europe of 7 million tonnes was forecast. Most exports to Eastern Europe take place in the second half of the crop year. Exports to Japan are predicted to remain at the same level as 1976/77 because of the Thailand crop failure. For these reasons extrapolations of the first quarter exports is likely to give too low a crop year total. General market comment similarly seemed to concur with USDA's forecast of crop year exports of 44.5 million tonnes.

The export ratio on USDA's estimates is 59.8, but on the extrapolation of domestic use, found to be the best procedure in building the models, the export ratio is 64.2. By the end of January the dollar had declined even further in value to under 83% of its pre 1971 value. On this basis Fig. 10.7 gives a prediction of 238 cents for the price over April. Allowing for the ± 5% range for errors and daily fluctuations this is identical to the September forecast, based on the loan support system. The price forecast for June is

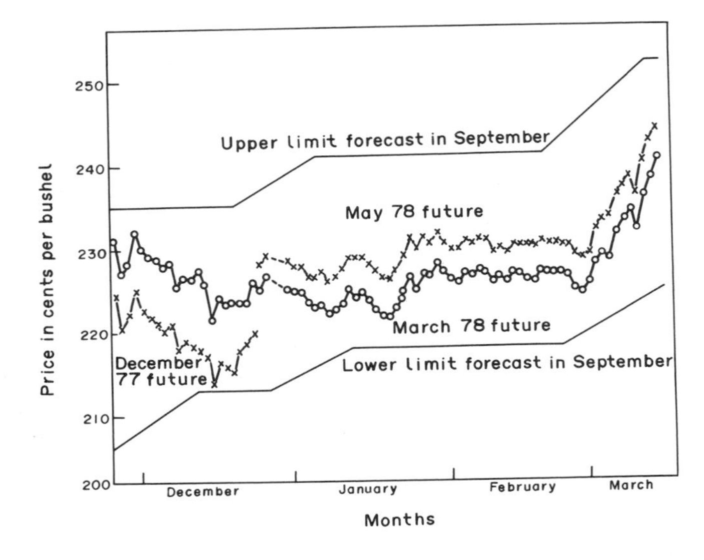

FIG. 10.10 Forecasts and daily prices on Chicago over November 1977 to March 1978

265 cents per bushel, note higher than the loan support rate September forecast. The export ratio for USDA's estimates gives a price of 246 cents, virtually identical to the September forecast. Thus the forecasts in February for the prices in April and June are the same as in September, but with the addition that prices over June could fluctuate as high as 278 cents.

The daily prices on Chicago over the period mid September to the beginning of March are shown in Figs. 10.9 and 10.10. The ranges forecast at the beginning of September for future prices are marked on the figures. Thus they show immediately how good the forecasts were over this six month period.

March

At the beginning of March it was announced that a severe drought had reduced the Brazilian soyabean crop. World soyabean prices soared from \$5.80 per bushel to reach \$7.30 by the end of March. Chicago maize prices rose in sympathy to touch 258 cents by the end of March, a rise of 16% over the month. A similar set of models for soyabeans suggested that prices would remain around the levels reached at the end of March through until early July. Since maize is predicted to rise to within the range 245–278 cents, depending on the domestic usage, the only forecast possible is that prices will remain around the levels reached at the end of March, 248–258 cents, over April and May possibly increasing further over June.

May

The April first stocks report showed that U.S. domestic use had fallen strongly in the second quarter of the crop year to 16% below the first quarter usage. This is a return to the pattern of the early seventies, rather than 1975/76 and 1976/77. An extrapolation of this behaviour implies a crop year feed use of only 97.5 million, compared to the forecast 101 million tonnes in February. This gives an export ratio of 61.0. Allowing for a further decline in the value of the dollar gives a forecast price range for June, the end of the old crop year, of 248–274 cents. Note this is higher than the forecast based on the loan support system, made in September.

June

Although there is considerable uncertainty regarding farmers' planting intentions, most forecasters are predicting a planted acreage of around 78 million acres. This implies that about 67 million acres will be harvested for

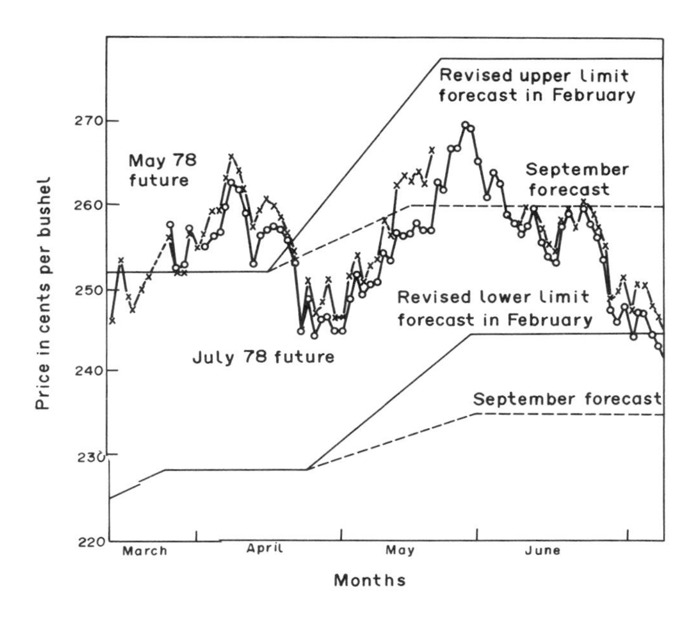

FIG. 10.11 Forecasts and daily prices on Chicago over March to July 1978

grain. Assuming a continuation of the trends in yield of the last 3 years gives a crop estimate of 153.3 million tonnes, with an error of ± 5 million tonnes. This gives a total supply of U.S. maize of around 181.8 million tonnes.

Assuming that the 1978/79 usages are the same as 1977/78 gives an Export Ratio of 62.9. Allowing for the decline in the dollar, Fig. 10.6 predicts prices will be in the range 212–235 cents over October and November. Note that this implies a strong fall in prices compared to current levels. Thus purchases for the late summer and autumn months should be delayed if possible.

USDA predicted that the new grain crop in the U.S.S.R. will be up by 5 million tonnes most likely. Argentina and South Africa have had good crops. The maize crop in Thailand should return to more normal levels for 1978/79. Hence 1978/79 U.S. exports should be slightly below the current crop year. Allowing for an increase in U.S. domestic use similar to the growth for 1976/77 to 1977/78 plus the lower exports gives an export ratio similar to the 62.9 derived from last year's usages. Thus little difference would be made to the price forecasts based on the 62.9 value for the export ratio given above.

The actual prices together with the various price forecasts for the period March to July 1978 are given in Fig. 10.11.

July

USDA's first crop estimate for 1978/79 gave a total supply of 185.9 million tonnes. Their first forecasts of domestic use and exports were 114.3 million and 44.4 million tonnes respectively. This gives an export ratio of 62.0, virtually identical to the one calculated in June.

August

USDA's August crop estimate showed an increase of 9 million tonnes on the preliminary July figure. Domestic use for 1978/79 was forecast at a higher level, compared to the July value, of 116.1 million tonnes. The export ratio value is now reduced to 57.6. Allowing for the further decline in the value of the dollar gives a mean forecast for October/November of 202 cents per bushel. The loan support rate was raised to 210 cents per bushel for 1978/79. The price forecasts from the models are thus well below the levels that are implied by this new loan rate. Hence unless there is a crop disaster prices over 1978/79 will be determined by the loan support rate. The price forecasts, together with the actual price over the year, for 1978/79 were shown earlier at the beginning of this chapter in Fig. 10.1.

11

Models for the Tactical Buying Problem

11.1 INTRODUCTION

This chapter reviews some of the approaches and models that have been put forward, or have been used for day-to-day decision-making in the commodity purchasing area. A major difficulty is that a model giving an efficient workable policy for a company is most unlikely to be publicised, because of the competitive advantage to the company concerned. Thus published work would be expected to be sparse, so that the small portion seen clearly could hide important and substantial amounts of work. Obviously specific *ad hoc* purchasing rules can be devised by observing the idiosyncrasies of the daily price movements of a particular commodity. Many of the unpublished policies adopted by companies, if in fact they do exist, are likely to be of this form. The emphasis of this chapter, however, is entirely on general models and general methods that can be applied to the day-to-day purchasing problems for any fluctuating price raw material. Only in this way can experience for one material in one purchasing situation be passed on to assist with other materials and other situations. It is this author's view and experience that the scarcity of published work reflects a real lack of work on the development of formal models for making purchasing decisions. This chapter therefore aims to give a comprehensive review of the current situation and form the basis for anyone who wishes to develop practical operational models for specific problems or carry out further research.

The order in which the various approaches are discussed in the chapter is somewhat historical since this naturally reflects the increasing sophistication and formal mathematical basis of the purchasing policies put forward at differing times. Most of the methods put forward are 'heuristic' purchasing policies. This means that either based on experience or an examination of the situation, a purchasing policy is devised that is asserted to be a good appropriate policy to use. The policies are generally based on some hypothesis about the process that generates daily prices over time. This hypothesis itself may be an enlightened guess or rigorously tested by high powered statistical analysis. The particular hypothesis concerning the process generating future prices then has to be converted into a specific price forecasting system to enable estimates or forecasts of the likely levels of future prices to be made at any time. The specific form that the asserted purchasing policy takes then depends upon some particular property of the price forecasting system devised. Only one approach has tried to model directly the day-to-day buying decision problem. Because this problem is a time sequential decision-making situation, the natural way to formulate it is in dynamic programming form. (The dynamic programming method was described earlier in Chapter 7 on the linseed oil delivery problem.) The dynamic programming model can be applied to a wide range of commodity purchasing situations that occur in practice. Again assumptions have to be made about the process generating prices and hence the appropriate price forecasting situation. The models can be solved for many alternative price forecasting systems to give a purchasing policy which is 'optimal' for the objectives and assumptions stated.

Only two of the approaches are from the very start directly based on the nonstationary of commodity price movements, that of the Chartists and Taylor's Price Trend Model. The underlying motive behind the approach taken by price chartists is that commodity price movements are so complex and capable of taking so many different forms that any method must be qualitative rather than quantitative. The approach concentrates on looking at graphs or pictures of price movements over time. Chartists then try to classify and catalogue all the different patterns of price movements that have ever occurred over the past. Such price patterns are identified for several weeks of price movement rather than days. No attempt is made to explain why a particular pattern of price movement occurred, in fact in the view of most chartists this is impossible and unnecessary. The next stages are to examine whether any one particular pattern of price movement always precedes another and more importantly whether a particular pattern of price movement can be identified at a very early stage in its development so that prices at the end of that particular pattern can be predicted. Unfortunately the verdict on their success in achieving these aims must be that they have failed. Some chartists do well on some patterns or for some periods of time, but no general successful procedure has yet been devised.

The final approach due to Taylor (1978), described at the end of the chapter, is based on the new statistical model proposed to explain daily commodity price movements, which was described in the appendix to Chapter 2. The model explicitly contains a trend term. A trend is assumed to persist until new information enters the market to produce some new trend. Such discrete changes are assumed to occur randomly and with low probability. The new trend value is assumed to be generated random from some probability distribution. The size of the purchase to be made at any time is then made proportional to the confidence that the current trend factor is significantly positive and will persist for some time. A drawback with this model is that it requires sophisticated analysis over several years of past prices to calculate the details of the statistical forecasting model. More importantly as yet it is only appropriate to the Type I continuous consumption continuous supply buying situation.

In the other four approaches discussed, Carlson and Corrigan, Morris (1959), Fabian *et al.* (1959) and Kingsman (1969), an initial purchasing policy is first derived on the assumption that future prices are generated from a constant probability distribution. The various policies are discussed and compared for the constant price distribution situation. It is shown that for different reasons both the policies put forward by Morris and of Fabian *et al.* are inappropriate for the real commodity purchasing situation. The Carlson and Corrigan policy is relatively simple to calculate but limited mainly to the Type I buying problem as described in Chapter 4. The decision price break policy derived as the solution to the dynamic programming model by Kingsman is slightly more complicated but can be extended to cover the variety of purchasing problems that occur in practice.

It is shown that the policy advocated by Carlson and Corrigan is unfortunately entirely unsuitable for the nonstationary price situation where there are short- and medium-term trends over time as well as changes in the size of the fluctuation about the general price levels. For successful practical use this policy requires an econometric/behavioural price model, as described in Chapters 8, 9 and 10, to give those times when prices can be expected to be relatively stable over the immediate future.

The decision price break model put forward by Kingsman extends easily to the nonstationary price situation. For the model to be an 'optimal' solution to the purchasing problem there should be no serial correlation between successive price changes or the future price distributions. Undoubtedly some serial correlation exists. In such cases the decision price break model becomes a good heuristic policy. As will be shown via case studies in the following two chapters, using an adaptive exponential smoothing model to forecast future prices and their distributions, Kingsman's decision price break model gives a good practical performance that will enable improvements in purchasing efficiencies to be made in practice.

In conclusion then it is recommended that in any study the Kingsman decision price-break model and the Taylor model should be considered, together with the Carlson and Corrigan model if the circumstances are right. No other general commodity purchasing model has yet been published that is worthy of consideration.

11.2 PRICE CHART PATTERNS

As mentioned in Chapter 1, some research has been carried out by those who work within the commodity markets on ways of forecasting prices. Although much of the details are kept hidden it is known that the preference is for 'technical' rather than 'fundamental' analysis, i.e. the study of the commodity market itself, the prices, the daily price spread, the volume of business etc. rather than the external factors that affect the supply of and demand for the commodity. The basic technique of 'technical' analysis is to plot charts (graphs) of daily prices and then examine them subjectively to try to pick out consistent patterns of price movement that repeat themselves. For this reason its practitioners are collectively known as 'chartists'.

This approach is based primarily on the Dow Theory originally developed for stock market price movements. As a major ('primary') uptrend of the commodity price proceeds there are numerous intermediate ('secondary') downwards reactions, each of which retraces a substantial proportion of the preceding rise. After each reaction, the price recovers and goes on to surpass the previous high price. Dow theorists keep on the alert for a recovery which falls short of the previous high. If, following such an abortive recovery, a downward reaction pierces the low point reached by the last previous reaction, evidence is at hand that the market has gone into a major ('primary') down-trend. This is illustrated conceptually in Fig. 11.1.

The trouble with this method is that there is no statistically significant way of evaluating it. For in addition to ascending and descending tops and bottoms, which are often referred to as 'channels' and which are amenable to reasonably precise definitions, chartists refer to 'heads and shoulders'

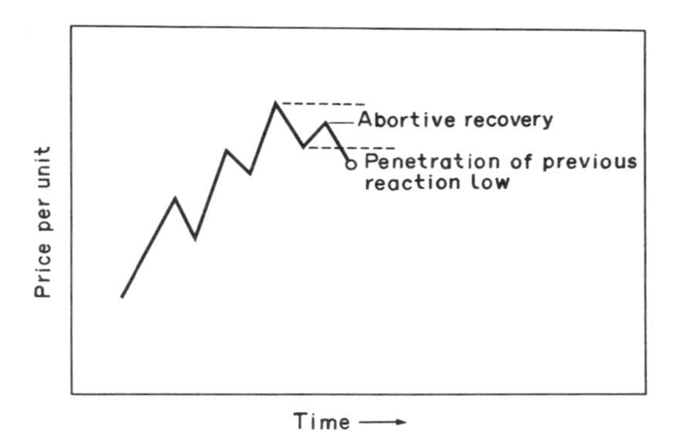

FIG. 11.1 Conceptual illustration of the Dow Theory

formations, 'triangles', 'rectangles', 'flags and pennants' and a list of other configurations with equally exotic names but with quite imprecise definitions. A half dozen chartist analysts looking at the same chart will rarely give anything near a unanimous interpretation. We therefore end up testing the chartist practitioner rather than the method. Unfortunately these exotic patterns are usually only clear well after that period of price movement has been completed. They can rarely be specified at an early stage of the development of the price pattern.

Although chartism is frequently described as simplistic and unsound, it should be remembered that no quantitative statistical processes have been developed that are generally accepted as describing accurately such unstable continuing changing time series as commodity prices. The exotic patterns proposed are esssentially attempts to derive complex models, which necessarily have to be descriptive. The real weakness of the method is the imprecision of the definition of the patterns, so that on the same data they are not repeatable from one practitioner to another.

A development of the chartist approach is the use of 'point and figure' charts. Point and figure analysts attempt to measure the price trends directly by charting the significant price changes only. The analyst chooses some minimum threshold level for the changes he will consider. Changes less than this are ignored. On the other hand if the price changes by three times this level, three entries will be made on the chart. Entries are made in the same column until the direction of prices reverses by more than the threshold level, whence entries are made in the next column, and continued in that column until prices reverse again. A further new column is then started.

The point and figure analyst examines the chart to discover areas of 'congestion', also called 'upside resistance areas', 'downside support areas', 'tops', 'bases', 'lateral trends', etc. Essentially these are narrow bands of price fluctuation, indicating a standoff between relative supply and demand pressures. Point and figure charting, by condensing many months of price fluctuations within a limited space, is designed to reveal such congestion areas. It is claimed that a break out of price from a congestion area very often seems to be symptomatic of a new trend.

Having detected what appears to be a new trend, a break out from the congestion area, the point and figure chartist measures the width of the band of congestion by counting horizontally the number of columns covered. If the price break out has indicated an upward trend, the number of columns multiplied by the given value of the minimum price change considered is added to the price level of the congestion area. The resulting price is said to be the forecast of the high price towards which the new upward trend is heading.

Little theoretical justification for this method has been given, only claims that on balance that the technique produces good results. Indeed the proof or disproof of the claim is an extremely difficult task. Usually several different price forecasts can be read from the same chart, depending on which price threshold level is chosen for the horizontal chart. In most cases the choice of the price threshold for the congestion area is not clear cut, and analysts use several levels to arrive at a range of possibilities. Using different levels for the minimum threshold price change often produces radically different forecasts of the new levels which prices can be expected to achieve.

Even if the claims made for the technique are valid, it suffers from a serious drawback, which makes it well nigh useless as regards assisting in commodity purchasing problems. There is no time dimension attached to the point and figure chart. No estimate is made of the time the price will take to move to its potential new high level nor how long it is likely to stay there. Thus no guide at all is given to the purchasing manager as to how much he should buy now and how many days' or weeks' future requirements he should cover. These methods are designed almost entirely for speculators, who are only interested in knowing whether prices are high or low and hence whether to sell or buy. Speculators have no particular interest in the time factor of how long prices will take to move from a low to a high and back to a low again. Virtually all these techniques come from the U.S. where speculating and gambling in the commodity markets by individuals and companies is alleged to be as common as, if not more than, such speculation in the stock market in the U.K.

Two useful reviews of chartists' methods from an outside objective viewpoint are given in the books by Cohen and Zinbarg, and Tewles.

11.3 MOVING AVERAGE AND SIMPLE PROBABILITY MODELS

In one of the first published papers on this area Carlson and Corrigan put forward a purchasing policy, for the continuous usage and supply situation, based on the use of long run moving average prices and the levels of the fluctuations of the actual prices about these moving averages. In addition for raw materials with strong seasonality in their price movements over the year they advocated a model based on this seasonality. For the general purchasing situation they found that 12 to 24 month moving averages of past prices (updated monthly) were an acceptable measure of the current general price level to be expected in most cases. This average they called an 'indicator'. If the price quoted is higher than the indicator value, then this implies that that price is high and the buyers should only make a purchase to meet current needs. If the price quoted is less than the indicator value, then it is a low price and the buyer should build up his stocks to some preassigned level to meet future consumption requirements. The indicator value is thus used as an initial reference point for determining whether price offers are high or low.

The stock to carry at any price level is directly related to the price spread, the variations of prices above and below the indicator value. In a study of the price spread, the indicator was plotted on a graph covering several years of the spot price performance. The frequency distribution of the deviations of the prices over each month about their monthly indicator value was then obtained. Considering only deviations when the price is below the indicator value this distribution might take the form shown in Fig. 11.2. This shows, for example, that the price is less than the indicator value but higher than \$2 below the indicator value for about 70% of the time that it is below the indicator value. It was assumed that price change intervals of \$0.50 were proper in the market for the commodity considered.

Figure 11.2 forms the basis of the proposed purchasing policy. If the price is higher than the indicator value the purchaser should buy no more than required to maintain his stocks at the minimum safety stock levels required to keep his manufacturing processes operative. Carlson and Corrigan argue that purchases to raise the stocks to cover future consumption should only be made when prices are lower than the indicator value. Furthermore, since the price is less than \$2 below the indicator value on 70% of the occasions that it is below the indicator value and hence the

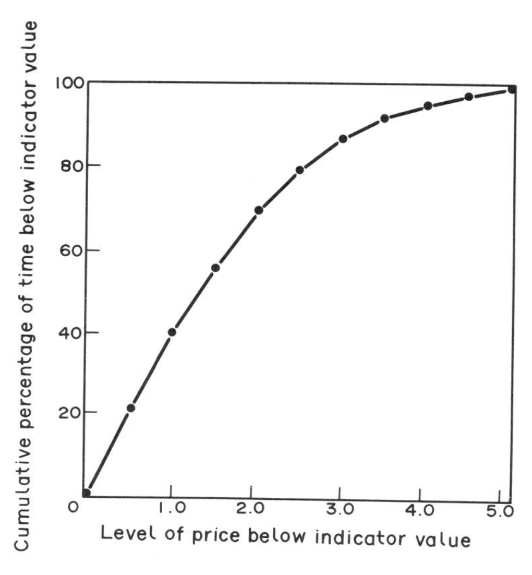

FIG. 11.2 Carlson and Corrigan's price spread distribution

occasions at which the buyer will voluntarily consider making purchases, the buyer should purchase sufficient to raise his stocks to cover 70% of his maximum 'purchase and stock cover' when the price offered is $2 below the current indicator value. Since their frequency distribution was derived from grouped data the $2.00 value is replaced by prices in the range $1.50–2.00 below the indicator value. Similarly for all other prices. The policy is then as given in Table 11.1, where as an illustration it is assumed that the maximum stock to have at any time is 50 weeks' consumption.

The next stage is to determine what should be the maximum and minimum values for the 'purchase and stock cover' to be held at any time for the particular purchasing problems. Carlson and Corrigan used a simple evaluation approach over past prices. If there are no inventory holding charges, then they say that the unit savings relative to the historical average price increase as the maximum cover increases and thereafter tend to level off as they approach the point where all purchases would have been made at the historical minimum price. Where there is a positive carrying charge the savings increase initially, then decrease to zero and become negative, since the cost of carrying the average inventory at larger levels of the maximum cover exceeds the savings made by purchasing at favourable prices. Thus for each level of the carrying charge there is a unique optimal value for the maximum purchase stock cover. This is shown conceptually in Fig. 11.3.

This policy, it is claimed, guarantees that the average purchasing cost will always be less than the historic average price.

There are practical problems for each particular situation in determining what should be regarded as the minimum price change to be considered, i.e., how should prices be grouped together to derive the price spread distribution. Should it be $0.50 as in Carlson and Corrigan's example or

TABLE 11.1 Carlson and Corrigan's moving average and price spread purchasing policy

Price range below indicator level	Level to raise stocks as %age maximum cover	Level to raise stocks in week's consumption
0.01–0.50	21	11.5
0.51–1.00	40	20
1.01–1.50	56	28
1.51–2.00	70	35
2.01–2.50	80	40
2.51–3.00	88	44
3.01–3.50	93	46.5
3.51–4.00	96	48
4.01–4.50	98	49
4.51–	100	50

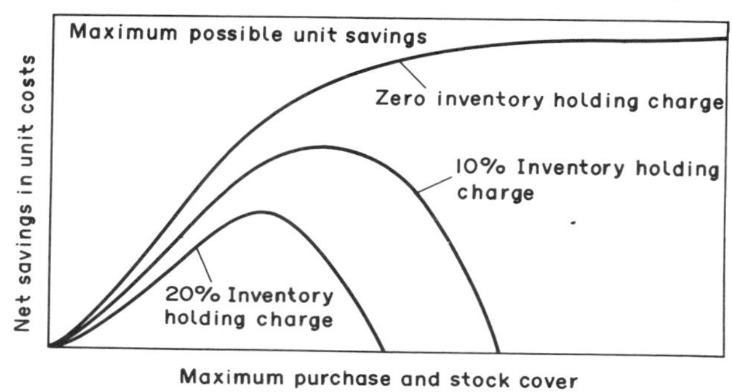

FIG. 11.3 Determination of the maximum purchase and stock cover for the Carlson and Corrigan moving average price spread model

$0.10, $1.00 or $5.00? Obviously alternatives should be evaluated. A similar problem applies to determining what should be the minimum purchase lot size. For the reasons discussed in early chapters this should be defined in day's or week's consumption. Since price offers usually occur daily then the purchase and stock cover, the commitment, should be in terms of days rather than weeks or half weeks as in Carlson and Corrigan's example. Some modification is necessary if a day's consumption is less than the minimum quantity that can be purchased at a particular price offer.

As we have seen the proposed policy allows the stocks, and hence purchase, to cover several weeks future consumption. The policy basically assumes that prices over future months will be randomly distributed above and below the current 12 or 24 months moving average indicator price level. This means that the current indicator value is the best forecast of the general price level over the next year. This is a very strong assumption indeed. The usefulness of the policy depends entirely on the validity of this assumption. Measuring how high or low prices are in terms of simple probabilities of their likely occurrence relative to the indicator value via the price spread distribution is based on the concept that over the long run prices are randomly generated from a constant probability distribution. If the price spread distributions change from month-to-month then such a simple probability measure of how high or low are prices is no longer possible.

Long run monthly moving average values are well known to lag significantly behind the actual price changes that occur from month-to-month. So if there are significant medium terms trends in prices over several months the forecasting errors become very large indeed. This can be seen in Fig. 11.4 which shows the average monthly cash corn (maize) prices on the Chicago market from October 1970 to September 1977. As

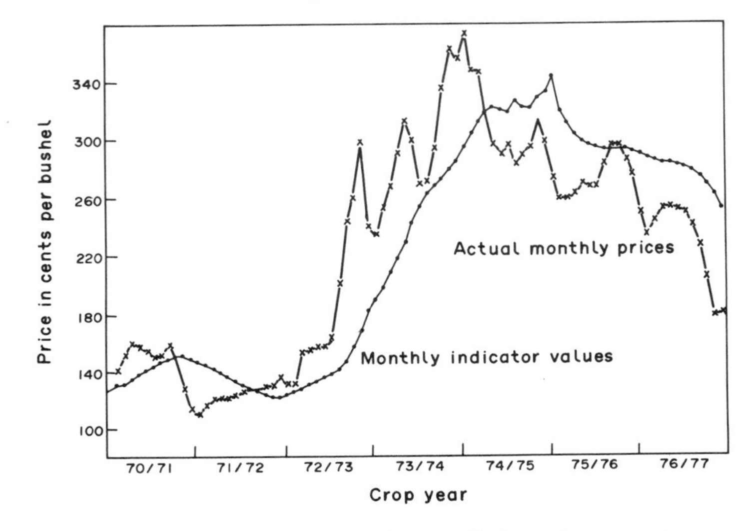

FIG. 11.4 12 Month moving average forecast (indicator) compared to actual monthly prices

can be seen the 12 month moving averages (the indicator values) lag well behind the actual price changes. The troughs and peaks of the indicator values are 12 months behind the troughs and peaks of the actual prices. As is clear from Fig. 11.4 the use of monthly moving average values will only be appropriate for those situations where there are no real changes over time in the price levels and where the prices generally fluctuate above and below a relatively constant price level for long periods of time, i.e. the stationary demand situation.

The Carlson and Corrigan policy was applied to purchasing maize requirements over the crop years 1974/75, 1975/76 and 1976/77. (This commodity and time period was used because it is the period chosen to illustrate the validity of the price forecasts made by the econometric and behavioural model derived for maize in the preceding chapter. The daily cash prices over each month were converted into a deviation from their month's indicator value for the four earlier crop years, 1970/71 to 1973/74 inclusive. This gave the price spread distribution as shown in Fig. 11.5. Note there were 994 days of market prices over this 4 year period. Prices are grouped into \$5 intervals. The prices ranged from \$41 below the indicator value to over \$150 above it. Figure 11.5 does not appear to follow any well-known standard probability distribution. However the Carlson and Corrigan purchasing policy is based only on those prices less than the indicator value. The relevant part of price spread distribution is shown in Fig. 11.6. This is exactly equivalent to Fig. 11.2 showing Carlson and

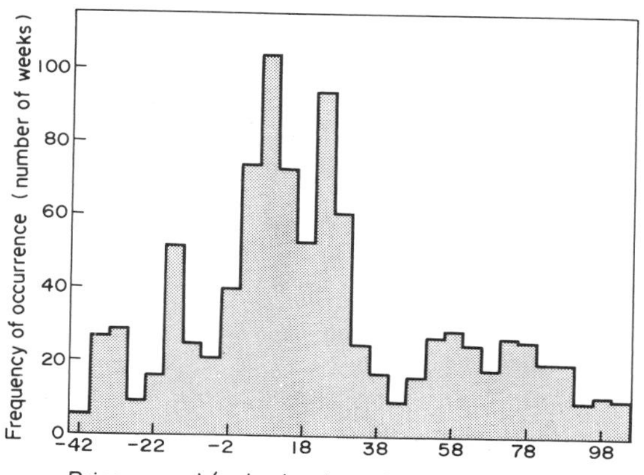

FIG. 11.5 Histogram of the differences between the actual monthly prices and the 12 month moving average value for maize over the crop years 1970/71 to 1973/74 (grouped data)

Corrigan's example. Only 22.5% of the total daily prices over the 4 year period were below their month's indicator value.

A purchasing policy for maize can be devised using Fig. 11.6 in the same way as Fig. 11.2 was used for the illustrative example. It was assumed that

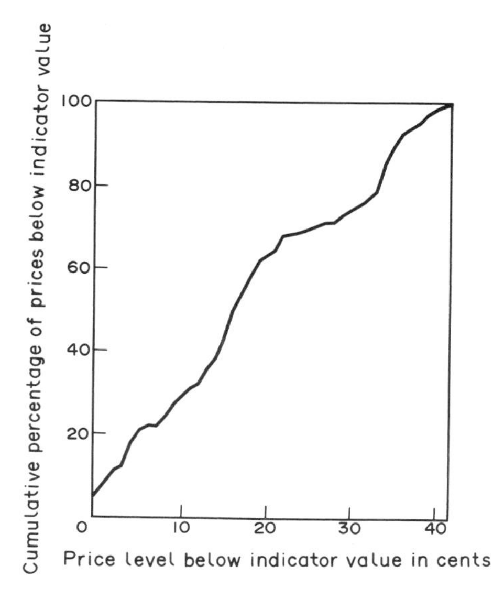

FIG. 11.6 Price spread distribution for monthly maize prices for the crop years 1970/71 to 1973/74

the maximum purchase and stock cover at any time was 12 months consumption. For simplicity all months were assumed to be of equal length of 4.25 weeks and that the weekly demand was constant over the period. It was also assumed that any size of purchase could be made. Analysis showed that using the policy nothing would have been bought in advance for the 1974/75 crop year consumption over 1973/74. Purchasing starts in October 1974 with the minimum safety stocks allowed. If no purchases are made voluntarily, because prices are high, then every month 4.25 weeks consumption must be purchased to maintain the level of the safety stocks. Such a purchase is assumed to be made at the average monthly price.

The purchases that would have been made are shown in Table 11.2, together with the average price paid per unit purchased. Sometimes several purchases are made in a month. Those months where 4.25 weeks consumption were purchased at the average monthly price are marked with an asterisk. The average prices per unit paid for each year's consumption are given at the bottom of the table and compared with the historic average

TABLE 11.2 Performance of the Carlson and Corrigan moving average and price spread policy in buying maize for usage from October 1974 to September 1977. (The prices in the table are in cents per bushel)

	1974/75		1975/76		1976/77	
Months	Purchases weeks usage	Price paid per unit	Purchases weeks usage	Price paid per unit	Purchases weeks usage	Price paid per unit
Oct	4.25*	374	4.25	292	9.5	249
Nov	4.25*	348	4.25	253	0.0	—
Dec	4.25*	347	4.25	271	0.0	—
Jan	40.3	305	4.25	255	0.0	—
Feb	14.95	286	0.0	—	0.0	—
Mar	4.25	270	0.0	—	0.0	—
Apr	0.6	283	0.2	264	0.0	—
May	7.85	282	0.0	—	0.0	—
June	4.25	282	0.0	—	0.0	—
July	4.25	281	0.0	—	0.0	—
Aug	0.0	—	8.16	276	0.0	—
Sep	8.5	293	20.4	265	0.0	—

Average price paid for the units consumed in the year shown

	1974/75	1975/76	1976/77
Carlson and Corrigan Policy	317.8	285.6	262.4
Hand-to-mouth Policy	312.3	275.0	230.5

yearly prices also. It is seen that for none of the 3 years did the Carlson and Corrigan policy give an average unit purchase price below the historic average annual prices. The performance for the 1974/75 crop year, obviously from Fig. 11.4 a difficult year with large price variations, was reasonable in the circumstances, giving a loss of 5.6 cents per bushel (1.8% of the average annual price). The performance then worsens over the other 2 years, a loss of 10 cents per bushel (3.7%) in 1975/76 and a loss of 31.9 cents per bushel (13.8%) over 1976/77. This example shows that contrary to the claim made, the proposed policy does not guarantee an average unit purchase below the historic average price.

More appropriate forecasting procedures could be devised to improve the estimate of the indicator value, by, for example, giving more weight to the more recent months. However the actual price movements shown in Fig. 11.4 demonstrate clearly that the assumption that futures prices over several months into the future can be described by the price spread distribution about the current month's indicator value is completely invalid. A trend component needs to be included in the prediction of the indicator value. Different price spread distributions could be derived for different future months. However if this is done there is no longer a simple method based on the probability of occurrence of different prices with which to derive the purchasing policy. The policy as stated is thus not appropriate to materials whose prices fluctuate by large amounts and where the general price level changes significantly over time. Essentially the policy ignores the existence of any short or medium trends in prices.

The basic methodological weakness of the Carlson and Corrigan approach is that they ignored the structure of the commodity purchasing decision as discussed in Chapter 4. They have combined together the strategic buying period, the active buying period and the tactical day-to-day buying policy. This leads to confusion, since the different aspects of these stages of the purchasing decision have not been appreciated. If a successful econometric/behavioural model for price behaviour can be constructed, as described in the immediately preceding chapters, then the Carlson and Corrigan policy may have a role to play. Using such an econometric model to derive the active buying period will also give the total amounts to be purchased over the active buying, say the next month or two. The model may also predict that prices will be relatively stable over the next month or over the next 2 months. The assumption of the Carlson and Corrigan approach of stationarity in both the price spread distributions and the general price level will then be reasonably valid. The approach can then become the basis of a day-to-day tactical buying policy.

The Carlson and Corrigan policy can thus be recommended for short-term tactical buying decisions, provided other methods have been devised which show that the prices over the short-term future will be relatively stable. It has the advantage that the calculations to work out the policy are simple and can be carried out by the buyer himself using only pencil and paper methods. If a trend is predicted for the short-term price movements, particularly a downward trend, then some of the potential gains will not be realised. Provided the predicted trend is small, then the simplicity of the policy may overcome the extra gains that could be achieved by more complex purchasing policies.

The detailed specification of the Carlson and Corrigan policy for a stable price situation over the active buying period, the next month or two, is as follows.

Let μ_t, be the current forecast average price level or indicator value at time t, let $\phi(u)$ be the distribution of the price deviations about the indicator value assumed constant for all time, M and m the maximum and minimum values of the purchase and stock cover. Note M is defined by the active buying period analysis of Chapter 4. Let P be a price offer. Then the policy can be generally expressed as follows:

1. $P \geqslant \mu_t$: then if S is current stock level before a purchase, (defined in days or weeks consumption) the amount to purchase, x, is given by

$$
\begin{aligned}
X &= m - S \qquad && \text{if } S \leqslant m \\
&= 0 && \text{if } S > m
\end{aligned}
$$

2. $P < \mu_t$: then derive $q = P - \mu_t$ and

$$y = \int_{q}^{\mu_t} \phi(u)du \Big/ \int_{-\infty}^{\mu_t} \phi(u)du$$

(to allow for nonsymmetric distributions), then the amount to purchase, x, is given by

$$x = \text{maximum } (m-S,\ yM-S)$$

Carlson and Corrigan recognised to some extent the inadequacies of moving average models. Some of the large variations between the indicator values and the actual prices may be due to persistent seasonal price movements. They suggest that for commodities with a periodic or seasonal price pattern, an efficient purchasing policy can be based on the periodicity. This policy will give a detailed specification of the purchases to be made week by week over the year. The example they give is illustrated in Fig. 11.7, which shows the price of a commodity over a 52 week period. The average price over the period was \$50 per unit with maximum and minimum prices of \$58 and \$42 per unit respectively. In the example the

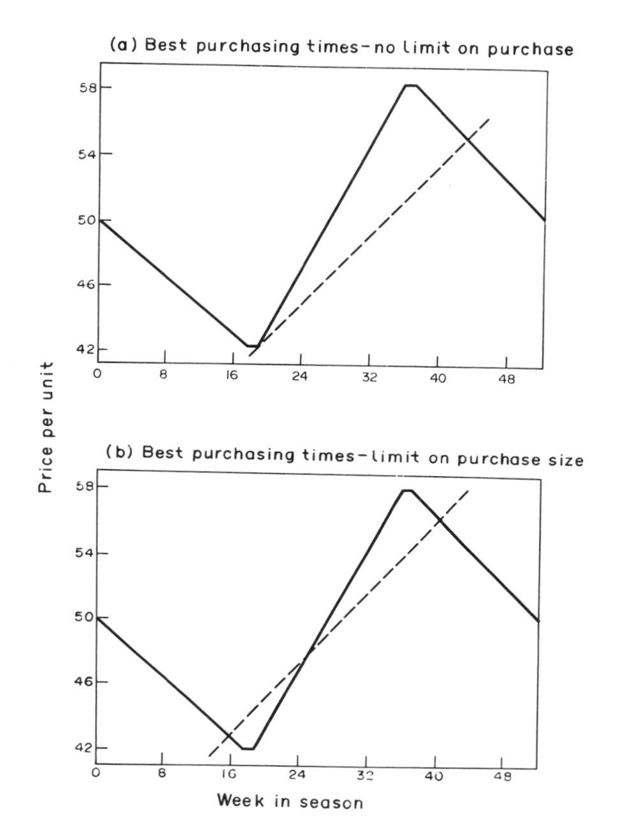

FIG. 11.7 Annual buying strategy based on seasonal price movements

inventory carrying charges are $0.5 per unit per week. The broken line, tangential to the price curve at week 18, shows the effective cost of purchasing at a price of $42 in week 18 and storing the commodity for use in some future week. Thus if stored and used in week 32, the total cost is $42, price paid, plus a total storage charge of $7 per unit, hence an effective cost of $49. The price that would have to be paid in week 32 is $56. Thus a saving of $7 is achieved by buying week 18 and storing for use. In fact, for all weeks such that the broken line is less than the price curve, it is cheaper to buy in week 18 and store until needed rather than buy in the market. This will be true for all weeks 19 to 43 inclusive.

Thus the purchasing policy can be specified as

(a) Buy each week's requirements in the market as needed for weeks 1–17 inc.

(b) In week 18 buy 26 weeks' requirements to cover needs for weeks 18–43 inc.

(c) From week 44 to the end of the year buy weekly in the market.

Assuming this characteristic seasonal pattern is repeated year after year this policy can be continued at an annual cost of $2,496.5 (purchase cost of $2,334 plus $162.5 in carrying charges) compared to a cost of $2,600 if requirements are purchased each week, assuming only one unit is consumed each week. The saving is approximately 4%.

This policy assumed an unlimited market. It can easily be applied to a restricted market in exactly the same way. This restriction may be imposed by the market or by the purchasing manager as a policy constraint.

The example given by Carlson and Corrigan considers a constraint of a maximum purchase of two and a half week requirements in any week. Figure 11.7b shows the price curve with a broken carrying cost line intersecting it at weeks 14, 24 and 40. The purchasing policy in this restricted market supply case is,

(a) Buy each week's requirements in the market for weeks 1–13 inclusive.

(b) From week 14 to week 24 inclusive, buy two and a half weeks' supply each week. There is now sufficient inventory to last until week 40.

(c) From week 41 to the end of the year buy weekly in the market.

The total annual cost, assuming one unit consumed per week, of this policy is $2,532.68 ($2,394.44 in purchasing costs and $138.24 in inventory charges). This restriction on the amount that can be bought each week has increased costs by $36.18. It still gives a saving of $67.32 per year, 2.6%, compared to a back-to-back policy. It is seen that with a well-defined seasonal or periodic price pattern, even with a relatively high annual carrying charge, an efficient purchasing policy can be devised.

Obviously in practice the short-term price fluctuation about the general trend of price movements will rather obscure sharp seasonal changes as postulated in Fig. 11.7. A policy specified in such precise detail as given above obviously cannot be determined. However, provided it is interpreted as giving the levels that the stock cover should aim to achieve at varying monthly or quarterly intervals of the year, it becomes a feasible possibility. For Fig. 11.7a for example the strategic policy could take the form of maintaining the stock cover near its minimum level, with several small purchases, over weeks 1 to 15 and 44 to 52. During weeks 15–20 purchases should be made at as low a price as possible, of a sufficient quantity to raise the stock cover to 24 weeks consumption by the end of week 20. This will cover consumption needs from weeks 20 to 43.

The use of the purchasing policy derived by Carlson and Corrigan depends entirely on the existence of such strong seasonal or periodic patterns and their continual occurrence year after year. If the yearly low occurs 4 weeks earlier or 4 weeks later than given in Fig. 11.7, most of the advantages of the policy will be lost. It is very doubtful that a persistent single seasonal pattern of price movements exists for any commodity in a strong enough form for such a simple policy to be useful. The lack of consistent seasonality was illustrated for cocoa in Chapter 4 in discussing the concept and estimation of the length of the active buying period, see Figs. 4.4 and 4.5. What was even worse was that in two successive years the minimum price of one year occurred at the same time of the year as the maximum price in the other year. Other examples of different seasonal price patterns from year to year were given for maize in Chapter 10. It seems more likely that there are a small number of alternative seasonal price patterns that can occur depending on the relative state of supply and demand factors. Furthermore even for the same seasonal pattern the peaks or troughs of the annual prices can differ by 1 or 2 months from one year to the next. Thus unfortunately it must be concluded that this Carlson and Corrigan seasonal model is most unlikely to be of any assistance in improving purchasing performance in practice.

11.4 'DOLLAR AVERAGING' AS A PURCHASING POLICY

William T. Morris discusses some simple models to derive efficient policies for purchasing a given requirement before a known deadline. He is basically considering the problems of a person who speculates in shares and securities. However, he notes in passing that the sample problem also faces an industrial purchasing officer buying leather, scrap iron or grain.

The main emphasis of his paper is on the policy of dollar averaging but he first of all discusses a single procurement strategy and finds the optimal form this should take by a simple reinterpretation of a model given by Simon. This policy states that all requirements are bought on the same day at the same price offer. Having obtained the form of the policy he then goes on to show that the single procurement strategy is the optimal policy. The policy says that if on day k (i.e. k days to go before the deadline) the price quoted, P, is below or equal to some price P_k, set by the buyer, then buy, otherwise wait until the next day and determine a new policy price P_{k-1}. If the price quoted on day $k-1$ is less than P_{k-1} then buy; if not wait until day $k-2$ etc. The policy price P_k should be set equal to the minimum expected price (following an optimal policy) over the next $k-1$ days to the deadline. This is essentially a problem in dynamic programming and is a special case of the general purchasing model discussed later in this chapter.

Morris then goes on to analyse the policy of 'dollar averaging' which is in common use with investors in the stock market, particularly with investors in unit trusts. The policy of 'dollar averaging' involves an equal money expenditure per time period regardless of price. The purpose of this policy is to free the investor from having to forecast future price movements and to average down the mean price of the stock held, as more and more purchases are made. If an equal amount of money is spent in each time period to buy as much as possible then more units of material are purchased in those time periods when the price is low than in those time periods when the price is high. The lower the price the more units that are purchased. Obviously the average price paid per unit of material purchased over a period will be lower than the average market price over the period.

An expression for the average price paid per unit can be derived in the following way. Consider a series of purchasing opportunities where there is one price offer and one purchasing opportunity in each time period. The 'averaging policy' is to spend £d in each time period. Let $\phi_j(p)$ be the probability distribution giving the relative frequencies of occurrence of particular price offers, p, in the jth time period. Then the expected number of units that will be purchased in the jth time period will be

$$s_j = \int_0^\infty \frac{d}{p}\,\phi_j(p)dp$$

Over the whole N time periods available the expected total number of units purchased will be

$$S = \sum_{j=1}^{N} s_j = \sum_{j=1}^{N} \int_0^\infty \frac{d}{p}\,\phi_j(p)dp$$

The amount of money spent in total is Nd. Thus the expected unit purchase cost over the N time periods will be

$$C = \frac{Nd}{S} = \frac{Nd}{\sum_{j=1}^{N} \int_0^\infty \frac{d}{p}\,\phi_j(p)dp}$$

$$= \frac{N}{\sum_{j=1}^{N} \int_0^\infty \frac{1}{P}\,\phi_j(p)dp}$$

i.e. independent of the fixed expenditure per time period of £d.

Under the simple assumption that a single constant probability distribution, $\phi(p)$, describes the relative frequency of occurrences of all future prices, the above expression simplifies to

$$C = \frac{1}{E\left(\frac{1}{p}\right)}$$

The expression $E\left(\frac{1}{p}\right)$ denotes the average value that will occur for $\frac{1}{p}$ over the long run. It is a well-known statistical result that if $E(p)$ is the average long run price offer then

$$C = \frac{1}{E\left(\frac{1}{p}\right)} < E(p)$$

If for example future prices randomly fluctuate between constant upper and lower price limits a and b, i.e. prices follow a rectangular distribution, then

$$\text{Average price offer} \quad = \quad E(p) = \frac{b+a}{2}$$

Average unit purchase cost

$$= \frac{1}{E\left(\frac{1}{p}\right)} = \frac{b-a}{\int_a^b \frac{1}{p}\, dp} = \frac{b-a}{\log_e (b/a)}$$

$$\text{Expected number of units purchased} = \frac{Nd \log_e (b/a)}{(b-a)}$$

Thus we can now calculate the relative savings of a dollar averaging policy compared to the average price offer for differing relative values for b to a. These are given in Table 11.3 for b/a up to a value of 4. It can be seen that the higher the range of variation for prices, given by the higher the ratio of b to a, the greater are the potential savings in purchasing cost for a 'dollar averaging' policy.

The final column of Table 11.3 shows the average number of units in the

TABLE 11.3 A comparison of dollar averaging with hand-to-mouth buying.

b/a	$E(p)$	C	Expected number of units purchased
2	$1.5a$	$1.44a$	$0.69Nd/a$
3	$2.0a$	$1.82a$	$0.55Nd/a$
4	$2.5a$	$2.16a$	$0.46Nd/a$

long run that will be bought over N time periods or buying opportunities. The higher the possible range of price variation, i.e. the relative values of b to a, the less the number of units purchased in total. This follows because the higher the ratio of b to a, the higher the general price level and hence the fewer units purchased.

All the values given in the table are the averages that will occur over the long run. For a small number of time periods the calculated averages will vary above and below these values. The average unit purchase cost may be higher or lower than the value given in the table, but it will always be lower than the average price offer over the time periods considered. More importantly, the number of units purchased over the time periods will vary above and below the long run expected number given in the final column of the table.

The investor in the stock market is not greatly concerned if he buys rather fewer or rather more units than he expected. For him dollar averaging has the attraction of not requiring price forecasts and is very useful if his money for investment is available in regular amounts, say regular savings from his salary. This specifies the value d of his expenditure in each time period. These factors explain its attractiveness and popularity with investors in the stock market.

However purchasing managers are concerned with a somewhat different problem. They have a known requirement, Q, which they wish to buy at the lowest possible average unit cost over N time periods. To apply dollar averaging they must first estimate d, the amount of money to spend at each buying opportunity via the formula

$$Q = \frac{Nd \log_e b/a}{(b - a)}$$

The variations in the number of units purchased over a relatively small number of time periods above and below their expected average value, means that in practice they will find they have bought rather more or rather less than their requirement. They thus will either have surplus stocks which may have to be sold off at a loss or be forced to make extra purchases over the last two or three time periods when prices by chance may be high. Thus even in the case of a stationary probability distribution for future prices, dollar averaging does not guarantee that the objectives of the purchasing manager are met.

In the more general case where short- and medium-term trends in price exist and where different probability distributions should be forecast for prices in different time periods, the amount of money, d, to spend should be calculated by the following formula

$$Q = \sum_{j=1}^{N} \int_{0}^{\infty} \frac{d}{p} \phi_j(p)\, dp$$

The problem just mentioned of variations in the actual amounts purchased above and below the expected average level become even more intense. The overwhelming constraint is that over the N time periods an amount Q in total must eventually be purchased whatever the price. This is the critical difference between a company buyer and a stock market investor. The dangers of this approach are obvious. The formula shows that the calculation of the value for d depends on good forecasts of the price movements, represented by $\phi_j(p)$. If d is too high because of bad forecasts, then they will buy all the requirements too early, when prices may fall to lower levels later; whilst if d is too low they will not have bought enough and have to buy the remainder on the last day perhaps at a premium price. This policy complicates the purchasing manager's first objective, that of always having sufficient in stock to cover the needs of his manufacturing process. The inevitable errors in the forecasts of $\phi_j(p)$ complicate the problem and increase the uncertainty rather than simplify the situation for the buyer. Thus dollar averaging is a policy that is not relevant to the problems of industrial purchasing managers.

Of interest is that dollar averaging could be applied with the amount purchased at a price proportional to $1/p^2$ rather than $1/p$, or $1/p^3$, $1/p^4$ etc. Extension of this idea of buying d/p^n units at the price p leads directly to the single purchase strategy of making only one purchase at the forecast lowest price over the whole N time periods (buying opportunities). This is the same strategy described at the start of this section and is proved later via dynamic programming models to be the best strategy to follow.

11.5 EARLY DYNAMIC PROGRAMMING MODELS

As we have seen the commodity purchasing problem is a continuous sequential decision-making situation. The form of the optimal policy will depend on the amount of information about the process which the decision maker is willing to consider. In general the process can be examined from two different points of view. Firstly only the average (or expected) development of the process need be considered and the model formulated in terms of the average state of the system. On the other hand the system can be observed and the information which becomes available with the passage of time regarding the particular states which it occupies can be used. By doing so a policy of a sequential nature can be formulated. It should be noted that whilst the policy is to be determined in advance, the future actions are not determined in advance but depend entirely on the development of the process. The times at which actions are taken to modify the system (e.g. make a purchase of a certain amount) are determined by what actually occurs in the system and not fixed initially on the basis of expected performance. This second approach naturally leads to the formulation of a dynamic programming model of the purchasing problem.

Fabian *et al.* is the first published work on formulating the general purchasing problem in dynamic programming form. The price and raw material demand or usage are initially considered as stochastic variables, both with stationary probability distributions. The objective was to minimise the average cost per period. It was shown analytically that the policy function relating price and stock level tended to a limiting form as the number of future buying opportunities increased. However it was found that the cost and magnitude of the computations required to solve the analytical model were too high for use in the practical problem.

The authors then adopted the first approach mentioned, of removing the stochastic nature of the problem to formulate policies in terms of the expected state of the system. The price forecasts were interpreted as prices that will occur with certainty and the mean usage rate was substituted for the usage rate distribution. The determination of the optimal purchase amount in any situation then becomes trivial. The most important parts of the practical system are the forecasting procedures, particularly for price. In their case only one price offer occurred for each month. Forecasting had only to be made for three months ahead, i.e. for three prices and three usage rates. In those circumstances the authors were able to develop an adequate method for price forecasting, which enabled a considerable saving over previous practice to be achieved.

Treating demand for raw material as deterministic is often an adequate approximation, since raw material demand is a controllable variable often manipulated in the production plan. However in most practical commodity purchasing situations it is completely wrong to treat price as a deterministic variable. As we have seen in the case illustrations of Chapter 3 there are always far more price offers than three at which to consider actively making purchases, typically between 20 and 100 price offers. With such large numbers of alternative price offers at which to make purchases, it is important to take account of the price fluctuations about the mean forecasts. Since most commodities are quoted daily, with large daily fluctuations and certainly large fluctuations about the monthly average price, a single monthly forecast price, as in Fabian *et al*'s case, does not provide the buyer with much assistance in choosing which are the low prices at which to make purchases. Furthermore short- and medium-term trends and seasonal price movements appear to exist in commodity prices which nullify the stationarity assumption for the probability distributions of their model. In most cases, it is the price variance about the trend as well as the trend itself that will determine the purchasing policy not the trend alone. Thus the approach advocated by Fabian *et al.* for their specific problem is not appropriate to the general commodity purchasing problem.

11.6 THE DYNAMIC PROGRAMMING MODEL FOR STOCHASTIC PRICES AND KNOWN DEMAND

This author has extensively analysed the dynamic programming model for the case of stochastically varying prices and many alternative opportunities to buy the materials required. A large range of different purchasing situations have been considered, some of which are discussed in this chapter. One overall result is, that for the case of known deterministic demand, the optimal purchasing policy, the solution to the dynamic programming model, always takes a basic general form. The objective of the policy is always taken as that of minimising the expected costs of buying the total usage requirements over a specified number of periods. Since demand is assumed to be known exactly, this is the same as minimising the expected price paid for each unit of material required.

The general form of the optimal purchasing policy is illustrated initially for a relatively simple Type 1 purchasing situation, see Chapter 4, section 4.6. This is the continuous consumption and supply problem where the future prices are assumed to be randomly generated from a known stationary price distribution. Various further assumptions have to be made to derive the dynamic programming model for this purchasing situation. These assumptions together with the model and its solution are given in an appendix to this chapter. Amongst these assumptions are that consumption occurs in each time period, new supply can be delivered in each period with no limitation on the purchase and supply quantities and that there is only one price offer in each period at which to make a purchase. For simplicity it is assumed that a purchase at a price offer in a period is delivered instantaneously so that it can be used to meet consumption needs in that period. If this assumption is invalid then some minimum safety stock level must be maintained at all times to cover consumption needs over the delivery lead time. Many of the assumptions made in formulating this initial model can be relaxed to extend it to cover the various practical purchasing problems that occur, as shown in later parts of this chapter.

A constraint that the maximum purchase and stock cover at any time should not exceed the consumption needs over M cumulative periods ahead is assumed to have been derived from the considerations of the strategic buying period as discussed in Chapter 4. As mentioned earlier the objective is to purchase the total consumption requirements over these M periods at the lowest expected total cost. The M period limit on future purchases applies to every period of time. Thus the purchasing policy applies to each successive price offer over time. The decision of interest is to decide what to buy now at a price offer to minimise the expected costs over M cumulative periods including the present one. At the next price offer the policy is applied in the same way based on the stocks and undelivered purchases at the start of that period to decide at that time what purchases should be made, etc.

The 'optimal' purchasing policy derived from the dynamic programming model takes the following form:

(i) The policy is to purchase at any price offer in a period a quantity of material which will raise the stock level to cover the consumption needs of a number of successive complete periods ahead. This stock level is called the 'target' stock level. Since it is defined in terms of cumulative time periods forward, including the present one, consumption in each of the periods can take any value. It is unnecessary to assume that consumption is the same in each period.

(ii) The optimal purchasing policy is defined in terms of a set of decision price breaks P_2, P_3, ... up to P_M and associated target stock levels that should be held at particular prices. These associated stock levels are those to achieve after a purchase at the current price offer. The purchase quantity is then the difference between the target stock level indicated for that price offer and the current stock in hand. If the current stock level exceeds the target stock level indicated then no purchase should be made. Note that in this context stock includes the physical stock in hand plus any purchases made earlier in time which have not yet been delivered.

Let p be the current price offer and S the current stock level then the policy is

Price	Target stock (after purchase)	Purchase (quantity)
$p \leqslant P_M$	Total demand over next M periods including present period $- D_M$	Max $\{0, (D_m - S)\}$
$P_{j+1} < p \leqslant P_j$	Total demand over next j periods including present period $- D_j$	Max $\{0, (D_j - S)\}$
$j = M-1, M-2, \ldots, 3, 2$		
$P_2 < p$	Demand for the present period only $- D_1$.	Mas $\{0, (D_1 - S)\}$

(iii) The decision price breaks are derived from the following recurrence relations.

$$P_2 = \int_0^{\infty} q\phi(q)dq - c$$

$$P_k = \int_0^{P_{k-1}} q\phi(q)dq + \int_{P_{k-1}}^{\infty} P_{k-1}\,\phi(q)dq - c \text{ for } k = 3, 4, \ldots M \tag{11.1}$$

Note that c is the cost incurred in holding one unit of material in stock for one complete time period. The major change to extend the policy to cover other purchasing situations occurs to the form of the recurrence relations for the decision price breaks, not to parts (i) and (ii) of the policy.

The general decision price break P_k is the price at which the purchaser is indifferent, with regard to his expected costs, as to whether he should cover the cumulative consumption demands for the next $k-1$ or k periods ahead, including the present period. As is shown in the appendix and confirmed more simply later in the chapter, the decision price break P_k is the lowest price that the purchaser expects to pay over the intervening $(k-1)$ periods for each unit of consumption in the kth period ahead in time, if it is not bought now at the price offer p.

If there is a forward or futures market for the raw material so that delivery can be made just prior to its time of consumption rather than at the time of purchase and if payment for the material is made after delivery, then the value of c (the storage unit incurred in his own premises by the purchaser) in the recurrence relations for the decision price breaks will be zero. If the delivery time is dependent on the time of purchase rather than usage then the purchaser will incur a cost for storage on his own premises until he uses the material. In such cases it follows that the decision price breaks will fall to zero as k is increased and become negative for larger values of k, since P_k is a monotonically decreasing function of k. For

$$P_k - P_{k-1} = \int_0^{P_{k-1}} q\,\phi(q)dq + P_{k-1}\int_{P_{k-1}}^{\infty} \phi(q)dq - c - P_{k-1}$$

$$= -\int_0^{P_{k-1}} (P_{k-1})\,\phi(q)dq - c \leqslant 0$$

i.e. $P_k \leqslant P_{k-1}$

This result is true whatever the value for c. However, if c is greater than zero then there will be a value of k, say $\hat{N}$, such that

$$P_{\hat{N}} \geqslant 0 \geqslant P_{\hat{N}+1}$$

The price offer made in any period is always positive, so that $P_{\hat{N}+1}$ (which is negative) will always be less than any price offer that can be made. Hence it will never be right to buy any more material than will cover $\hat{N}$ cumulative periods demands. If M, the number of periods or buying opportunities from the strategic buying period, is larger than $\hat{N}$ then the active buying period should be $\hat{N}$ periods long. Decision price breaks for the optimal purchasing policy only need to be calculated for $k = 2, 3, \ldots$ up to $\hat{N}$. This situation applies whether c includes all the costs associated with carrying stocks of materials or is based on merely the interest cost on the capital investment in raw material stocks.

The effect of different values for c on the decision price breaks for the simple case when $\phi(q)$ is a rectangular distribution over the range £0 – 200

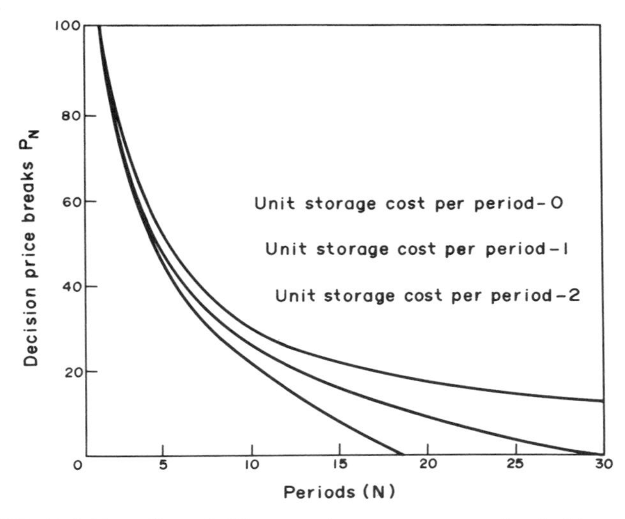

FIG. 11.8 The effect of different unit storage costs on the decision price breaks for a rectangular distribution

is shown in Fig. 11.8. It is seen from the figure that if c = £1 (1% of the mean price level) then $\hat{N}$ is 27, whilst if c = £2 (2% of the mean price level) then the value of $\hat{N}$ is 17.

In many situations there is effectively an absolute lower limit on the prices that will be accepted in the market for a raw material, say $P_{\min}$. This can arise because Governments commonly subsidise agriculture by guaranteeing to buy up the surplus production at some specified minimum price, which can be quite high. The Common Agricultural Policy operated by the countries of the EEC is an example of this situation as is the loan support system for agriculture of the U.S. government, (already mentioned for maize in Chapter 10). There will then be a value N^* such that

$$P_{N^*} \geqslant P_{\min} \geqslant P_{N^*+1}.$$

This value of N^* will be the length of the active buying period (purchase and stock cover) since the purchaser cannot buy at a price less than $P_{\min}$.

If the raw material is perishable with a limited storage life then this will determine the maximum period of time over which the material can be stored. Provided a 'first in—first out' system is used to transfer raw material stocks into the production process, this shelf life sets the length of the active buying period and the maximum number of decision price breaks that need to be calculated.

The decision price break policy for simplified purchasing situations can be derived more directly. For simplicity it is assumed that the storage cost is zero. If there are only two periods in which he can make a purchase, the buyer has to consider whether to buy his requirements today or tomorrow. If he does not buy today and waits until tomorrow then he would expect to pay tomorrow's average price offer, i.e. the average value of whatever probability distribution is used to generate prices. The price he expects to pay is the price he would pay on average if the same situation occurred many many times. This is the average, or mean value, of the price probability distribution. Let his average price be denoted by P_2. The buyer will obviously not purchase his requirements today if the price offered is higher than P_2, since he expects to make a purchase at the price P_2 tomorrow. Thus his policy will be to buy today if the price offer is less than P_2 and wait until tomorrow if it is higher than P_2. If it exactly equals P_2 it doesn't really matter what he does in terms of his expected costs, since he either pays P_2 now or expects to pay P_2 tomorrow.

Before the first day's price offer is made, the unit price he expects to pay for the requirements to be bought over either of the 2 days is the weighted average given by:

(Average price offer over all those prices less than P_2) × (Probability first day's price offer is less than P_2 so that he buys on first day) +

(Average price offer on second day which is P_2)×(Probability first day's price offer is greater than P_2 so that he waits to buy tomorrow).

Let this expression be denoted by P_3. Consider the situation when there are 3 days to make the purchase. Then if he does not make a purchase on the first day he will make the purchase over the next 2 days and as just shown he would expect to pay a price P_3. Hence the best policy to follow to give the lowest expected price over the 3 days is to buy on the first day only if the price offer is less than P_3. If it is higher he waits since over the following two days he expects to pay a lower price, i.e. P_3. Over 3 days or buying opportunities, prior to the first price offer, the price he expects to pay is given by the same formula as for buying over 2 days but with P_3 replacing P_2. Let the expected price paid over 3 days be denoted by P_4. This procedure can be repeated continuously until it generates the lowest price that he would expect to pay over the longest period of time allowed by the strategic or active purchase and stock cover policy. The weighted average formula given above in words is exactly the same as the recurrence relations (11.1) for calculating the decision price breaks in mathematical terms. So the P_2, P_3, P_4,..., etc., are the decision price breaks given earlier.

An illustrative example

Consider a purchasing manager buying the commodity whose price history is illustrated in Fig. 11.9 with a continuous requirement of 10 tons per day to fulfill the production needs. Let the maximum purchase and stock cover, because of limited storage, be 100 tons covering 2 weeks

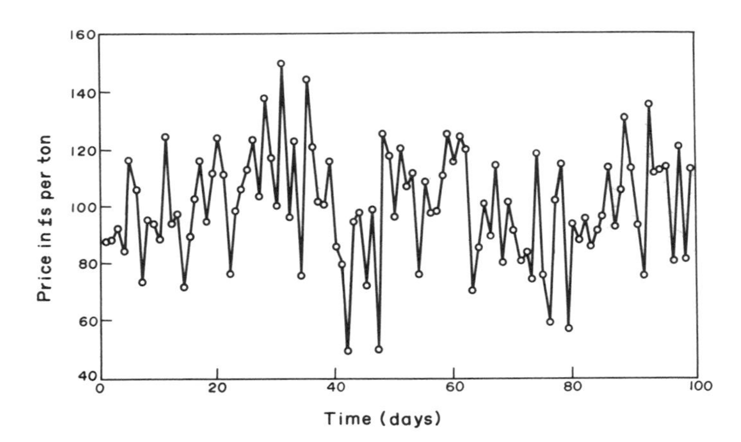

FIG. 11.9 Commodity price history

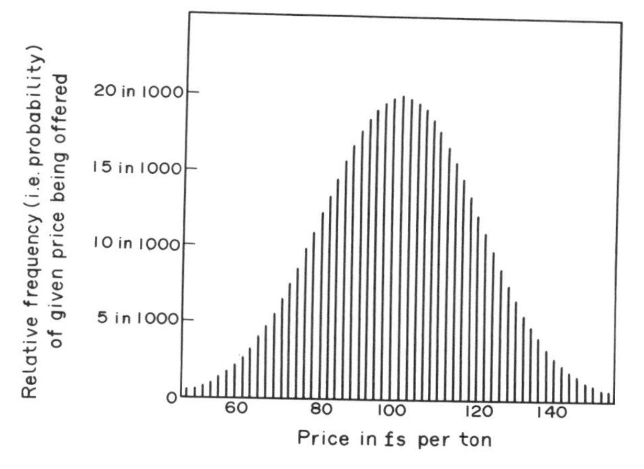

FIG. 11.10 Histogram of the relative frequency of occurrence of the different prices for the illustrative problem

consumption. The prices shown in Fig. 11.9 fluctuate above and below the calculated average of £100 per ton, with no prices above £155 per ton or below £45 per ton. These prices are in fact randomly generated from the normal distribution shown in Fig. 11.10, which has a mean of £100 and a standard deviation of £20. This distribution then gives the forecasts of the future price offers to be expected. This distribution should then be substituted for $\phi(q)$ in the recurrence relations (11.1) for the calculation of

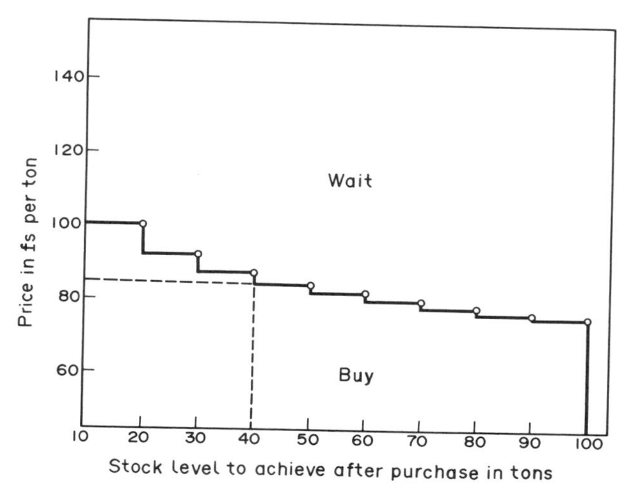

FIG. 11.11 The purchasing policy for the illustrative problem. (The 'o's indicate the decision price breaks)

decision price breaks. The DP purchasing policy based on these decision price breaks will take the form given in Fig. 11.11, where the 'o's show the successive decision price breaks. It should be noted that since the maximum stock that can be carried is 100 tons, 10 days consumption, there is no need to calculate more than ten decision price breaks. The buyer will not be able to cover consumption more than ten days into the future.

On any particular day, he should take the price offer made to him and draw a horizontal line on Fig. 11.11 through that price offer to see where it meets the purchasing policy, which is the stepped line drawn through the 'o's. He should then read off the corresponding point on the horizontal scale which indicates the stock level which he should have after his purchase at that price offer. For example, let the price offer be £85. A horizontal line drawn through the £85 point meets the purchasing policy line at a stock level of 40 tons. If the buyer already has 20 tons in stock or on order, then he should make a purchase of 20 tons so as to raise the stock level after purchase to 40 tons. If he had 40 tons or more already in stock, then he should not make a purchase.

The purchasing policy illustrated in Fig. 11.11 can now be applied to purchasing this commodity over the next 2 months or so, 50 days. For simplicity it is assumed that delivery of new material is instantaneous, i.e. an order placed at today's price offer is delivered immediately ready for use today. If this were not so the actual stocks at all times must be maintained above some minimum safety stock level to cover consumption over the delivery lead time. To avoid biasing the results because of good or bad initial prices it is assumed that there are 50 tons in stock at the start of the next 50 days and that the purchasing manager must end the 50 days with 50 tons in stock, if necessary making an unwanted purchase at the final price offer. To achieve this the effective consumption on the final day must be taken as 60 tons, the normal 10 tons for usage plus the 50 tons of required closing stock. The purchasing policy is defined in terms of the number of days future consumption to cover at a price offer. Since 10 tons are required each day the number of days to cover at a price offer are multiplied by 10, to give the target stock level in tons. However, towards the end of the period, if the price offer implies that number of days to be covered includes the final day of the 50 day period, then the target stock in tons to have at that price will be the number of days to cover times 10 plus 50 tons extra on the final day to give the desired closing stock, or 100 tons if this is lower because of the maximum stock that can be carried at any time.

The performance of the purchasing policy can be evaluated by working successively through the prices over the 50 day period to determine what purchase decision is recommended for each day. The result of this is shown in Table 11.4. The price on the first day was £96. Figure 11.11 shows that at this price there should be 20 tons in stock. Since there is already more stock

TABLE 11.4 The performance of the decision price break policy on the simulated illustrative problem

Weeks	Prices	Target stock	Opening stock	Purchase	Closing stock	Weeks	Prices	Target stock	Opening stock	Purchase	Closing stock
1	96.0	20	50	0	40	26	71.1	100	80	20	90
2	108.4	10	40	0	30	27	100.0	10	90	0	80
3	148.3	10	30	0	20	28	111.5	10	80	0	70
4	105.2	10	20	0	10	29	102.2	10	70	0	60
5	92.9	20	10	10	10	30	104.7	10	60	0	50
6	48.9	100	10	90	90	31	75.2	100	50	50	90
7	113.3	10	90	0	80	32	81.4	60	90	0	80
8	101.5	10	80	0	70	33	86.6	40	80	0	70
9	72.7	100	70	30	90	34	112.8	10	70	0	60
10	136.7	10	90	0	80	35	150.0	10	60	0	50
11	106.2	10	80	0	70	36	117.9	10	50	0	40
12	115.4	10	70	0	60	37	98.6	20	40	0	30
13	80.9	60	60	0	50	38	117.8	10	30	0	20
14	98.2	20	50	0	40	39	123.4	10	20	0	10
15	103.0	10	40	0	30	40	102.6	10	10	0	0
16	86.8	40	30	10	30	41	74.6	100	0	100	90
17	124.6	10	30	0	20	42	85.5	40	90	0	80
18	77.7	80	20	60	70	43	135.8	10	80	0	70
19	111.0	10	70	0	60	44	96.2	20	70	0	60
20	114.9	10	60	0	50	45	112.0	10	60	0	50
21	93.4	20	50	0	40	46	103.1	10	50	0	40
22	74.7	100	40	60	90	47	116.0	10	40	0	30
23	58.2	100	90	10	90	48	83.8*	80*	30	50	70
24	71.1	100	90	10	90	49	126.8	10	70	0	60
25	100.4	10	90	0	80	50	96.8	20	60	0	50

*At this price the stock cover should be 5 days consumption. Since the demand on the last day of the period is 60 tons in order to give a closing stock of 50 tons, the target stock will be 80 tons.

than this, the opening stock of 50 tons, no purchase should be made on day 1. On day 1, 10 tons are used in the production process so that the opening stocks on day 2 are 40 tons. The price of £108.4 on day 2 from Fig. 11.11 shows that the desired stock after purchase should only be 10 tons. Since there is already more in stock than this, no purchase should be made. Day 3 and day 4 will be similar. On day 5 the price offer is £92.9. Figure 11.11 shows that 20 tons should be in stock at this price. Since the opening stocks are only 10 tons, an amount of 10 tons should be purchased to bring the stocks up to the desired level. Since 10 tons is used by production on day 5, the opening stock for day 6 will be 10 tons. The price offer on day 6, £48.9, is very low indeed so that Fig. 11.11 shows that the desired stock after a purchase should be the maximum possible value of 100 tons. Since the opening stock is only 10 tons, a purchase of 90 tons should be made to bring the stocks up to the desired level. The evaluation then continues as shown in Table 11.4. On day 48 the price offer is £83.8 which via Fig. 11.11 gives a stock cover level of 5 day's consumption. The total desired stock for day 48 will be the sum of the consumption for day 48, day 49, day 50, day 51 and day 52. The consumption on day 48 is 10 tons, on day 49 is 10 tons, but on day 50 is 60 tons, in order to give a closing stock at the end of 50 tons. The consumptions for days 51 and 52 are zero. Hence the desired stock level after purchase will be 80 tons. Since there is an opening stock of 30 tons a purchase of 50 tons should be made on day 48 at a unit price of £83.8.

With this policy the average price per ton paid for the consumption requirements over the 50 days is £71.3 per ton. This performance is achieved with no knowledge of the prices that will be offered each day, other than that they will be generated randomly from the frequency histogram of Fig. 11.10. If the buyer had followed a back-to-back policy of merely buying each day's requirement of 10 tons on that day's price offer, the average cost per ton would have been the average price offer over the period of £99.8. Thus the decision price break purchasing policy would have saved £28.5 per ton over the back-to-back policy. With hindsight knowledge of all the actual prices on each of the 50 days, the very best the buyer could have done would have been to have bought his requirements at an average cost of £66.5 per ton. The purchasing efficiency, as defined in Chapter 5 to measure purchasing performance is

$$\left(\frac{\text{Average price} - \text{Average unit cost paid}}{\text{Average price} - \text{Hindsight minimum unit cost}}\right)$$

$$\text{i.e.} \quad \frac{99.8 - 71.3}{99.8 - 66.5} = \frac{28.5}{33.3} = 85.6\%$$

Thus even with a very large uncertainty in the future prices that are offered, but with knowledge of the upper and lower limits containing future price movement and some measure of the relative frequency with which particular prices will be offered, it is possible to devise a purchasing policy which will give a purchase cost close to the absolute minimum and with a large saving over a back-to-back policy.

The Carlson and Corrigan policy was also evaluated over this same set of 50 prices under the same conditions. The average cost per unit purchased was £73.5 per ton. This is £2.2 per ton more than the decision price break policy of the dynamic programming model. The savings on the back-to-back policy for the Carlson and Corrigan policy are 7.5% less than for the decision price break policy. So for the situation of a stationary known price distribution the Carlson and Corrigan policy performs very well. Because of its easier calculation process it might be preferred to the decision price breaks approach. The acid test of the best choice will be how the decision price break policy performs on nonstationary prices where the Carlson and Corrigan policy failed utterly. As we shall see later in this chapter and in the two case studies in succeeding chapters the decision price break policy is easily modified to cover nonstationary prices. The case studies show that it gives very good purchasing performances indeed. Hence it is a better and more practical policy to use than that put forward by Carlson and Corrigan.

One other point is that purely by chance there are slightly more lower prices in the 50 days of the evaluation of Table 11.4 than would be expected over any 50 day period in the long run. The results quoted are thus slightly better than would be obtained in the long run. It can be shown theoretically, by treating the changing stock situation from day-to-day as a Markov process, that over the very long run the average cost per unit paid with the decision price break policy would be £73.3 for prices randomly generated from the frequency histograms of Fig. 11.10. The hindsight unit cost would be slightly higher so that the long run purchasing efficiency would be just over 80%. In this illustrative example the distribution generating prices was known exactly. In practice we can only estimate the distribution. A purchasing efficiency of 80% for a 10 day planning horizon is thus likely to be the absolute maximum level that can be achieved. It is impossible to do better than this. In practice the purchasing performance would be less good. What levels can be achieved in practice will have to be determined through actual case studies as described in Chapters 12 and 13. One other interesting point is that by the same theoretical process the average long run performance of the decision price break policy for a 20 day maximum stock cover would be £67.6 per ton. Hence increasing the maximum stock cover from 10 to 20 days' consumption would reduce costs by a further £5.7 per ton.

11.7 SEVERAL PRICE OFFERS DURING A PERIOD

The model so far has assumed only one opportunity to buy in each period. A period is essentially defined as the interval of time during which only one delivery of material is made from the supplier. In many practical situations several different price offers may be made during the 'consumption or delivery' period. Many commodity markets have dealing periods several times during a day. Thus a scrap metal dealer in nonferrous materials may offer the same scrap at different prices over the day although he may only deliver physically once per day. In other cases delivery may be on a weekly or monthly basis although prices are offered daily. The extension of the previous model to cover several price offers in any period is trivial. It is assumed that the material for use in any period must be bought on or before the last buying opportunity in the previous period. Suppose that there are ten buying opportunities in each period and that as before the optimal purchasing policy will be that which minimises the costs of purchasing the requirements over N 'consumption' periods. In terms of the model formulated in the previous section, this implies a maximum purchase and stock cover of $10N$ periods with a demand characteristic that the only nonzero demands are d_1, d_{11}, d_{21}, d_{31}, etc., up to d_{10N-9} (where again the backward numbering convention is used). All other demands are zero. Substituting these values in the policy rules and recurrence equations (11.1) gives the optimal purchasing policy for this particular buying situation. The extension to any number of buying opportunities in each period immediately follows.

11.8 PURCHASING A SINGLE LOT BEFORE A GIVEN DEADLINE

A further simplification of the model covers the Type II problem of Chapter 4, of purchasing a specified amount of a commodity before a given deadline with N opportunities to make a purchase, the problem discussed by Morris.

The model for the Type I continuous consumption and supply situation gives the policy for deciding what purchase to make at a price offer to minimise the expected cost of purchasing all the requirements over M periods. This is a continuous 'rolling' limit always covering M periods forward in time at each successive price offer. In this Type II situation there is an absolute deadline by which the single lot must be purchased. At the first price offer there will be N periods over which to buy the lot, at the second offer only $N-1$ periods, at the third only $N-2$ periods etc. Assuming a stationary probability distribution for future prices then the decision price breaks P_2, P_3, ..., up to P_N will be required at the first price offer, but only P_2, P_3, ... up to P_{N-1} are required at the second price offer, only P_2, P_3, ... up to P_{N-2} at the third price offer etc.

Furthermore, since the single lot only needs to be bought by the deadline it follows that the demand in each time period is zero except at the final price offer. Referring to the policy given at the start of section 11.6 it follows that D_j, the total demand over the next j periods will be zero unless it includes the period of the final price offer. Hence at the first price offer D_j is zero for all values of j except for $j = N$. At the second price offer D_j will be zero except for $j = N - 1$. The policy of section 11.6 stated that if the price offer was between the decision price breaks P_j and P_{j+1} the purchase quantity should be such as to raise the stocks to the value D_j. Obviously at the first price offer, since D_j is zero except if $j = N$, it follows that no purchase will be made if the price offer is greater than P_N. If the price offer is less than P_N a purchase will be made to raise the stock level to cover the single lot required, say d. Hence, if S_N is the opening stock, the purchase quantity will be

$$x = \text{Max } \{0, (d - S_N)\}$$

The opening stock at the start of the second period, S_{N-1}, will be the opening stock at the start of the initial period plus the purchase made in the first period. Hence

$$S_{N-1} = S_N + \text{Max } \{0,(d - S_N)\} \qquad \text{if the first price offer} \leqslant P_N$$

$$S_{N-1} = S_N \qquad \text{if the first price offer} > P_N$$

At the second buying opportunity, the purchase quantity x is given as

$$x = \text{Max } \{0,(d - S_{N-1})\} \qquad \text{if the second price offer} \leqslant P_{N-1}$$

$$x = 0 \qquad \text{if the second price offer} > P_{N-1}$$

and similarly for all successive buying opportunities.

At the start of the N periods prior to the deadline the buyer obviously has nothing in stock so that S_N is zero. The purchase at the first price will thus be either d or nothing, depending on whether the price offer is less than or greater than P_N. If a purchase d is made at the first price offer then S_{N-1} will be d, so that all further purchases at successive offers must be zero. If no purchase is made at the first price offer then S_{N-1} is zero. Hence the purchase at the second price offer can be either d or 0. Continuation of this argument for the second and succeeding periods shows that only one single purchase of the total lot required will be made. The policy reduces to the simple form that, if the price offer when k periods remain to the deadline is less than or equal to P_k buy the total amount required, otherwise wait. The policy starts with $k=N$, $k=N-1$, $k=N-2$, ..., etc. Whenever the first price offer less than or equal to its associated decision price break occurs the purchase should be made. The decision price breaks are calculated by exactly the same recurrence relations as given in equations (11.1).

11.9 INTERPRETATION OF THE PRICE BREAKS FOR PURCHASING A SINGLE LOT

An interpretation of the price breaks given by the recurrence relation of equations (11.1) can be obtained from this simple case as follows. Since the lot is bought in one single purchase we can, without loss of generality, take d_1 as equal to one. For simplicity the storage cost is taken as zero.

Let C_k be the expected cost of purchasing this unit when there are k opportunities before the deadline. If $\phi(q)$ is the probability density function of future prices, then using the policy just defined above of a single purchase to be made if the price offer is less than P_k, it follows that

$$C_k = \int_0^{P_k} q\phi(q)dq + C_{k-1} \int_{P_k}^{\infty} \phi(q)dq$$

The minimum value of C_k with respect to P_k is derived from the equation

$$\frac{dC_k}{dP_k} = P_k\, \phi(P_k) - C_{k-1}\, \phi(P_k) = 0$$

Generally $\phi(P_k) \neq 0$; hence the minimum value of C_k w.r.t. P_k is given by

$$P_k = C_{k-1}$$

i.e. The price break P_k is the expected cost of purchasing each unit of the amount required over the next $k-1$ buying opportunities before the deadline, if we do not buy it this period. This reinforces the interpretation of the decision price breaks given in the appendix to this chapter describing the initial dynamic programming model.

11.10 PRICE DISCOUNTS

The policies can be adapted to deal with the case of price discounts for larger quantities than the buyer wishes to buy initially. The buyer compares the discount price with the decision price breaks to determine how many periods ahead he should cover with a purchase at the discount price. He then converts this into the actual tonnage he should buy at this new price. If the tonnage is sufficient to obtain the individual discount then the larger purchase at the discount price should be made. If the policy indicates that the discount offer should be refused, then the interpretation of the price breaks can be used to determine the expected cost of purchasing the extra quantity required to obtain the discount. This cost can be compared with the discount price to obtain the extra opportunity cost incurred if the discount offer is accepted. One problem that arises generally is that the probability distributions for future prices can only be estimated; they cannot be known with certainty, particularly for nonstationary distributions obtained from statistical forecasting systems. The uncertainties in the forecasts of future prices give some risk that the price the purchaser can expect to pay for future requirements may be higher than that calculated from the decision price breaks based on the current forecast of future price movements. The opportunity cost of accepting the discount offer can be weighed against the possible extra costs from the risk of errors in the forecasts of future prices. This may lead to accepting the discount offer in many cases where the original policy suggested its rejection.

11.11 EXTENSION OF THE INITIAL MODEL TO NONSTATIONARY PRICE DISTRIBUTIONS

A critical assumption of the initial dynamic programming model of section 11.6 is that the future prices that will be offered to the buyer are all assumed to be randomly generated from the same fixed probability distribution. This means that the distribution remains unchanged, whatever the actual values of the successive prices offered. In general, as the various illustrations of Chapter 2 have shown, such a stable probability distribution is likely to exist only rarely and for limited periods of time. Raw materials prices are highly unstable with regard to level, trend and variance.

The dynamic programming model given in the appendix can easily be extended to the situation of nonstationary prices, irrespective of whether the successive prices are correlated or not. In the nonstationary price situation different price distributions will be required to generate the prices to be expected at different future buying opportunities. Let $\phi_k(q)$ be the probability distribution for the prices to be expected in the kth period. The only change necessary to the model in the appendix is to replace $\phi(q)$, the previously assumed constant price distribution, by $\phi_k(q)$.

The problems in the nonstationary price situation arise in deriving a solution to the model to give the optimal purchasing policy in a reasonably simple form which can be worked out in practice for particular situations. For an 'exact' solution for practical use, based on the decision price breaks, certain conditions on the nature of the prices and the probability distributions must exist. These conditions concern the presence or not of serial correlations between successive prices. More exactly, in order to derive a 'simple' purchasing policy, the distributions for the prices at each of the future buying opportunities must be assumed to remain unchanged whatever the price offers made between now and that buying opportunity. For example, for a ten period maximum cover as in the illustrative example of section 11.6, the tenth period hence is the furthest ahead for which we can consider making a purchase today. The question is whether the forecast of the probability distribution for the prices in the tenth period that we made at the beginning of today remains constant for the next 10 days, or whether it should be changed by the next ten price offers. We can of course choose only to revise our forecasts and hence our purchasing policy at infrequent intervals. The condition is then satisfied. Subjective forecasts made by buyers will usually have this property, as will forecasts generated from econometric models of market price behaviour. In both cases forecast revisions will not usually be made more frequently than once per month. If no other price forecasting system is used, then the forecasts of the daily prices over a month will remain unchanged within that month whatever happens to the daily prices.

The different distributions of the prices on different days, as estimated prior to each price offer, are the current forecasts of the underlying process generating prices. The price break solution is then 'optimal' for these forecasts. This is only an 'optimal' solution to the real buying problem if the current forecasts, without modification, are an exact description of the way prices are generated.

If the forecasts of the future distributions are derived via statistical forecasting methods, then we cannot pretend that the pricebreak model is an 'optimal' solution to the tactical buying problem. Statistical forecasts are explicitly obtained as a weighted average of the previous forecasts and the current price offer. The assumption of independence between the future probability distributions and the intervening future price offers is immediately broken, since we will update the forecasts and hence the decision price breaks at regular intervals. In the ten period maximum cover example we will update our forecasts ten times before we reach the tenth period hence and have to make a purchase, if we have not done so earlier. We could update the forecasts and hence the price breaks immediately after today's price offer is received, prior to making a purchase decision. In practice there is often insufficient time to do this. Furthermore such a procedure would only give us the correct decision regarding buying some material today or tomorrow. It does not give us the correct decision comparing today, tomorrow and the day after tomorrow etc.

The problems that such serial correlation between successive prices cause, can be seen as follows. Consider firstly a two-period problem with a known price offer today of p and a probability distribution $\phi_1(q)$ for tomorrow's price q. The decision whether to buy tomorrow's requirements today at the price offer p or wait until tomorrow to buy them at whatever price is offered depends on whether p is less than or greater than

$$\int_0^\infty q\phi_1(q)dq,$$

the expected price for tomorrow. This is the exact solution to the dynamic programming model for the two-period problem. If $\phi_1(q)$ is independent of today's price, p, then

$$\int_0^\infty q\phi_1(q)dq$$

is merely the current forecast mean price, m say, made before today's price offer is known. If tomorrow's price depends on today's price offer p, then

$$\int_0^\infty q\phi_1(q)dq$$

will be a function of p, say $\mu(p)$. Generally the current forecast mean, m, will not be a solution to the equation

$$p - \mu(p) = 0$$

Let the solution to this equation be p_2. In general p_2 will depend on the values of the forecasting parameters estimated after yesterday's price as well as today's price offer. Assuming for simplicity, as is likely to be the case for most statistical forecasting systems, that $\mu(p)$ is a monotonic function of p such that

$$p - \mu(p) \leqslant 0 \text{ if } p \leqslant p_2$$

$$p - \mu(p) > 0 \text{ if } p > p_2$$

then the solution, to the dynamic programming model, giving the lowest expected purchasing costs will be to buy tomorrow's requirements at the price offer p if it is less than the value p_2, otherwise wait until tomorrow and buy at whatever price is offered.

In the case of a three-period situation then the decision whether to buy the requirements for the third period (period 3) at today's price offer p depends on whether p is less than or greater than $\psi(p)$, where

$$\psi(p) = \left\{ \int_0^{p_2} q\phi_1(q)dq + p_2 \int_{p_2}^{\infty} \phi_2(q)dq \right\}$$

and $\phi_2(q)$ is the current forecast probability distribution for the unknown price q that will occur in the second period (period 2). This again is the exact solution to the dynamic programming model for the three-period situation. Assuming as earlier that $\psi(p)$ is a monotonic function of p, then the solution p_3 to the equation

$$p - \psi(p) = 0$$

will be the decision price break determining whether to buy the requirements for period 3 at today's price offer or to wait and buy them at the price offers in either period 2 or period 3 using the two-period policy just derived.

The decision price break p_2, used to calculate the function $\psi(p)$ in this three-period situation is the price break that will determine whether the requirements for period 3 are bought in period 2 or left until the end in period 3. If $\phi_3(t)$ is the distribution for the unknown price offer t in period 3, then the solution to the two-period problem above showed that p_2 is the solution to the equation

$$p - \int_0^{\infty} t\phi_3(t)dt = 0$$

The distribution $\phi_3(t)$ for the unknown future price offer in period 3 will depend upon the unknown price offer q in period 2 plus the forecasting parameter values at the start of period 2. Assuming the forecasting system is updated each period after each price offer, these parameter values will depend upon the forecasting parameter values at the start of the period 1, the first period, and the price offer p made in this first period. Thus the price break p_2 depends upon both today's price offer p and the unknown price offer q that will be made in period 2. Hence the calculation of the function $\psi(p)$ requires the evaluation of a complex repeated integration process. No simple closed form can be determined for $\psi(p)$.

In the theoretical sense the general model for nonindependent price distributions can be solved. A series of recurrence relations can be formally written down. However, they require the calculation of repeated integrals. This must be performed numerically for each particular situation and the specific starting values for the forecasting parameters. It is an extremely difficult and complex process. The complexity of such a procedure rules out the possibility of a practical solution for online use.

The only possible approach to derive solutions for practical operations is to ignore the dependence of successive prices. Let us assume that the forecast price distributions are independent for the three-period situation above and let m_2 and m_3 be the mean constant forecast prices for periods 2 and 3. Then the solution to the complex integral equations reduces to the simple extension of the recurrence relation equation (11.1)

$$p_2 = m_3 = \int_0^\infty q\phi_3(q)dq$$

$$p_3 = \int_0^{m_3} q\phi_2(q)dq + m_3 \int_{m_3}^\infty \phi_2(q)dq.$$

Exactly similar results will be derived for a four-period example, five-period, six-period, etc.

Thus if we either assume independence between successive future price distributions or act as if they are independent we will obtain a modified decision price break policy for our purchasing problem for the case of nonstationary future prices in the same form as derived for the simplified model in section 11.6. The only change to the policy given there is to the recurrence relations to calculate the decision price breaks, equations (11.1). Let the forecasts of the probability distributions for future prices be $\phi_M(q)$, $\phi_{M-1}(q)$, ..., $\phi_1(q)$ over the successive M periods of the maximum stock cover. The price breaks will now need a double suffix and become P_{kj}. This is the price to be used for the period k buying opportunities remaining to the end of the M period forward, from now, when we consider raising the stock to cover the cumulative demands over the next j periods. The recurrence relations for the price breaks take the form

$$P_{kj} = \int_0^{P_{k-1,j-1}} q\phi_{k-1}(q)dq + P_{k-1,j-1} \int_{P_{k-1,j-1}}^{\infty} \phi_{k-1}(q)dq - c$$

for $j = (2, \ldots, k)$ and $k = (2,3, \ldots, M)$ (11.2)

where c is the unit storage cost per period. The initial conditions are that P_{k1} = '∞' for all k as no shortages are allowed. As before we apply the recurrence relation successively to obtain the complete set of price breaks. As we move forward through time, it will be necessary to recalculate the price breaks each period since the price distributions may have changed, following a revision to the forecasts. The price break P_{Mj} is equivalent to the price break P_j in the policy for the simplified model of section 11.6.

A modified decision price break policy has been derived as a practical solution to the purchasing problem for the case of nonstationary prices.

This is an heuristic or approximate policy. We can only say that this approach offers a reasonable policy that is likely to be of the right form and give reasonable purchasing costs. We cannot assert that it will be an 'optimal' policy. How well it will perform depends on the type of forecasting system we use, how accurately forecasts can be made for a reasonable time into the future and whether generally changes in the forecasting parameters from one updating to the next are small. For agricultural materials in Type 2 purchasing situations, market behavioural price determination models (see Chapters 8 to 10) can be used to indicate short active buying periods, as discussed in the appendix to Chapter 4. Over such periods prices and their distributions are more likely to be relatively stable. This gives some confidence that the modified price break model should perform moderately well in reducing purchasing costs. However the model is unlikely to give as good a performance for the continuous Type 1 situation as the illustrative example of section 11.6 suggested.

11.12 RESTRICTED SUPPLY OR RESTRICTED PURCHASE QUANTITY SITUATIONS

In certain practical buying situations there are restrictions on the amount of raw material that can be purchased at any time or at a particular price offer. This is particularly so when the market is made up of a large number of small suppliers or producers. A buyer requiring 5000 tons of a cereal for consumption in some month may have to make ten purchases each of 500 tons, at different buying opportunities or prices from different suppliers, because no one supplier has more than 500 tons available. Examples of such restricted supply situations in the U.K. are English barley and natural rubber. Buying barley is an example of the Type II situation, purchasing a known amount over several opportunities prior to a known deadline, whilst buying natural rubber is an example of the modified Type I problem discussed in section 11.7. Other markets are specifically organised in the form of auctions to sell materials in many successive small lots, for example tea and tobacco. In such auction markets a buyer has to bid for and purchase many of these small separate lots in order to obtain his total requirement.

In many situations buyers prefer, whenever possible, to buy their requirements in several sub lots rather than one single large purchase, see Chapter 3. In such cases the buyer sets the supply restriction himself. The multilot rather than a single purchase strategy was briefly discussed in Chapter 4. Such a choice in practice will be relevant mainly to Type II purchasing problems. In the standard Type I problem of continuous consumption and supply, see section 11.6, purchases will be made at many opportunities and different prices since something must be in stock ready to meet each period's consumption. There is thus some 'price averaging' already occurring.

If in a Type II problem a buyer either has to, or decides to purchase his requirements in ten sub lots, he must try to pick out the ten lowest prices over the active buying period rather than the single lowest price. As well as anticipating future price increases or decreases, the policy for such situations must anticipate coming up against the supply restrictions on future occasions when he might prefer to buy more. Thus purchases will have to be made earlier at higher prices than would be the case with a single lot strategy. Clearly a multilot purchase strategy will theoretically lead to higher long run average purchase costs than a single purchase strategy.

The single lot strategy for the Type II problem is a case of 'putting all of one's eggs in one basket' by buying the whole requirement when one thinks the lowest price offer over the whole buying period has occurred. The motivation behind the deliberate choice of a multilot strategy is the difficulty buyers find in picking out this lowest price offer. It also recognises that some few bad purchases are almost inevitable and it is an attempt to mitigate them by averaging their price effects down with two or three other purchases.

A real justification for seriously considering a multilot strategy is the nonstationary nature and volatility of raw material prices. Abrupt changes to the price movements can occur part way through the buying period due to some unexpected event happening. A 'good purchase' made at an earlier time, when prices were predicted to rise continuously over the buying period, can then turn into a poor or bad purchase if an unexpected or unpredictable event causes prices to fall strongly. Such 'mistakes' will be rare, but may be quite costly if they do occur. It may be worth paying slightly extra on the long-term average purchasing costs to avoid the rare occasions when a bad short-term purchasing performance occurs. Since it is virtually impossible to model theoretically in a statistical manner the relative frequencies with which such rare events (abrupt reversals of the price trend) may occur and also model their consequent effects on purchasing costs for a general commodity purchasing situation, the value of the multilot rather than the single purchase strategy will have to be assessed by an evaluation of the alternative strategies for the specific commodity of interest. Such a comparison is made at the end of the cereals buying case study described in Chapter 12. It is seen that, provided it is possible, purchasing a known amount prior to a deadline in 2 or 3 sublots, rather than one single amount, does make sense sometimes in practice, because of the sharp changes in price movements that can occur, particularly over the old crop/new crop transition period for annual crops.

A policy for the multilot Type II purchasing problem can be derived relatively easily as follows. For simplicity it is assumed that each sublot consists of one ton only and that all future prices are generated from the same probability distribution $\phi(p)$. There are in total N buying opportunities prior to the deadline at which M sublots must be purchased. No more than one sublot can be bought at the price offer at each buying opportunity. The results of the earlier models show that either one whole sublot or nothing should be purchased at any time. The extension of the previous results indicates that the policy will be based on a set of decision price breaks Q_{ij}. If there are j lots still to be purchased and i buying opportunities remaining to the deadline then a purchase of a sublot should be made if the price offer is less than or equal to Q_{ij}, otherwise nothing should be purchased and the buyer should await the next price offer. Let C_{ij} denote the expected cost of buying j lots over i buying opportunities.

If one lot is purchased now at the price offer p, the expected cost of purchasing all j lots will be the current price offer p plus the expected cost of buying $(j-1)$ lots over the remaining $(i-1)$ opportunities, i.e. $(p + C_{i-1,j-1})$. If no purchase is made at this current price offer, the buyer has to purchase j lots over $(i-1)$ opportunities and his expected cost will thus be $C_{i-1,j}$. Hence the expected cost of buying j lots over i opportunities, C_{ij}, will be given as

$$C_{ij} = \int_{0}^{Q_{ij}} 1.(p+C_{i-1,j-1})\ \phi(p)dp + C_{i-1,j} \int_{Q_{ij}}^{\infty} \phi(p)dp \qquad (11.3)$$

The minimum value of C_{ij} with respect to the decision price break Q_{ij} occurs when

$$Q_{ij} = C_{i-1,j} - C_{i-1,j-1} \qquad (11.4)$$

By definition of the problem whenever $i=j$, i.e. the number of opportunities left equals the number of lots still to be purchased, a purchase of one lot must be made at each opportunity. Mathematically this gives the boundary conditions

$$i \leqslant j\ Q_{ij} = \text{‘}\infty\text{’}$$

The Q_{ij} for all i and j up to $i = N$ and $j = N$ can be obtained by iterating between the two equations (11.3) and (11.4).

If there are no restrictions on the purchase quantity then $j = 1$ and this model reduces to that already derived for the Type II problem in section 11.8.

If the future price distributions are nonstationary then the decision price break formulae can easily be modified as was done in equation (11.2) for the 'single purchase' type of problem. The dynamic programming model given in the appendix can easily be extended to cover restrictions on the purchase quantity at any time for the modified Type I problem of section 11.7. Provided that the restriction on the purchase quantity can be expressed as some complete number of future periods' demands, as is usually the case, a decision price break model with a 'simple' closed form for the recurrence relations equations can be derived, see Kingsman.

11.13 THE STOCHASTIC DEMAND SITUATION

In some commodity purchasing situations it may be inappropriate to treat demand for the commodity in future periods as known deterministic values. Future demands may fluctuate wildly from period to period around some constant level. Hence future demands must be treated as stochastic variables generated from various probability distributions. In such situations it is not possible to plan to meet all the possible demands that can occur. The consequences and penalty costs of being unable to satisfy some consumption demands at times must be explicitly taken into account in the planning of future purchases.

The basic dynamic programming model of the appendix can be extended to cover the stochastic demand situation, see Fabian *et al.* and Kingsman. The model can be solved in the theoretical mathematical sense. Unfortunately the solution does not take a simple form which can be easily worked out for particular situations as in the previous cases. No simple decision price breaks can be derived to give the best purchasing policy to minimise expected costs. The solution requires an iterative procedure between two rather complex integral equations. The equations must be solved by numerical methods for each particular situation for the specific probability distributions for prices and demand at each point in time. The computational procedure is complex and lengthy, even for short buying periods of only ten buying opportunities for example. For the practical situation with many buying opportunities, where the price forecasts, and hence the probability distributions for future prices, must be updated after each price offer, the model just cannot be solved in a sufficiently short period of time at reasonable computational cost to be used as the basis of a purchasing policy for practical use.

Kingsman has considered various approximation approaches to the problem which might give implementable solutions without incurring excessive extra costs above the theoretical optimal solution. It was found that a reasonable approach was to consider the stochastic prices and the stochastic demand separately.

Firstly the decision price break purchasing policies of sections 11.6 and 11.11 should be used considering only the future fluctuating prices to determine how many future periods' demands should be covered at the current price offer. Let the current price offer be p and the number of cumulative periods to be covered at this price be N. Also let c_1 be the storage cost per unit of material per time period. The stochastic demand aspect is then covered subsequently by a further approximation method. The N periods whose demands must be covered by a purchase at the current price offer are considered as one single period. Let the probability distribution for this total demand over the N periods be denoted by $\psi(x)$. For small values of N $\psi(x)$ can be derived by a convolution of the demands for the individual periods, alternatively appealing to the Central Limit Theorem for values of N greater than five, say, $\psi(x)$ will be a normal distribution. This normal distribution will have a mean value and variance equal to the sum of the mean values and variances of the individual periods' demands. The cost of holding one unit of stock over the whole period whose demand distribution is $\psi(x)$ will now be Nc_1. Let the penalty costs for not being able to meet demands be incurred as a cost of c_2 for each unit of demand unsatisfied. Then the amount to have in stock, say z, at the price offer p will be the value of z minimising the expected costs expression

$$pz + \int_0^z Nc_1(z-x)\psi(x)dx + \int_z^\infty \left\{ \tfrac{1}{2}\, \frac{Nc_1 z^2}{x} + c_2(x-z) \right\} \psi(x)dx$$

The right value for z is then given as the solution to the equation

$$p = c_2 \int_z^\infty \psi(x)dx + Nc_1 \int_0^z \psi(x)dx - Nc_1 \int_z^\infty \frac{z}{x}\, \psi(x)dx.$$

With the assumption of normality for the distribution $\psi(x)$, a table of values or a nomogram can be derived to give the solution to this equation for particular values for the mean and variance of $\psi(x)$.

The equation above has the property that as p decreases in value, z will increase. Thus in this stochastic demand situation if a purchase has already been made to cover N periods' demands and then subsequently a lower price is offered which implies covering the same N periods' demands, the value of z at the second price will be higher. Hence a second purchase or even a third, etc., may be made to cover the demands over the same number of periods if lower prices occur after the initial purchase. Remember in the original deterministic demand situation that only one purchase was made to cover N periods' demands. However low later prices were, no further purchases for these N periods' demands could be made.

This and other approximation methods work since, as in the classical economic order quantity models, the total expected costs in the theoretical optimal purchasing policy for a particular price offer are relatively insensitive to moderate changes in the stock levels to be held at that price above and below the optimal level. In view of this and the problem of estimating the precise values for the mean and variance of $\psi(x)$, the cost parameters c_1 and c_2 there is little practical value in calculating the exact theoretical solution to the total problem.

11.14 TAYLOR'S PRICE TREND MODEL

One final approach to the tactical buying problem should be mentioned. It is a quite recent development proposed by Taylor. It is based directly on the new model for daily commodity price movements presented initially in Taylor and Kingsman and in rather more detail in Taylor. This new model was briefly described in the appendix to Chapter 2. The proposed buying policy is an heuristic method in the same sense as the Carlson and Corrigan method, since it is not derived from some explicit model of the commodity purchasing problem. It appears to be a very promising approach. It is the only formal method based directly on the nonstationary nature of price movements, on the trends in daily prices that can be detected. Some results of applying it to copper are given in Taylor. It has not yet been applied directly to a practical purchasing situation within a company. Thus judgement on its potential usage and its practical value to some extent must be reserved as yet. One drawback in its current state is that it is only applicable to the simple straightforward Type I purchasing situation, the continuous consumption and supply case. How it should be modified to apply to other Type I situations and the Type II situation discussed for the decision price break model is obviously a matter for further research.

The new model for commodity price movements states that there is a trend present in daily prices, say μ_t at time t. There is a constant probability p that no discrete change to the trend occurs on any day. There is thus a constant probability $1-p$ that new information enters the market on any day and causes a discrete step change in the trend value. The new trend that arises after the new information is received is a random variable with a distribution $f(\mu_t)$. The model states also that the variance of the new trend distribution is time dependent and directly proportional to the variance of the residual errors.

The trend term each day can be estimated as some form of exponentially weighted past values allowing for the time dependence of the variance of the daily price changes. The calculated trend $\hat{\mu}_t$ is only an estimate of the real value of the underlying trend. How accurate this estimate is depends on the time dependent variance of the trend distribution $f(\mu_t)$. Obviously purchases should only be made now at time t to meet future consumption needs if the real daily trend is positive and likely to persist for some time. The purchasing rule is then to make the size of the purchase directly dependent on the strength of the buyer's belief that the real trend is definitely positive and will stay positive for an appreciable period of time. A measure of this strength of belief is the direct ratio of the calculated trend to the calculated standard deviation, i.e. $\hat{\mu}_t/\hat{\nu}_t$. Consider the situation when this ratio takes a high value such as 1.0, i.e. the calculated trend is equal to the calculated standard deviation. Since a major change in trend will only occur on a relatively small proportion of days, $1-p$, and since the trend is calculated as an exponentially weighted moving average of the daily price changes it will require quite a large number of days to elapse before the calculated trend falls to zero. Furthermore many succeeding days of the new negative trend will be required to overcome the total increase in prices over the earlier period of continuous positive trends, and hence bring a day with a price level below the current price offer. Thus it is worthwhile buying a large amount of material at the current price to cover consumption over many future periods.

Unfortunately, because of the time dependence in the variance of the price changes, it is not possible to estimate statistically for this new model the number of buying opportunities that can be expected to elapse before the level of prices falls below today's price offer for any particular value of $\hat{\mu}_t/\hat{\nu}_t$. Thus the purchasing policy is defined as follows:

'The stock level to achieve after a purchase at time t is $k(\hat{\mu}_t/\hat{\nu}_t)$, where k is some constant of proportionality to be determined.'

Because all of the earlier research described in this chapter has shown that the stock must be defined in terms of complete periods' consumptions the above expression should be rounded to the nearest integer. As is usual there will be maximum and minimum values for the stock level given by the strategic or active buying period considerations for each particular purchasing problem.

To avoid biasing the results the constant k should not be chosen retrospectively by evaluations over past prices, which have been used to derive the parameters of the new price forecasting model. Taylor recommends that several series of 'synthetic commodity prices' should be generated which include the time dependent properties of the prices of the commodity of interest. The best value for k is then obtained by evaluation over these synthetic series. For example, for the daily purchasing of copper over 1966–1974 Taylor found that the best value for k was 80.

One procedure for updating the parameters of the system is as follows. Let x_t be today's price and x_{t-1} be yesterday's price. Then

1. Calculate the daily price change $z_t = (x_t - x_{t-1})$

2. Calculate the mean absolute deviation of the daily price changes

$$s_t = 0.9\, s_{t-1} + 0.1\, |z_t|$$

The standard deviation of the daily price changes is 1.33 s_t since the distribution of price changes is leptokurtic rather than normal.

3. The updated trend estimate $\hat{\mu}_t$ is then

$$\hat{\mu}_t = \frac{s_t}{s_{t-1}} \left(q\hat{\mu}_{t-1} + (p-q)\, z_t\right).$$

The parameter p has already been defined as the probability no new information enters the market to change the trend on any day. The parameter q is merely a weighting factor.

4. Calculate the standard deviation of the daily trend distribution

$$\hat{\nu}_t = \gamma\, (1.33\, s_t)$$

where γ is a further constant of proportionality.

The parameters p and γ must be estimated from a long series of prices so as to minimise the sum of squared differences between the estimated and theoretical autocorrelation functions. The weighting factor q is then given as the solution in the range 0 to 1.0 to the quadratic equation

$$q^2 - q\left(\frac{(1+p^2) + \gamma^2(1-p^2)}{p}\right) + 1 = 0.$$

It is a slight drawback to the use of this model that a lengthy series of past daily prices, preferably 1500 or more, i.e. at least 6 years of data, is required to obtain good estimates for the parameter values. However, Taylor has estimated the parameters for the new model of price movements, but not the purchasing rules, for several different commodity price series so that it is possible to make a reasonably close estimate of the values for any new commodity series. It is thus only necessary to check these values over a smaller period of past prices.

An idea of the value of his proposed purchasing rule is that when it was applied to the daily purchasing of copper on the LME settlement cash prices over 1969 to 1975 it gave a reduction of 2.1% over a back-to-back policy of buying daily at the average market price.

Taylor's purchasing rule is to make the size of the purchase linearly proportional to the strength of the belief in the existence of positive daily trends, i.e. $\hat{\mu}_t/\hat{\nu}_t$. Obviously other functional forms may be appropriate and should be evaluated as part of any research study. This may be particularly relevant if and when the approach is applied to the Type II purchasing problem. However, the results quoted for copper in the previous paragraph show that it is potentially a very valuable new approach.

Appendix

Solution to the Initial Dynamic Programming Model

The basic assumptions used to derive an initial simplified dynamic programming model to give a purchasing policy are as follows:

1. Time is divided into successive periods which contain one price offer, which in practice can be months, weeks or days etc.
2. A 'planning horizon' of N periods is arbitrarily chosen. The decision criteria will be to minimise the expected cost of purchasing the total consumption requirements for these N periods.
3. Demand is known for all periods of the planning horizon. Let the demands be d_N, d_{N-1}, ..., d_2, d_1, where d_j is the demand in the period when j buying opportunities remain before the end of the planning horizon. (Note that in common with the convention often used in dynamic programming methods the periods are numbered backwards from the end of the planning horizon. Thus the rth period forward in calendar time is the $(N-r+1)$th period from the end of the planning horizon and has demand d_{N-r+1}. Conversely the jth period from the end of the planning horizon is the $(N-j+1)$th period forward in time from the beginning of the planning horizon. Consider a planning horizon where $N = 40$. The 10th price offer is thus the 31st offer from the end of the planning horizon.)
4. No shortages are allowed. Hence at the beginning of each period the buyer must already have in stock, purchased in previous periods, or must purchase at this opportunity sufficient to meet his demands for this period. For simplicity it is assumed that delivery of purchase orders is instantaneous.
5. The storage cost is c per unit per period.
6. The buyer is offered a price quotation for the commodity at the beginning of each period. If he does not buy at that price offer, then he must wait until the beginning of the next period before he can make a further purchase.
7. There are no supply restrictions on the amount that the buyer can purchase at any price offer. He can buy as little or as much as he wishes.

8. It is assumed that the prices in successive periods are independent random variables generated from some stationary probability density function $\phi(q)$.
9. Stock once bought cannot be resold at any later date.

The model is developed for a Type I purchasing situation, continuous supply and continuous consumption, since as was demonstrated in Chapter 4, the Type II situation of purchasing a known amount before a known deadline, is merely a special near degenerate case of the Type I problem. Many of the assumptions made to derive this initial model and its solution can be relaxed to extend it to more general situations.

At the beginning of each period the state of the system can be described by the stock on hand, S, the current price offer quoted to the buyer, p, and the number of periods for which a purchasing decision is to be made.

With k opportunities remaining before the end of the planning horizon, the expected cost of a k period optimal policy consists of

(a) the cost of buying an amount x in this kth period
(b) the inventory holding cost for the stock and purchase over the kth period
(c) the expected cost of a $k-1$ period policy starting with the derived opening stock at the start of the next period and the unknown price offer q that will be made in that period.

The usage of the commodity in the kth period d_k is assumed to take place uniformly over the period so that the holding cost (b) is given by

$$c(x + S - \tfrac{1}{2}d_k)$$

The cost of the optimal one period policy for the last period with an initial stock S is

$$F_1(S,p) = \min_{x+S\geq d_1} \{px + c(x + S - \tfrac{1}{2}d_1)\} \tag{11.5}$$

And the cost of the k period purchasing policy with an initial stock S and a price offer p is $F_k(S,p)$, given by Bellman's Principle of Optimality (see Chapter 7) as follows

$$F_k(S,p) = \min_{\substack{x+S\geq d_k \\ x\geq 0}} \left\{ px + c(x + S - \tfrac{1}{2}d_k) + \int_0^\infty F_{k-1}(S + x - d_k,q)\, \phi(q)dq \right\} \tag{11.6}$$

The restriction $x + S \geqslant d_k$ ensures that there is sufficient material available to meet the current period's demand, i.e. no shortages are allowed. Since demand is deterministic and known an upper limit on the purchase quantity is

$$x + S \leqslant \sum_{i=1}^{k} d_i.$$

We require the decision rules relating the amount to purchase, x, to the price p for $k = N$, since we wish to determine the optimal policy to minimise the expected cost of purchasing all the requirements over the N period planning horizon.

The solution is obtained by solving the functional equations (11.5) and (11.6) using the method of induction. The form derived for the function $F_N(S,p)$, the expected total purchasing costs over the N period planning horizon is as follows:

Let $P_2, P_3, \ldots, P_k, P_{k+1}, \ldots P_N$ be the decision price breaks as defined in equation (11.1) in section 11.6.

Let D_j be the cumulative demands over the first j periods of the planning horizon so that

$$D_j = d_N + d_{N-1} + \ldots + d_{N-j+1} = \sum_{i=N-j+1}^{N} d_i.$$

Hence the cumulative demands from the initial period of the planning horizon up to and including the jth period prior to the end of the planning horizon will be

$$D_{N-j+1} = \sum_{i=j}^{N} d_i = d_N + d_{N-1} + \ldots + d_{j+1} + d_j.$$

Let X_i be the amount of the demand in the ith period prior to the end of the planning horizon not covered by the stock in hand S at the start of the N period planning horizon. It follows that

$$X_i = d_i \qquad \text{if } S \leqslant D_{N-i}$$

$$X_i = D_{N-i+1} - S \qquad \text{if } D_{N-i} \leqslant S \leqslant D_{N-i+1}$$

$$X_i = 0 \qquad \text{if } D_{N-i+1} \leqslant S$$

Using the variables just defined

If $p \leqslant P_N$

$$F_N(S,p) = p \text{ Max } \{0, (D_n - S)\}$$

If $P_{k+1} < p \leqslant P_k$ (for k = N−1, N−2, ... 2)

$$F_N(S,p) = p \text{ Max } \{0, (D_k - S)\} + \sum_{i=1}^{N-k} P_{N-i+1} X_i \qquad (11.7)$$

If $P > P_2$

$$F_N(S,p) = p \text{ Max } \{0, (d_N - S)\} + \sum_{i=1}^{N-1} P_{N-i+1} X_i$$

The policy described in section 11.6 is obtained by interpreting this mathematical solution in terms of the original purchasing problem and expressing it in words rather than mathematical symbols.

Since X_i is the demand in the ith period prior to the end of the planning horizon not covered by the initial stock in hand S, it follows that the cumulative demand over the first k periods of the planning horizon not covered by the initial stock in hand is $(X_N + X_{N-1} + X_{N-2} + \ldots + X_{N-k+1})$. However, this must be equal to either zero or the cumulative demand over the first k periods of the planning horizon less the initial stock in hand, whichever is the larger value. It follows that

$$\text{Max } \{0, (D_k - S)\} =$$

$$X_N + X_{N-1} + X_{N-2} + \ldots + X_{N-k+1} = \sum_{i=N-k+1}^{N} X_i$$

Hence the functional expression for $F_N(S,p)$, equation (11.5), for $P_{k+1} \leqslant p \leqslant P_k$ can be rewritten as

$$F_N(S,p) = p \sum_{i=N-k+1}^{N} X_i + \sum_{i=1}^{N-k} P_{N-i+1} X_i \qquad (11.6)$$

This expression shows that if $i \leqslant k$ (i.e. $P_i \geqslant P_k \geqslant p$) the X_i occurs in the first term on the left-hand side of the equation with a coefficient of p, whilst if $i>k$ (i.e. $P_i \leqslant P_k \leqslant p$) the X_i occurs in the second term on the left-hand side with a coefficient of P_{N-i+1}. Hence the requirements for period i not covered by the initial stock in hand are either all purchased at the opening price offer p ($\leqslant P_{N-i+1}$) or at the price P_{N-i+1} over later periods. Note that $N-i+1$ is the number of buying opportunities at which to make purchases for the requirements of the ith period prior to the end of the planning horizon. Thus P_{N-i+1} can be interpreted as the expected minimum cost using the policy of buying one unit over the remaining $N-i$ buying opportunities before its usage if it is not bought at the initial price offer p. The solution to the dynamic programming model directly gives the interpretation of the decision price breaks derived only in an indirect fashion in section 11.8.

As discussed in Chapter 4, the practical purchasing problem contains constraints imposed on the freedom of action given to the buyers, either based on the company's overall policies or by an assessment of the buyers' ability to forecast prices and purchase well over differing periods of time ahead. This led to the concept of a strategic and active buying period or purchase and stock cover. This imposes the limitation that at any time the buyer cannot make purchases for future requirements more than a certain specified number of periods ahead of the current time. As discussed in Chapter 4 this constraint is almost always in terms of time, periods ahead, rather than in absolute tons. Let the maximum buying period be M periods long or alternatively the maximum purchase and stock cover be the next M periods' cumulative demand. This constraint means that in solving the functional equations for the dynamic programming model, equations (11.5) and (11.6) there is an extra constraint to be added at each stage that

$$x + S \leqslant \text{next } M \text{ periods cumulative demands.}$$

The impact of this constraint on the solution and the purchasing policy is that the decision price breaks only need to be calculated for up to P_M, since the purchase and stock cover can not exceed M periods demands. The policy will be the same for prices above P_M, but for $p \leqslant P_M$, no matter how low the price offered might be, the policy is to raise the stock level after a purchase to cover the cumulative demands over the next M periods, including the current period, only.

The solution to the model derived applies for any value for N, the number of periods in the planning horizon over which the expected purchasing costs are minimised. N can be as small or as large as we please. Indeed N can be allowed to tend to infinity. If we are only concerned with a fixed planning horizon, i.e. we are not interested in purchasing requirements for usage in any period further ahead than N periods forward, then as we move forward through time we will need the policy for $N-1$ for the next price offer, for $N-2$ at the one after that, $N-3$ at the following one etc., down to $N=2$ at the last but one buying opportunity and price offer. This fixed horizon approach means that it is assumed that the need for the material of interest will suddenly end after N periods. However, in general in the continuous consumption and supply situation the buyer has to act as if the need for the material continues for ever. Hence it implies that N must be allowed to tend to infinity.

As the length of the planning horizon N becomes extremely large relative to the length of the maximum buying period M, then we have exactly the same policy applying to every buying opportunity. We only need to calculate the decision price breaks up to P_M. The situation then becomes the same as assuming we have an M period 'rolling planning horizon'. At each price offer at successive buying opportunities over time, we use the policy as modified to give the purchasing decision at that price offer which minimises the expected costs of purchasing the requirements over the M periods from that offer into the future.

12

Practical Implementation of the Decision Price Break Policy Case Study 1—Day-to-Day Cereals Buying

12.1 THE GENERAL PROBLEM

The validation of the decision price break model on the example of the simulated prices, generated from a known normal distribution, showed its value for giving low purchasing costs. However such good results depend on knowing the underlying probability distributions that describe the price movements. Even if prices are stationary, the forecast in practice of the relative price frequencies will not be as accurate as in the simulated example, so that a lower purchasing efficiency can be expected. Although on some occasions for short periods commodity prices do appear to randomly fluctuate about some constant level, more usually, as we have seen in numerous examples, the price series is highly nonstationary with regard to level, trend and variance. In the previous chapter it was shown that the modified simple decision price break policy of equation 11.7 is only an approximation to the 'optimal' policy for nonstationary prices. The crucial question is whether forecasts of the future underlying price movements and price fluctuations can be made that are sufficiently accurate to enable the simple decision price break policy to make good purchasing decisions in practice. Included in this, is to determine for what periods of time ahead the assumption of stationary price movements about the current forecasts remain sufficiently valid before the nonstationary effects outweigh good purchasing decisions. Since some commodity prices are apparently more volatile and nonstationary than others this time period may differ from commodity to commodity.

In the derivation of the purchasing model no constraints were placed on the way the forecasts should be generated. The only requirement was that forecasts be provided as relative frequencies of particular prices being offered at any buying opportunity. The forecasts may be derived from one of the many forms of statistical forecasting models that exist, whose output is in the required format directly. Alternatively they may be provided entirely by a buyer from a subjective assessment of all the market factors based on his experience and knowledge of the market. In this case some processing of the forecasts into the appropriate format of a probability distribution will undoubtedly be required. The forecasts could also be the output of a complex econometric behavioural model of the particular market derived from relationships between prices and various production and consumption factors, such as briefly described for cocoa in Chapter 4 and discussed more extensively in Chapters 8, 9 and 10. Any of these methods can be used or indeed any appropriate combination of them. All that is required, is that the forecasts be presented in the form of probability distributions for the prices to be expected at each future buying opportunity over the buying period.

This chapter and the next describe two case studies of the derivation of appropriate price forecasting systems and their use within the simple decision price break policy for two practical buying situations, firstly for wheat and maize and secondly for copper. The cereals buying study was carried out in a major U.K. animal feeds manufacturing company and the copper buying study for a relatively small producer of specialist copper cable. The two studies are described in some detail generally in the way the studies were actually performed, although invariably hindsight smooths out some of the rough edges, blanks out unsuccessful paths of development and gives a firmer and more systematic justification for the various steps in the study than was usually formally worked out at the time. This approach is used because this author feels that newcomers to the field can often learn more from what was actually done than from a text book hindsight approach describing what should have been done. Inevitably circumstances change over the course of a study both in a planned and an unexpected manner. The problems caused by the latter can really only be properly appreciated by using a semihistorical approach to a description of the study.

As discussed in Chapter 2, classical time series analysis has been singularly unsuccessful in detecting systematic movements in commodity prices. This means that *a priori* no particular statistical forecasting system will provide good future price forecasts. What we are seeking to establish is whether a forecasting system can be determined that, combined with the decision price break model, will in practice give good purchasing decisions. The usual forecasting criterion of minimum squared errors is inappropriate. The impact of over or under forecasts on the purchasing decisions may be significantly different. Indeed it may be appropriate to deliberately introduce bias into the forecasts. Furthermore even with an unbiased forecast the normality assumption for the distribution of forecast errors may be invalid. We should not be surprised if the smoothing constants, for example, in the statistical forecasting systems take vastly different values to those usually found when they are applied to sales forecasting problems. It will also be necessary to examine in detail the idiosyncracies of each forecasting system, the way it reacts to different short-term changes such as changes in trend, ramps, plateaux, isolated high or low prices etc.

Because *a priori* the price forecasts will often be bad forecasts, we need to examine the detailed performance of the forecasting/purchasing system with more care than usual. We must check whether 'good purchasing decisions' are made deliberately rather than by accident, as for example buying always at the end or always at the start of the buying periods for each consumption requirement. It is this type of analysis and detailed discussion of the results that will lead us to the best overall system and not consideration of the classical statistical goodness of fit tests etc., for statistical forecasting. There may indeed be specific properties of the price series, insignificant on a statistical analysis, that in the combined forecasting/purchasing system can be used to advantage. To some extent informed judgement rather than objective statistical testing is needed to determine which forecasting system to try or what modifications to make to it. Hopefully the detailed discussion of the two case studies illustrates what is required.

Initially it was thought that because of all of these problems the buyer's subjective views would form the best basis for developing forecasts. Unfortunately for a variety of reasons, discussed in the first case study, this was found not to be the case. It was then found that a particular statistical forecasting system, based only on past prices, was appropriate for both the cereals and the copper study for making good decisions for buying periods of around 2 or 3 months. In both cases the adaptive exponential smoothing method of Trigg and Leach was superior to all others tried. The value of this system is that it is designed to respond quickly to significant changes in level and trend of the price movements. In view of the extreme volatility of prices, with the benefit of hindsight it could be argued that this should have been apparent. However, judging by the reviews of various forecasting systems that appear in the literature at periodic intervals there is no universal acceptance of what are the best forecasting systems for particular types of time series. Suffice it to say here, that at the start of the two case studies there was no evidence or grounds for confidence that any statistical forecasting system could provide adequate price forecasts.

The two case studies show that the decision price break policy using price forecasts from a statistical forecasting system only, updated regularly, can lead to significant reductions in purchasing costs. The results of the cereals buying study described in this chapter may overestimate possibly the potential savings, since the model parameters were 'optimised' during the course of the study. For a variety of reasons, unconnected with the technical value of the work, it was not possible to obtain the real price data for a further period to see how the policy performed over succeeding years. However, in the copper buying study, the decision price break policy was applied to a long period of prices beyond the original study. The results were up to expectations and became the major basis of the company's day-to-day copper buying policy. Since the same price forecasting method is used in the two case studies and since copper prices are more volatile and unstable than those of cereals, it is reasonable to conclude that the cereals policy would perform almost as well in later years as in the actual study period.

It should be noted that in neither of the two case studies are the price forecasts derived from a combination of the outputs of an econometric behavioural model of the market and linear statistical forecasting systems. This is because these case studies were historically the first practical problems to be studied by the author. It was only later that the author developed the approach to medium-term price forecasting described in Chapter 8. The day-to-day cereals buying study was performed in 1967 and 1968 whilst the maize price forecasting model of Chapter 10 was developed in 1975 and 1976.

12.2 THE CEREALS BUYING SITUATION

As described in Chapter 3 for the animal feeds manufacturer LFF, consumption of cereals with respect to purchase and delivery is on a monthly or half monthly basis, mainly because the vast bulk are imported into the U.K. by sea. Delivery is specified by company strategy to occur in the month or half month prior to its intended usage. Delivery can occur at any time in the month or half month specified and is entirely at the discretion of the seller. The animal feed company's strategy was that a purchase for delivery in any month must have been made by the end of the previous calendar month. As discussed in Chapter 4, each delivery month can be considered as a distinct commodity purchasing problem. Each problem is of the Type II form of Chapter 4.6, of purchasing a known amount of material over a finite buying period before some specified deadline. Any remaining amounts of material required must be purchased at this deadline, if they have not all been bought earlier in the buying period.

In the study it was assumed that a delivery of cereals would occur in every month. The amounts to be delivered were derived using the methods described in Chapter 6 and by an interchange of information between the chief buyer and the formulation manager, see Chapter 3, section 3.3. The actual amounts to be delivered each month do not affect the day-to-day purchasing problem of deciding when or at what price to make purchases.

The study therefore assumed that each month's requirements were the same value. The objective was thus to devise a tactical purchasing policy and price forecasting system that bought each month's requirements at the lowest purchase cost. Furthermore purchases were made on a futures price basis and payment was made at the end of the month when delivery occurred. Hence there is no need to include any storage cost in the decision price break model.

Because of the large price fluctuations in both the short- and long-term that can occur, see Chapter 2, it was initially thought that the subjective judgements of the buyer, with his continuous contact with the market, would be the major source for price forecasting. The form of forecasts used initially, and for some substantial part of the study, reflect the solution adopted for converting the buyer's forecast into the probability distributions required as input to the decision price break policy. Further work showed that a linear statistical forecasting system could give purchasing results at least as good as the best buyer in the company and better than several others. Hopefully a combination of the buyer's judgement with the model recommendations would lead to synergy and a further improvement.

The company had several branch factories in different parts of the U.K. Each factory had a branch buyer who was executively responsible to the branch factory manager. The branch buyers had the individual responsibility for purchasing most of the raw materials used by the branch factory. In addition the branch buyers were functionally responsible to a central chief buyer. The chief buyer on a regular basis sent out recommendations on future price movements to the branch buyers. The branch buyers were expected to follow the central chief buyer's advice, unless there were important, unusual factors present affecting the different local markets where the branch buyers actually purchased their materials.

The chief buyer's recommendations and forecasts were issued in the form of a written market survey for all the commodities the company needed to purchase. It was sent as useful information to other relevant senior managers as well as the branch buyers, because the raw material costs were such an important part of the company's economic viability. This survey was entirely qualitative. However it did give a record of the chief buyer's forecasts as made at the time rather than with hindsight. It provided the basis upon which purchasing decisions were made by the branch buyers. It was therefore information which could form the basis of a feasibility study of the decision price break model.

12.3 INITIAL FEASIBILITY STUDY

The chief buyer was accustomed to making his forecasts for future price movements in general qualitative terms that a 'bull' or 'bear' market was likely to occur, i.e. prices would increase or decrease. The market terms 'bull' or 'bear' merely indicate the direction of the price movement but say nothing of its likely magnitude. Discussions were held with the chief buyer to agree a quantitative interpretation of these qualitative forecasts. A 'bull' or 'bear' market was to be interpreted as an increase or decrease of the same fixed amount for all the periods covered in the initial study. The forecast of the final price at the end of the buying period, the deadline for a purchase, would be the current price plus this increase or decrease, for the 'bull' or 'bear' market respectively, or would be the same as the current price if the buyer predicted a steady market. The straight line joining the initial current price to this final forecast price then gave his estimate of the most likely price movements.

The chief buyer recognised that prices would actually fluctuate up and down rather than move smoothly in a particular direction and that his final forecast price was overprecise. He indicated that his forecasts should be interpreted as meaning that the final price would be anywhere within fixed limits above and below his actual forecast. He thought these limits should be the same for a 'bull', a 'bear' or a steady market. These limits are the points C and D in Fig. 12.1. He implicitly assumed that prices would be symmetric about his forecast trend. The prices over the first few days of the buying period would also fluctuate about his forecast trend but not by as much as for the later price offers. Lower limits were therefore specified for the initial price, the points A and B. The lines AC and BD of Fig. 12.1 thus become his estimate of the forecast envelope of the upper and lower limits containing the future price movements.

As is the case for most buyers he had little knowledge of probability theory or statistics. He could not therefore refine his forecasts further,

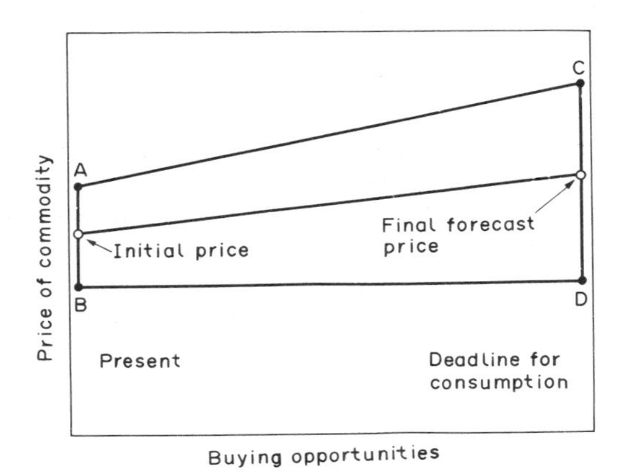

FIG. 12.1 The forecast envelope for the central buyer's forecasts

based on his judgement and experience, to suggest a probability distribution for the frequency with which the prices between his upper and lower limits would occur. He was not prepared to commit himself to the view that prices near to his mean forecast trend were more likely than prices near to the upper and lower limits. As far as he was concerned all prices within these limits were equally likely to occur. Since he had not provided quantitative forecasts previously there was no method of evaluating the distribution of the prices about his forecast trend. In these circumstances the prices each day were assumed to be subject to a rectangular distribution between the upper and lower forecast limits for that day. After some experience in providing this type of forecast it may be that other probability distributions to describe the price variation within the limits will be found to be more appropriate.

The detailed formulae, for calculating the values of the decision price breaks of the theoretical purchasing policy for this simple forecasting system, are given in an appendix to this chapter. This policy, using the chief buyer's forecasts in the form described, was applied to the purchasing situation at the two major company factories over the previous year. This initial study was based on historical data. Although the buyer's forecasts for the previous 2 years were available, complete daily price data however had only been collected and stored for ten delivery positions for maize in factory A, and for three periods for both maize and wheat in factory B. Contrary to the stated usual policy of the company of having deliveries every period, no actual purchase had been recorded for one maize and one wheat delivery position in factory B. These positions were still included in the study of the applicability of the decision price break model to give a larger sample of results for factory B. There are thus sixteen separate purchasing decisions to be made.

For each monthly delivery position the relevant forecasts of future prices were substituted in the recurrence relation given at the end of section 11.10 in the previous chapter. This gave the numerical values of the decision price breaks to be used for each day of the buying period. Starting with the price offer on the first day of the buying period and continuing over all the days in the buying period, the prices were compared with that day's decision price break. If the price offer was larger than the decision price break for that day no purchase was made. The same comparison was then made for the following day. As soon as a price offer occurred that was less than or equal to the decision price break for that day, a purchase was made of the total amount required for that delivery position. No further purchases were made for that delivery position since the decision price break model showed that only one purchase of the total amount required should be made for each delivery position. The same exercise was carried out separately for all sixteen delivery positions. The results are given in

Table 12.1. All prices are given in shillings per ton. The shilling was in use prior to decimalisation of the U.K. currency in the early seventies and 20 shillings were equal to one pound sterling.

The first column of Table 12.1 shows the purchase price per ton that would have been paid if the decision price break policies had been used as just described. Remember that the decision price breaks are calculated from the chief buyer's own forecasts. The second column of the table shows the average price per ton that was actually paid for the material by the branch buyers. The third column shows the average price offer over the

TABLE 12.1 Cereal Buying Simulation

Maize Purchases for Factory A

Delivery position	Decision price break purchase price	Actual purchase cost	Average price (over Buying Period)	Lowest price (over Buying Period)
1	480.0	487.5	487.75	479.75
2	482.75	490.0	488.25	477.5
3	487.0	496.5	490.5	478.0
4	488.75	488.75	490.5	476.5
5	473.0	480.0	487.25	473.0
6	473.0	492.5	486.5	466.5
7	464.0	490.25	474.5	456.5
8	462.5	490.5	471.5	451.75
9	460.0	482.0	472.0	454.25
10	460.0	482.75	472.0	454.5

Maize Purchases for Factory B

Delivery position	Decision price break purchase price	Actual purchase cost	Average price (over Buying Period)	Lowest price (over Buying Period)
1	489.0	491.0	492.0	480.75
2	481.75	486.75	486.75	479.25
3	471.5	—	484.5	471.5

Wheat Purchases for Factory B

Delivery position	Decision price break purchase price	Actual purchase cost	Average price (over Buying Period)	Lowest price (over Buying Period)
1	508.5	—	507.5	503.5
2	511.0	503.5	512.5	503.5
3	488.5	494.0	505.0	486.0

buying period for each delivery position, the hand-to-mouth policy price. The fourth column shows the lowest price that was offered over each buying period. The purchasing efficiency values (as defined and discussed in Chapter 5) are shown in Table 12.2.

For maize the results show that, using the *same forecasts* as given to the buyers, the decision price break policy gives substantially lower cost purchases than the buyers of both factories actually made. The average purchasing efficiency of over 60% is a very good result. In factory A the policy would have saved 3% on the buyer's maize purchasing costs. Unfortunately the buyer in factory A paid more than the average price for almost all delivery positions. It should be noted that for maize the price with the theoretical policy is always below the average price. For wheat there was one bad purchase over the average price. This arose because

TABLE 12.2 Purchasing efficiencies

Maize Purchases for Factory A

Delivery position	Decision price break purchase	Actual purchase made
1	97	3
2	51	−16
3	28	−48
4	12.5	12.5
5	100	51
6	67.5	−30
7	58	−87.5
8	46	−96
9	68	−56
10	69	−61
Average	60	−33

Maize Purchases for Factory B

Delivery position	Decision price break purchase	Actual purchase made
1	27	9
2	67	0
3	100	—
Average	65	—

Wheat Purchases for Factory B

Delivery position	Decision price brake purchase	Actual purchase made
1	−25	—
2	17	100
3	87	58
Average	26	—

N.B. Purchasing Efficiency $= 100 \times \dfrac{\text{(Average Market Price} - \text{Actual Price Paid)}}{\text{(Average Market Price} - \text{Market Lowest Price)}}$

See Chapter 5.

prices moved completely contrary to the original price forecast. No revisions had been made historically to the forecasts so no revisions were made to the forecasts in the evaluation exercise.

The wheat results give only a moderate positive purchasing efficiency. The two recorded wheat purchases of the buyer in factory B are very good. If this is a true reflection of his wheat purchasing ability little improvement to his performance could be made. His maize purchasing results are rather less impressive, although they are below the average price offer over their buying periods.

12.4 MAIN STUDY USING CHIEF BUYER'S FORECASTS

The results of the feasibility study showed the potential usefulness of the theoretical decision price break purchasing policy using the chief buyer's forecasts. Certain comments and reservations about the results were, however, expressed by the chief buyer. The most important of these was on the validity of the price offers used. It was said that some of the prices recorded historically were not 'real' offers at which the company could have bought the amounts it required. They were either offers of small amounts from U.K. farmers, which would have provided only a very small fraction of the company's requirements (only applicable to wheat since maize is not grown in the U.K.), or more generally they were offers by shippers to test the market, which would have been withdrawn if the company had attempted to buy moderate to large amounts. These comments affect not only the theoretical policy costs but also the average and lowest prices over the buying period. Thus Table 12.2 might be giving a very biased measure of the buyers' purchasing efficiency. The central buyer also pointed out that if he had been aware that prices were moving contrary to his initial forecasts he would have modified them.

It was agreed that the study would continue on a real time basis making purchasing decisions 'on paper' each day alongside the buyer using the decision price break model. The study would be carried out for factory B which was considered to have the best buyer in the company. Because of the reservations expressed on the past price data, this buyer would record daily the lowest price offer for each delivery position which he believed to be a 'real' price offer. The chief buyer agreed to make specific *quantitative* forecasts of prices each time a particular delivery position could first be considered for a possible purchase by the buyer of factory B. The central buyer undertook to modify his forecasts whenever *he* thought that his original initial forecasts should be revised.

However the central buyer was only prepared to forecast the general trend of prices. He was not prepared to put any limits on the likely fluctuations about his trend nor indeed any limits on the likely errors in his trend forecast. Since this was the first time the chief buyer had made explicit quantitative forecasts, no information on the size or distribution of the likely forecast errors was available. Analysis of the data used in the preliminary study showed that the difference between the highest and lowest price offers over the buying periods had been 20 shillings per ton or more and had always been less than 50 shillings per ton, for both wheat and maize. Accordingly it was decided to simulate the theoretical policies for the two cases of limits of ± 10 shillings and ± 25 shillings per ton about the buyer's forecast of the final price over the buying period. These values then determine the points C and D of Fig. 12.1. As before the limits A and B of Fig. 12.1 were taken as ± 5 shillings per ton about the initial price offer. Since there was no historical data available to evaluate how accurate the central buyer's forecasts would be, two alternative probability distributions to describe the relative frequencies with which prices between the upper and lower limits would occur were used in the evaluation. These were the rectangular distribution as used in the feasibility study and a normal distribution. As mentioned earlier a normal distribution implies more confidence in the buyer's forecast of the overall trend of prices than a rectangular distribution, since there is a greater probability of prices occurring near to the mean value than further away.

The delivery positions for this exercise in factory B were always on a monthly basis, i.e. delivery in June, July or August etc. The branch buyer bought his maize and wheat through the local 'informal' cereals market. This local market was formed by the agents of the shippers of maize and wheat into the nearby port and the various local consumers of cereals. Many years ago there had been an active 'Corn Exchange' at the port where price fixing and trading took place openly. However this had declined and then closed down. Trading and price offers currently took place via telephone conversations between the agents and the potential buyers. The agents of the different sellers would normally contact the branch buyer in factory B with a price offer on every day. Thus the number of buying opportunities for any delivery position would be the same as the number of days, excluding weekends and public holidays, from the day the chief buyer said purchasing could begin for that position up to the first day of the calendar month of the actual delivery. In practice there could be a smaller number of actual offers since on some days no seller was willing to fix a firm price, perhaps because they had no unsold maize or wheat left at that time. On other occasions price offers continued into the delivery month. This arose when the seller had not sold all of the maize or wheat carried by a ship already sailing to the U.K. across the Atlantic for delivery late in the month. The policy was simulated on the basis that if a purchase had not been made earlier it must be made on the last estimated offer of the buying period or the first market day of the calendar month of delivery, whichever was the earlier, at whatever price was offered. The possibility of delaying purchases to buy material 'afloat' at lower prices was ignored.

The detailed forecasts made by the central buyer together with the number of buying opportunities (days with price offers) predicted to occur in each buying period are shown in Table 12.3. One important point to note, which was a major factor in determining the subsequent course of the

TABLE 12.3 Central buyer's price forecasts

	Maize			Wheat		
Delivery Month	Start price	Final price	Predicted offers	Start price	Final price	Predicted offers
June	501.7	501.7	50	493.6	493.6	45
July	499.0	499.0	60	515.0	515.0	50
August	491.7	491.7	60	490.0	490.0	25
September	481.7	481.7	60	486.0	480.0	40
October	479.2	479.2	55	468.0	450.0	60
November	486.7	450.0	90	458.5	445.0	75
December	470.5	450.0	70	458.5	450.0	85
January	459.2	452.5	100	438.5	450.0	75
February	459.2	455.0	100	438.5	450.0	85
March	459.2	455.0	100	438.5	450.0	80

Reforecasts after devaluation of sterling by 14% in November.

January	512.0	508.0	45	502.5	493.0	30
February	517.0	522.5	50	502.5	493.0	45
March	530.0	522.5	65	502.5	493.0	30

study, was that the central buyer did not revise his forecasts, except when a devaluation of the pound sterling relative to other currencies took place in mid November 1967. Sterling was devalued by 14% so that a step increase in the U.K. imported cereals prices occurred. Obviously a reforecast is required in these circumstances. Presumably no other modifications were made to the forecasts earlier or later because the central buyer felt that his original forecasts were as accurate as could be expected.

Some examples of the price movements that actually occurred over the buying periods for both wheat and maize are given in Figs. 12.2 to 12.6. The central buyer's mean forecasts are also drawn on the graphs together with the estimated upper and lower limits on the price fluctuations; the lines AC and BD for the wider limits of ± 25 shillings on the final forecast price and the lines 'AC and BD' for the narrower limits of ± 10 shillings. All the price offers that were made for each delivery month are marked on the graph sequentially with no gaps left for those days when no real price offers, in the opinion of the branch buyer, were made. The line CC'D'D indicates the time predicted for the final price, e.g. the 50th price offer for July maize in Fig. 12.2. In practice there were only 45 real price offers made for July maize up to the deadline of the first day of July. The final price on each figure is for the first day of the delivery month. For most delivery months price offers were made for several days into the delivery month itself. These were sometimes higher and sometimes lower than the immediately preceding prices. As stated earlier no purchases could be

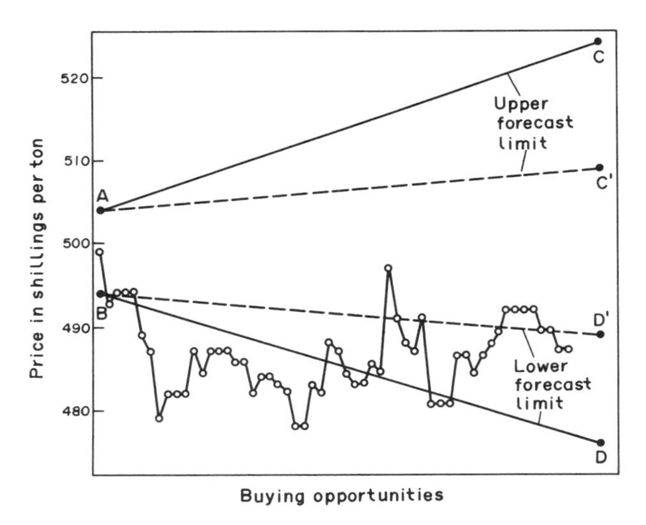

FIG. 12.2 The central buyer's forecast envelope and the actual daily prices over the buying period for July maize

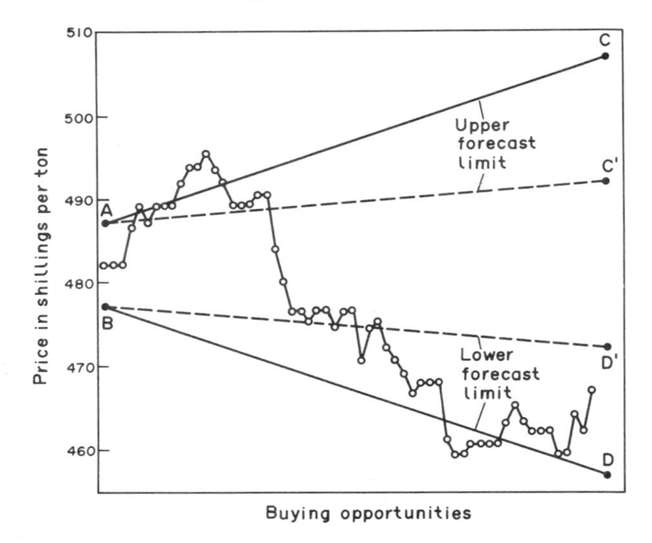

FIG. 12.3 The central buyer's forecast envelope and the actual daily prices over the buying period for September maize

made at any of these prices in the evaluation exercise. If a purchase had not been made earlier it had to be made at the price offered on the first day of the month of delivery.

One immediate result is that the predicted number of price offers was always larger than the number of price offers that actually occurred, by up

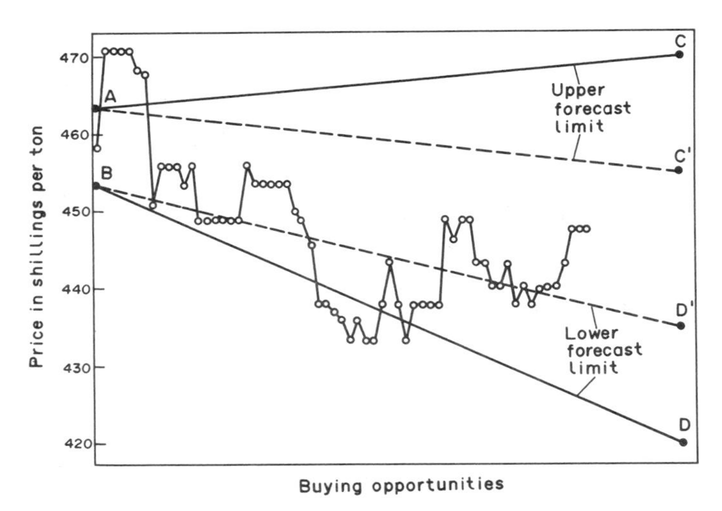

FIG. 12.4 The central buyer's forecast envelope and the actual daily prices over the buying period for November wheat

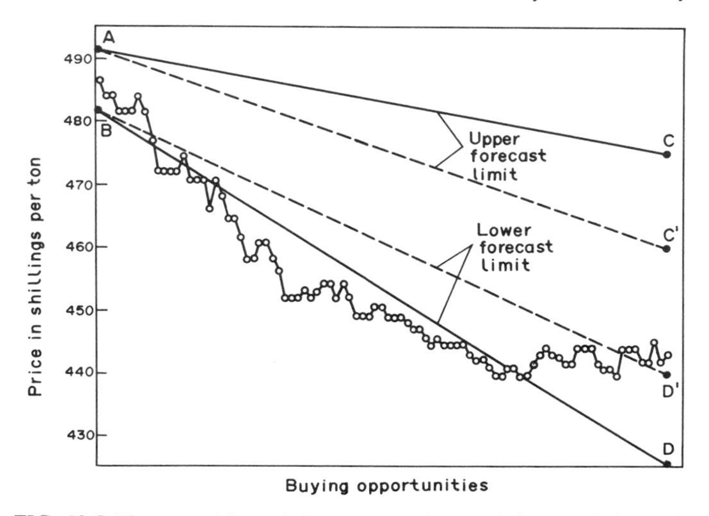

FIG. 12.5 The central buyer's forecast envelope and the actual daily prices over the buying period for November maize

to 14 offers or almost 20% for the case of November wheat for example, Fig. 12.4. Thus there are quite a large number of days in practice when no offer was made by the agents of the shippers or no 'real' price offer in the opinion of the branch buyer. Thus for future use the predicted number of

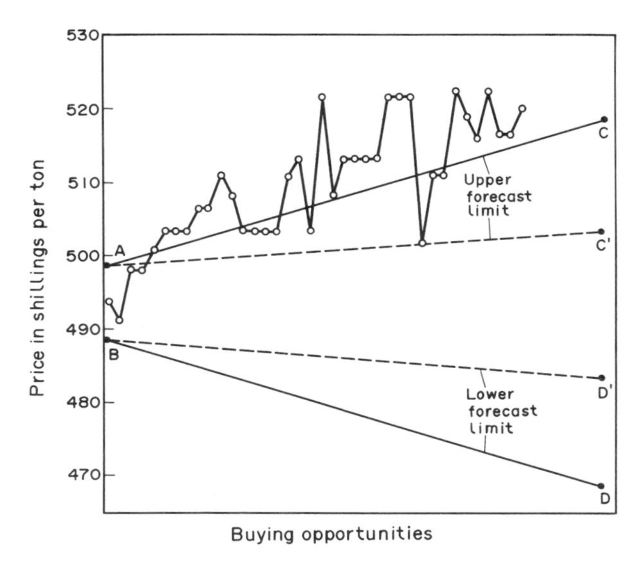

FIG. 12.6 The central buyer's forecast envelope and the actual daily prices over the buying period for June wheat

price offers that will occur over the buying period should be taken as about 10% less than the number of calendar days, excluding weekends and public holidays, from the time buying can commence to the first day of the month of delivery.

It had originally been intended to carry out a day-by-day real time simulation. However as often happens in such cases the daily prices were not provided on a regular daily basis by the buyer but in monthly batches at irregular intervals. It was however recognised by the buyers that an evaluation could be performed just as if it had occurred in real time. The first evaluation of the decision price break policies was carried out for the June to November delivery positions in early December 1967.

The devaluation of the pound sterling in November 1967 caused several complications to the study. Devaluation led to a steep increase in U.K. prices of about 14%. It also apparently caused unusual price fluctuations in the early weeks after devaluation not really related to the underlying real supply and demand circumstances, according to some of the buyers. Thus adjusting earlier prices by 14% would not give a consistent continuous price series. It was therefore decided to evaluate the buying performance for all those months for which delivery had to occur before devaluation i.e. June to November deliveries. December was ignored since it had prices both before and after devaluation. December prices had been on a downward-trend prior to devaluation, so that leaving a purchase until the end of the buying period was reasonable. Unfortunately the 14% increase after devaluation ruined this policy. A bad purchase prior to devaluation would have seemed a better decision. Thus no real conclusion about the efficacy of the theoretical policy or the buyer's usual performance could really be made from the results for the December delivery position.

The results of the exercise are shown in Table 12.4, together with the minimum and average price offers over the buying periods. The final column of the table shows the average price per ton paid by the buyer himself for his purchases. An indication of the overall performance for both maize and wheat is shown by calculating the average price per ton paid assuming equal amounts bought for each delivery month. These averages are shown in the last rows of the table for the maize and wheat sections. The best result for maize, a rectangular distribution with the wider limits, lies just about half way between the average and minimum prices, indicating a good performance. The purchasing efficiency value is 56% compared to 33% for the buyer himself. The best result for wheat is for the wider limits also but has a purchasing efficiency of only 17%, much worse than for maize. For wheat the buyer did not make purchases in all delivery months. For those months for which he did make purchases his average cost was much better than that of the best theoretical policy. The poor wheat results for the decision price break policy were to a large extent due to the very bad purchase for June delivery.

TABLE 12.4 Purchasing performance using the central buyer's forecasts only

Raw Material	Delivery Month	Price paid using price break policy: Narrow Limits C′,D′ = ± 10, Rectangular Distribution	Narrow Limits C′,D′ = ± 10, Normal Distribution	Wide Limits C,D = ± 25, Rectangular Distribution	Wide Limits C,D = ± 25, Normal Distribution	Minimum Price Offer	Average Price Offer	Actual Price Paid by the Company
MAIZE	June	494.2	494.2	479.25	479.25	479.25	488.5	484.0
	July	489.2	493.0	479.25	479.25	478.0	486.6	485.5
	August	479.25	479.25	470.6	474.2	464.25	482.75	476.4
	September	474.5	474.5	464.2	464.2	459.25	474.25	461.25
	October	466.6	466.6	461.6	461.8	449.0	457.0	454.25
	November	444.5	446.0	445.0	445.0	439.25	453.0	453.0
	Averages	474.7	475.6	466.7	467.3	461.5	473.7	469.1
WHEAT	June	520.0	520.0	520.0	520.0	491.0	511.5	501.0
	July	501.0	501.0	494.8	494.8	486.0	503.0	—
	August	476.0	476.0	473.5	473.5	473.5	485.75	—
	September	466.0	466.0	458.6	458.6	436.0	451.5	443.5
	October	438.6	438.6	433.6	433.6	433.5	447.75	442.5
	November	438.6	438.6	449.0	449.0	433.5	447.75	451.0
	Averages	473.4 (465.8)	473.4 (465.8)	471.6 (465.3)	471.6 (465.3)	458.9 (448.5)	474.5 (464.6)	— (459.5)

N.B. For wheat the figures in brackets in the last row are the average values taken over only those months when the buyer made purchases.

Examining the individual monthly results shows that for maize the rectangular distribution never gives higher purchase prices than the normal distribution. For wheat on the other hand, there is no difference between the prices at which purchases are made for the rectangular and normal distributions. The wider limits on the price fluctuations give lower purchase prices for maize for all months except November. However, for wheat, the wider limits are better for some months and the narrower limits for others.

The results obtained from this exercise for both wheat and maize are remarkably similar to those obtained in the initial feasibility study. A good purchasing performance is achieved for buying maize, producing significant savings below a back-to-back policy both in this exercise and the feasibility study. The maize performance is again generally better than that of the buyer. The overall purchasing efficiency at 56% is very similar to the just over 60% achieved in the feasibility study. The wheat purchasing performance is only moderate, again as happened in the feasibility study. For those months where the buyer did make a purchase he bought better than the decision price break policy, exactly as happened in the feasibility study.

Thus apparently the decision price break policy using the chief buyer's own forecasts gives a good purchasing policy for maize, although not for wheat. It is rather surprising that there is such a large difference between the performances for maize and wheat buying. They are both cereals used in animal feeds production, to some extent being competing products. Although November wheat price movements are very different to those of November maize, they are very similar to those of July and September maize. A more detailed analysis of the purchasing performance and the price forecasting performance would be useful to throw some light on why such differences have arisen between maize and wheat buying performances. It might also explain why the wider limits in the price fluctuations about the forecast trend are virtually always better than the narrower limits and hence provide the confidence in the model necessary for practical implementation to begin.

12.5 ANALYSIS OF THE MAIN STUDY RESULTS

For the first 5 months for maize and the first 3 months for wheat the central buyer forecast that prices over the buying period would fluctuate about the level of the initial price offer with no discernible trend. A small decrease in price over the buying period was predicted for September wheat and a large fall in price for October and November wheat and November maize. For maize the central buyer always overforecast both the final price offer and the prices over the second half of the buying period. On average the actual final price was overforecast by 17 shillings per ton, with the overforecasting error ranging from 5 to 27 shillings per ton. For wheat the forecasting performance was rather more variable. The forecast of the final price ranged from overforecasting by 44 shillings per ton to underforecasting by 26 shillings per ton. The forecast of the final price offer was very good for 2 of the 6 months. However, even in these two cases the prices over the second half of the buying period were well below the forecast trend.

Except for June wheat, most of the individual prices over the buying periods were below the forecast trend. June wheat was the one case where prices rose continuously over the buying period well above the central buyer's forecast. Indeed except for November wheat (Fig. 12.4) most prices were outside the narrower upper and lower price fluctuation limits (AC′ and BD′). In some cases, e.g. November maize (Fig. 12.5), the prices were below the wider lower price fluctuation limits, BD. Very large variations in prices occurred over the short-term. Even for November wheat, where the central buyer's forecast of the price was very accurate, prices both above the wider upper fluctuations limit, AC, and below the wider lower fluctuations limit, BD, occurred on several days.

Prices for July maize (Fig. 12.2) fell from the very beginning of the buying period to a lower level and then fluctuated about this lower level for most of the buying period. A purchase was made early in the buying period as soon as prices fell substantially below the forecast trend. The size of the fall required to trigger off a purchase depends more on the size of the limits on the price fluctuations about the trend, rather than the particular statistical distribution used to describe the relative frequencies of occurrence of prices within the upper and lower limits. With the narrower limits a purchase was made at the second price offer after the initial large fall in price, whilst with the wider limits the purchase was delayed for a further 6 days after three further moderate falls in price had occurred. It is clear, that if July maize prices had continued to fall over the second half of the buying period at the same rate as over the first half, the purchase price would have been very much higher than the average price over the buying period. This is precisely what happened for September wheat. The central buyer's forecast was that a very small decrease on the initial price offer of 486 shillings per ton to 480 shillings per ton would occur. Prices fell from the very start of the buying period and continued falling so that the final few prices were 45 shillings below the forecast final offer, almost twice as large an error as for any of the maize forecasts. An early purchase well below the forecast price offer was made but this was substantially above the later prices. In the event it was a rather bad purchase well above the average price offer. Substantially wider limits on the price fluctuations about the forecast trend than were used would have been required to give reasonable purchases for September wheat.

The September maize situation was quite similar to that of July. Prices initially rose above the forecast level and stayed there for about 4 weeks, see Fig. 12.3. They then began a continuous fall over the next 8 weeks, followed by 4 weeks of relatively stable prices. A purchase was made soon after prices fell below the forecast price level for the narrower limits, AC′ and BD′, but delayed for a further 2 weeks of falling prices for the wider limits, AC and BD, exactly as happened for July maize and September wheat.

For November maize and November wheat, Figs. 12.4 and 12.5 the central buyer forecast that prices would fall over the buying period. Whenever a falling trend or a zero trend exists for price forecasts made in the form illustrated in Fig. 12.1, the decision price breaks for every day of the buying period are always below the forecast final price offer. The wider the limits imposed on the price fluctuations about the forecast trend the further in value below the forecast final price offer are the decision price breaks. The decision price breaks however are always greater in value than the lower limit on the price fluctuations for the final price offer, point D′ in Fig. 12.1. These various properties are proved in the appendix to this chapter. If prices do in fact continue falling over the buying period then a purchase will invariably be left until the final few price offers and most likely until the final price offer itself. If however the minimum price offer occurs early in the buying period and is close to or smaller than the lower limit forecast for the final price offer, then a purchase will always be made early on in the buying period, during those days when the prices are falling to their minimum value.

November wheat prices fell much faster than predicted over the first half of the buying period but then rallied strongly back to the forecast final offer. The minimum price offer was smaller than the lower limit forecast for the narrower price fluctuations, the point D’ in Fig. 12.4, but significantly larger than the lower limit for the wider price fluctuations, the point D in Fig. 12.4. Hence with the narrower price fluctuation limits a good purchase was made during the fall to the minimum price half way through the buying period. With the wider limits on the price fluctuations a purchase was left until the actual final price offer. The rally from the minimum price offer half way through the buying period to the final offer was very large, so that leaving the purchase until the final offer was quite costly compared to the earlier price offers and above the average price offer over the buying period. November maize prices were similar to those of November wheat. However the rally from the minimum price to the final offer was very small so that little was lost by leaving the purchase to the final offer.

June wheat was one of the rare cases where prices continually increased over the buying period (see Fig. 12.6). The chief buyer’s forecast, which he never revised, was that prices would fluctuate about the initial offer of 493.6 shillings per ton. Prices in fact rose strongly above this forecast to reach 520 shillings per ton on the first day of June, the final price offer of the buying period. All the decision price breaks for the theoretical policy were below 490 shillings for both the wide and narrow error limits so that a purchase was left until the final price offer at the purchase deadline. The central buyer's forecasts, as for September wheat, were drastically wrong, although in his defence when prices fell from 521.5 shillings to 501 shillings on the 29th price offer this could have suggested that prices were now about to fall below the forecast price level. Hence no revision would be required to his original forecast of the final price. The factory buyer, obviously using his own forecasts, bought at this 29th price offer. Interestingly if the chief buyer had made a similar type of forecast, that prices would fluctuate about the initial price at the end of the previous week, one day earlier than he actually did, when the price was 501 shillings, a very good purchase would have been made with the narrow price fluctuation limits (but not the wide limits) at either of the first two price offers of 493.6 or 491 shillings per ton.

Although it was evaluated at a much later date than the initial exercise, it is useful to consider briefly what happened with February maize after the sterling devaluation since this was the only occasion that the central buyer forecast that prices would increase. His prediction of the final price offer in this situation was remarkably accurate. The price offers and the forecasts are shown in Fig. 12.7. In this instance prices initially increased at a faster

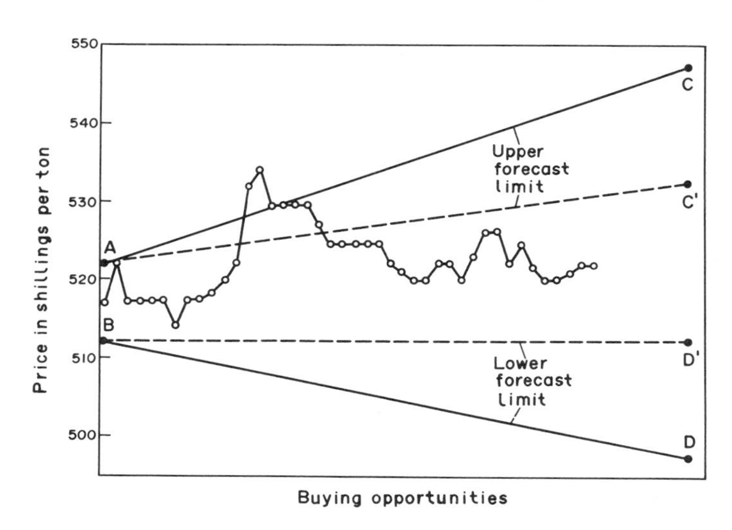

FIG. 12.7 The central buyer's forecast envelope and the actual daily prices over the buying period for February maize

rate than the forecast trend. They then fell back to the forecast trend from above and fluctuated only narrowly about it for the last third of the buying period. Since the decision price breaks of the theoretical policy were always well below the forecast trend no purchases were made until the purchase deadline on the first day of February.

June wheat and February maize, Figs. 12.6 and 12.7 were situations where price movements were generally increasing over the whole buying period. The prices were above the forecast trend of price movements over most of the buying period, i.e. the central buyer was *underforecasting* the path along which prices would move. The decision price breaks will always be below the mean forecast prices for each day. Exactly how far below will depend upon the size of the limits assumed for the fluctuations above and below the mean forecast trend. If, as in both instances, the prices begin to move upwards initially and then remain above the mean forecast path for most of the buying period, a purchase will inevitably be left until the final price offer or the last few price offers. Thus a higher price than the average price over the buying period will be paid. This occurs even where a relatively accurate price forecast is made, as for example for February maize. If prices initially move downwards for a short time, as would have happened with June wheat if the buying period has been started 1 or 2 days earlier, a very good purchase would be made at one of these early low prices if the error limits on the price variability are not too wide.

The size of the limits to be put on the fluctuations of prices above and below the forecast trend obviously depend on the accuracy of the forecasts of prices over the whole buying period. If the price forecasts are good, as for example November wheat, then the narrower limits give lower cost purchases. If the price forecasts, whilst generally giving the right direction in which prices will move are poor, such as for July and September maize, the wider limits give better purchases. If the forecasts are really bad, as for September wheat, then even wider limits than ± 25 shillings per ton would be required to give good purchases. Since the central buyer's forecasting performance is so variable and since the data sample is so small, and only covers a very small subset of the different types of price movements over a period that can occur, no objective assessment can be made. At this stage the ± 25 shillings per ton used in the evaluation would seem a reasonable compromise if the same method of forecasting is to be used in future.

Comparing the various under and overforecasting situations that occurred in the evaluation, it appears that the best results were obtained for the overforecasting case where the forecasting errors for prices over the second half of the buying period were not too excessive. The 2 months for which the most accurate forecasts were made in the sense of having most prices within the narrower fluctuations limits were November wheat, where a downward trend was predicted, and February maize where an upward trend was predicted. A much better purchasing result was obtained for the overforecasting situation, November wheat, than for the underforecasting situation, February maize. This result together with the earlier discussion in this section would strongly suggest that in general it will be better to bias price forecasts towards overforecasting rather than underforecasting future prices.

If the forecast of the general trend of prices is too low a purchase is virtually certain to be left until the final price offers at higher prices. If the forecast of the general trend is too high a purchase is likely to be made in the first half of the buying period. In the former case one will certainly pay a higher price than necessary, whilst in the latter case, provided the overforecasting error is not too large, a reasonable to good early purchase will be made although there is small possibility that even lower prices may occur later. Some of the risks of bad purchases arising from too large an error in the overforecasting situation can be averted by reforecasting at regular intervals or whenever the errors to date become too large.

12.6 CONCLUSIONS ON THE USE OF THE CHIEF BUYER'S FORECASTS

The conclusions from the above analysis are that a good purchasing policy has been derived for the situation where prices fall by a small amount over a period of up to 3 months. The central buyer appears able to make reasonable forecasts of future prices in such circumstances. The decision price break policy using the central buyer's forecasts then gives purchases at good prices usually well below the average price offer over the buying period. However the central buyer was unable to predict in advance the substantial increase in prices that took place for June wheat nor the very large fall in prices that occurred for September wheat. In both these cases the poor price forecasts led to the decision price break policy giving very bad purchases, for two entirely different reasons, as discussed in the previous section.

In general buyers will be faced with a mixture of strongly or weakly rising, strongly or weakly falling, or steady prices over the different buying periods. Any practical purchasing policy must be able to cope with all of these situations rather than a small subset of them. Alternatively some method is required that can anticipate in advance which of these situations is likely to occur.* At the time of the study no such method was available.

*The work described in Chapter 10 shows that this can be done successfully. The various patterns of price movement that occurred for the delivery months evaluated in the study are quite usual during the old crop/new crop transition months for maize when the new crop is expected to be much higher than the old crop was. However these results were only available several years after the study described in this chapter. It is apparent that the chief central buyer had no model of the old crop/new crop transition period in his mind when making his forecasts.

Even if it were, a price forecasting system and purchasing policy has only been devised to cope with generally steady prices or weakly falling prices. No real assistance can be offered for the other situations. The central buyer's current method of forecasting is clearly inadequate to properly cover all future situations. An obvious way of improving the forecasts is to introduce regular monitoring and updating of the forecasts whenever necessary. Another approach is to evaluate whether any of the standard statistical forecasting systems are of value.

It is very difficult with subjective forecasts, such as those made by the central buyer, to determine when the prices have moved significantly away from the forecast path and hence that a new forecast should be provided. Remember that the central buyer felt unable to put upper and lower limits on the size of the likely fluctuations about his overall forecast trend. Even if prices move markedly away from his forecast overall trend it does not necessarily mean that the forecast of the final price offer should be modified. The forecasts of the final price offer for both November maize and wheat were very accurate, yet in both cases prices fell below the wider forecast lower fluctuation limit on the prices, the line BD. Reforecasting, when prices fell below the lower fluctuations limit, to reduce the forecast final offer would have been a mistake in these months. Deciding when a forecast should be revised is thus in practice a rather more difficult problem than may be thought from an initial cursory view of the situation.

A further example of these difficulties can be seen by comparing the first 4 weeks of the price movements (15–20 price offers) for September maize and June wheat. The difficulties can only really be appreciated by covering up all prices after the first '20' price offers. These offers are shown in Fig. 12.8. The movements over the first '20' offers are remarkably similar for the two cases. Prices increased from the very start to above the upper fluctuation limit and then fell back slightly towards the mean forecast price level. Figure 12.8 would certainly suggest that the price forecasts for both

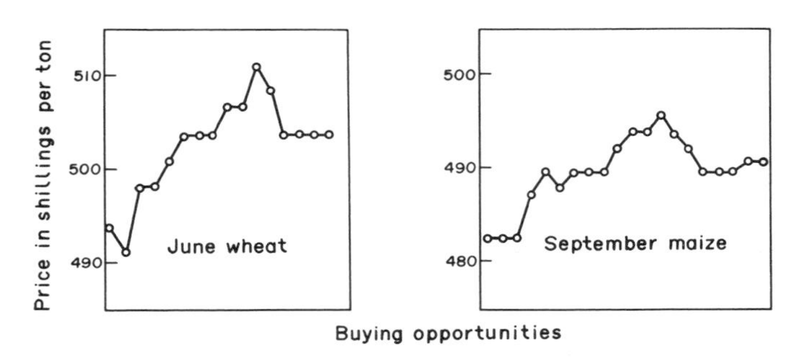

FIG. 12.8 An example to the difficulties in price forecasting

cases should be revised to predict an upward trend. This would obviously have been right for June wheat where the upward trend continued over the rest of the buying period, but not for September maize where prices fell drastically over the next few days.

The problem is to forecast the general trend of prices over many days into the future and not just to forecast the next price offer. The forecast of the overall trend of prices, i.e. of the path prices will follow over the buying period, depends upon the 'start price' used as well as on the forecast of the final price offer. It is rather arbitrary to use the price offer on the day buying can commence for a particular delivery month. This price offer could have been unusually high or low. It might have been better to have used the simple average of prices over the previous week or some weighted average of past prices as the start price. As mentioned earlier there is some slight evidence, the case of June wheat, that such a modification could have improved the purchases. If the forecasts are routinely updated at regular intervals then there is the opportunity to update the 'start price' as well as the forecast of the final offer. Such a revision to the start price could in fact change the direction of the forecast price trend. For example after 2 months for November wheat the 'start price', based on some past average, would probably have been just under 440 shillings per ton. If the forecast of the final price offer were unchanged, the forecast trend would give an upward trend over the expected remaining 40 price offers or so of the buying period. Furthermore at each subsequent updating there are fewer offers remaining in the buying period. There is thus a smaller period of time over which cumulative forecast errors can mount up. It would therefore seem reasonable to reduce the upper and lower limits on the fluctuations about the final offer. A combination of this change plus the updating of the start price could have improved some of the purchases e.g. November wheat although of course it might have worsened others.

Since the central buyer was unable to give a size to the limits on the movement away from his forecast trend which would indicate automatically that the reforecast is necessary, the only option is to decide at the very start to review the forecasts at regular intervals. A formal system must be set up to ensure that the central buyer and the research team automatically discuss, evaluate and review the forecasting performance to date at regular intervals.

An alternative approach is to change the method of price forecasting. Many statistical forecasting systems exist for predicting the future values of time series. Their properties have been extensively analysed. Particularly important is that control limits can be set in advance which will indicate whether the underlying time series has changed its behaviour and hence that the current price forecasts have become invalid. Manual intervention can then be made to modify the forecasts. It could be that this might be the best way for the central buyer to make forecasts. His forecasts at the beginning of the buying period could be used as the initial values for the parameters of the forecasting system. Routine updating would be made by the statistical forecasting system and the buyer required to intervene only when the forecasting system itself indicated that manual intervention to override the statistical forecasts was necessary.

It was thus decided that the study would change direction to evaluate alternative forms of statistical price forecasting systems to see if they would provide an input to the decision price break policy which would lead to good purchasing decisions for all the types of price movement that could occur.

12.7 SIMPLE EXPONENTIAL SMOOTHING FORECASTS

It was decided initially to evaluate the use of simple exponential smoothing forecasting methods for revising the forecasts at regular intervals. The forecasts were converted into the same form used earlier, i.e. Fig. 12.1, to provide the appropriate input for the decision price break policy. The central buyer's forecast of the final price offer was replaced by an exponentially smoothed price, as was the start price. The smoothed price level, $Z(t)$, at time t following a price offer $P(t)$ was derived via the formula

$$Z(t) = \alpha P(t) + (1-\alpha)\ Z(t-1)$$

where α is a smoothing constant, generally in the range 0.1–0.2. The exponential smoothing model may also incorporate a smoothed trend factor, $B(t)$, via the formula

$$B(t) = \alpha\{Z(t) - Z(t-1)\} + (1-\alpha)\ B(t-1)$$

With this model the forecast of the price L offers ahead is given by extrapolating the trend as

$$Z(t+\mathrm{L}) = Z(t) + \left(\frac{1-\alpha}{\alpha}\right) B(t) + L\ B(t)$$

The exponential smoothing model was updated after each price offer. This was not necessarily every calendar day for the reasons mentioned earlier. However the decision price breaks of the theoretical policy were only recalculated at fortnightly intervals, every 10 days.

At each revision the current value of $Z(t)$ was used as the start price and $Z(t+\mathrm{L})$ as the forecast final price, replacing the buyer's forecasts. L was the number of prices estimated to remain from the time of the revision of the policy to the end of the buying period. At each revision the initial error limits, A and B, on the start price were taken as plus or minus 5 shillings, as before. The error limits on the final forecast price, C and D were reduced at each successive revision of the theoretical buying policy from the starting value of ± 25 shillings per ton. The limits were recalculated as follows:

$$X = \frac{\text{Numbers of Offers Remaining from Now}}{\text{Original Estimated Total Number of Offers}} \times 25$$

This is then rounded up to the nearest multiple of 5, to give the limits C and D. Thus for example if X were 17 the limits C and D are taken as ± 20 shillings on the current forecast of the final price offer. The distribution of the price fluctuations about the trend was taken as normal, because the forecast errors of a simple exponential smoothing system should be normally distributed about the forecast trend.

Except for November maize no prices were recorded for any delivery position prior to the start of its buying period. It was thus not possible to 'run-in' the exponential smoothing forecasts before active buying started. At the start of each buying period the central buyer's forecasts were used as the input to the theoretical policy as described earlier. His start price and final forecast price were also used to give the initialising values $Z(0)$ and $B(0)$ for the smoothing formulae. The price fluctuation and error limits, C and D, were taken as ± 25 shillings about the buyer's forecast final price. After 10 days the first revision to the theoretical buying policy was made using the exponentially smoothed forecast price levels and trends.

At this point in time data became available on the remaining months up to March 1968. The December, January, February and March offers for both wheat and maize had been affected part way through the offer period by the effects of devaluation. As mentioned this introduced an immediate step change in the price levels. Some concern was expressed by the buyers as to making a fair comparison between their purchases and those indicated by the models. For example, if the models made an early purchase before devaluation, whilst the buyers held off since they expected a continual drop in prices, they would have bought after devaluation at a very much higher price. It was agreed that the study should consider only December offers prior to devaluation and January, February and March offers after devaluation and also only consider the buyer's purchases that were made in these periods, even if they had only bought part of each month's requirements at that price.

An interesting result of the exercise was that the best results for maize were obtained by ignoring the trend component in the exponential smoothing model, whilst the best results for wheat were with the trend adjusted model. In the zero trend model at each revision of the buying policy, the smoothed price level, $Z(t)$, was used for both the final forecast price and the start price. The results of the exercise for the best models for each cereal are shown in Table 12.5. Overall the theoretical policies could have given a saving of just over ½% on the actual price paid by the buyer for both wheat and maize. The purchasing efficiency achieved for maize by the theoretical policy was 58% and for wheat was 47%. The maximum potential savings below a back-to-back policy were small, 3% for maize and 3.5% for wheat. The company buyer's purchasing efficiencies were 35% and 29% for maize and wheat respectively, very creditable performances. The range 30–35% appears to be the level that a good buyer can achieve using only informal subjective methods.

For wheat the major difference between the results for trend corrected and zero trend models occurred for June delivery, where with the zero trend model a purchase was left to the very end of the buying period at a higher price of 520 shillings. The asterisks on the entries in Table 12.5 denote those months where a purchase was made compulsorily at the very last price offer of the buying period. Note that for wheat only one month was bought at the very last price offer. By contrast the zero trend wheat model left six of the months until their final offer. For maize there were 7 out of the 10 months left until the final price offer of the buying period. During the time period considered in the exercise, it was undoubtedly best to wait and buy maize in the last few offers for most months. Yet a doubt must persist as to whether the buying system with these forecasts was anticipating this situation, and hence deliberately holding off buying, or whether the forecasting system was grossly in error and purchases were made at the last price offer by default. The trend corrected model for maize was significantly worse, overall only just under the average back-to-back policy price. In general wheat had larger price changes over the buying periods than maize. In addition wheat had some months with significant overall upward price movements. The differences between the trend corrected and zero trend model for maize also probably reflect the nonsymmetric effects on the purchasing decisions of over and underforecasting errors. It could be that the trend corrected smoothing model for maize overreacts to short-term trends in prices and unwisely projects them over the whole buying period.

A satisfactory system has now been developed for wheat buying for the first time. There is some doubt and uncertainty about the value of the zero trend model for buying maize. Such a model does not buy well if prices continually increase over the buying period. Obviously periods of rising

TABLE 12.5 Purchasing performance using simple exponential smoothing to generate price forecasts for the decision price breaks

	MAIZE				WHEAT			
Delivery Month	Purchase price	Average price over buying period	Minimum price over buying period	Actual cost paid by company	Purchase price	Average price over buying period	Minimum price over buying period	Actual cost paid by company
June	485.4*	488.5	479.25	484	503.6	511.5	491	501
July	479.25	486.6	478.0	485	494.8	503.0	486	—
August	465.4*	482.75	464.25	476.4	476.0	485.75	473.5	—
September	467.0*	474.25	459.25	461.25	448.6	451.5	436.0	443.5
October	451.6*	457.0	449.0	454.25	438.6	447.75	433.5	442.5
November	443.4*	453.0	439.25	453	438.6	447.75	433.5	451.0
December	444.4*	450.75	440.75	445	438.6	447.5	433.5	448.0
January	522.0*	520.5	509.5	519.5	498.6	498.75	483.5	494.25
February	517.6	522.25	514.5	—	486.2	493.5	478.5	487.5
March	514.6	524.5	514.5	522	473.6	486.25	463.5	486.25
Averages	479.6	486.0	474.9	—	469.7	477.3	461.2	—
Averages over months when buyer made a purchase	474.6	482.0	470.5	478	465.8	473.0	456.7	469.2

* Denotes months when a purchase had to be made at the very last offer of the buying period

prices for maize will occur at some time in the future. Thus other forecasting systems were considered to see if performances as good as those of Table 12.5 or better could be achieved without an undue reliance on buying at the last price offer.

12.8 ADAPTIVE SMOOTHING FORECASTING METHODS

Many statistical forecasting procedures have been developed to attempt to overcome the problems of modelling and then forecasting time series. These include Brown's method of exponential smoothing (1962), Box Jenkins methods (1970), and adaptive exponential smoothing as developed by Trigg and Leach (1967). The Box Jenkins methods are undoubtedly superior to Brown's method since they encompass a more general class of model. However they require a rather complicated parameter estimation procedure and do not provide a simple method of updating the forecasting parameters to adapt them to changes in the price stream. Commodity prices in general have large fluctuations about significant short-term and medium-term trends. It is thus necessary to have a forecasting system which reacts quickly to real changes though suppressing the white noise in the system. The Trigg and Leach adaptive smoothing system has been advocated as a simple procedure for dealing with such problems. The model is again a linear model. At time t, the forecast L offers ahead, $F(t,t+L)$, is given in terms of the present smoothed price level, $Z(t)$, and the present estimate of trend, $B(t)$, as

$$F(t,t+L) = Z(t) + L\ B(t)$$

where $Z(t)$ and $B(t)$ are updated daily after each price offer using the one step ahead errors of the forecast via the formula

$$E(t) = P(t) - F(t-1,t)$$

$$Z(t) = Z(t-1) + B(t-1) + \alpha\ E(t)$$

$$B(t) = B(t-1) + \beta\ E(t)$$

Note that different smoothing constants are used for the smoothed price level and the trend. In the strict model α and β are related by the formula

$$\beta = 2 - \alpha - 2(1-\alpha)^{\frac{1}{2}}.$$

The point of the adaptive smoothing method of Trigg and Leach is that the smoothing constant α is itself updated continually. It is set equal to the numerical value of the ratio $S(t)/M(t)$; if the ratio is negative the negative sign is ignored since α must always be positive. The two factors in the ratio are updated continuously as follows.

$$S(t) = (1-\gamma)\, S(t-1) + \gamma\, E(t)$$

$$M(t) = (1-\gamma)\, M(t-1) + \gamma\, E(t) \qquad \text{if } E(t) \text{ is positive}$$

$$= (1-\gamma)\, M(t-1) - \gamma\, E(t) \qquad \text{if } E(t) \text{ is negative}$$

and γ is a constant parameter generally over the range 0.05 to 0.15. For this forecasting system prices are normally distributed about $F(t, t+L)$, with a variance $V(t,t+L)$, given by

$$V(t, t+L) = \{1.25\, M(t)\}^2\{1+\alpha(L-1)\}.$$

The magnitude of the adjustments to the parameters each day depends on the errors. $S(t)$ is the smoothed mean error and $M(t)$ the smoothed mean absolute error. Thus if there is a large price change in the same direction as the current trend of the price movement, $S(t)$ is increased proportionally more than $M(t)$, so that α and β are increased. Hence more weight is given to the more recent prices than usual in updating $Z(t)$ and $B(t)$. If prices are relatively steady or fluctuating randomly within the existing upper and lower limits then $S(t)$ does not change much relative to $M(t)$. The values of α and β remain small so that less weight is given to the most recent data. The system can thus be expected to cope with all the types of movement that occur in the wheat and maize prices. A further advantage is that limits can be set on allowable values for α, so that warning signals can be given if the prices and the forecasts move out of line too much. It is on these occasions that the buyer can use his skill and experience to modify the forecasts. The only drawback with this forecasting system is that, if the price series violently fluctuates up and down about a constant level the forecasts may oscillate too much. The system was originally designed for sales forecasting where there are far less violent random fluctuations than in commodity prices.

Again it was found that, although the price forecasts were updated after each price offer, it was sufficient to only recalculate the decision price break policy every fortnight. As in the previous exercise the central buyer's forecasts were used to calculate the decision price breaks for the first ten offers and for initialising the price forecasting system. This forecasting system was simulated with values of γ of 0.1 and 0.05. The results for both maize and wheat are shown in Table 12.6. It is interesting that $\gamma = 0.1$ gives the best results for maize whilst $\gamma = 0.05$ is best for wheat. No maize purchases and two wheat purchases are left until the final price offer in the buying period. The wrong values of γ would increase costs by just over

½% in both cases and give average purchasing costs higher than the buyer himself paid. In fact, the overall results for the best maize and wheat models are virtually identical to those achieved with the previous simple exponential smoothing model. However for maize the individual monthly prices paid are very different to those of the simple smoothing model, compare Table 12.6 to Table 12.5. Since few purchases are now made by default at the final price offer of the buying period, these adaptive smoothing models, which give a 1.5% saving on a back-to-back policy, are an acceptable satisfactory basis for a day-to-day purchasing policy.

The purchasing efficiencies for the policies together with that achieved by the buyer for each delivery month are shown in Table 12.7. Over all the delivery months the average purchasing efficiencies for the two materials were very similar, at 50% for maize and 47% for wheat. Note the consistency of the buyer's actual performance for maize, where every month was bought at lower than average price over the buying period. The theoretical policy made three purchases at prices slightly greater than the average price. The results in the table demonstrate an important point that in order to make overall 'savings' in purchasing costs it may be necessary

TABLE 12.6 Adaptive exponential smoothing model purchases

Delivery Month	MAIZE		WHEAT	
	γ=0.1	γ=0.05	γ=0.1	γ=0.05
June	479.25	486.5	511.0	501
July	480.25	481.5	506.0	498.5
August	486.50	479.25	476.0	476.0
September	459.25	484.25	453.5	438.5
October	454.25	453.0*	438.5	438.5
November	441.5	440.5	434.0	451.0*
December	442.75	442.75	448.5	438.5
January	523.25	519.5	498.5*	498.5*
February	524.5	522.0	486.25	483.5
March	514.5	527.25*	471.0	473.5
Averages	480.0	483.7	472.3	469.8
Averages over months when buyer made a purchase	475.0	479.3	467.6	465.4

* Denotes months when a purchase had to be made at the very last price offer of the buying period

on a few occasions to make a loss compared with a back-to-back policy. Comparing the buyer's wheat performance with his maize performance shows that a consistently positive purchasing efficiency for all individual purchase orders will be a rare event.

Intelligent manual overriding of the recommendations of the above system, particularly when the parameter α of the forecasting system is outside its allowable limits, should enable further reductions in cost to be made. Any forecasting system is always reacting to prices after they have occurred. The adaptive system perhaps reacts quicker than other alternative simple systems. So although it is updated daily it still lags behind the price movements. If prices peak at some maximum and the trend then reverses direction it will take the forecasting system some offers to catch up. The smoothed mean error from being high and positive, on the upward trend side, decreases to zero and then becomes large and negative on the downward trend side. The smoothed mean absolute error is relatively unaffected over such a period of movement. Thus α, which gives the proportion of the current forecasting error which is added to the smoothed price level etc, must initially fall to zero before increasing again at the very moment a reversal of the price trend occurs. If the price is on an increasing trend and falls for three successive offers, with this forecasting system the

TABLE 12.7 Purchase efficiencies

Delivery Month	MAIZE		WHEAT	
	Policy	Buyer	Policy	Buyer
June	100	47	51	51
July	74	19	27	—
August	−20	34	80	—
September	100	87	84	52
October	34	34	65	37
November	84	0	−23	−23
December	80	53	64	−5
January	−25	9	2	25
February	−29	—	67	40
March	100	25	56	0
Averages	50	—	47	—
Averages over months when buyer made a purchase	59	34	46	23

decision price break policy will very often indicate a purchase should be made at either the second or third of these offers, since the forecasting system still predicts an upward trend. The purchase for February maize (Fig. 12.7) is a good example of this situation. Prices were initially stable for some days at 517 shillings per ton, increased rapidly to 534 shillings and then stabilised for 4 days at 530 shillings. The price trend was now strongly positive. Prices fell to 527 shillings. Such a fall was within the day-to-day price fluctuations, estimated by the mean absolute error. The price level and trend factors were changed little for the reasons mentioned above. A strong upward trend is still forecast. The next day prices fell to 524.5 shillings per ton. Over the 2 days the fall was more than the day-to-day variation and the price forecast was still for prices to remain over 530 shillings per ton. A purchase was thus made. If the central buyer's forecasts have any merit, lower prices than this can be expected, so manual intervention to delay a purchase at this point is appropriate. Obviously in the opposite circumstances if prices fall rapidly to a minimum level with an abrupt rally thereafter, the delay in the forecasting systems adapting to this situation means that a purchase will never be recommended at that minimum price offer. Hence in general, manual intervention should be carefully considered whenever a period of rapid upward or downward price movement followed by an opposite price reaction occurs.

The buying system with an adaptive exponential smoothing forecasting model merely treats the prices as numbers. No account has been taken that the prices are for maize or wheat rather than some other raw material. There are presumably special features which the buyer could take advantage of to override or delay the recommendations of the automatic buying system. One particular feature is the finite crop year and that prices generally reach a peak in June. The influence of the new crop is thus generally to reduce prices over the summer months. For example it affected the prices of July maize 'afloat' during the first week or so of July. It also caused a steep drop of 15 shillings per ton half way through the August price offers. The theoretical model, with no knowledge of this possibility built in, had made an early purchase prior to this drop. A similar situation occurred with the July wheat prices. If the buyer could anticipate that the new crop would have this effect, he could override the earlier recommendation and delay purchases until the second half of the buying period. However the buyer himself, although he made a good purchase, bought before the new crop affected prices. It appears that the buyer's knowledge and experience, because it is qualitative rather than quantitative, is insufficient to be able to make a reasonable judgement of the likely effect of the forthcoming new crop. It was thoughts and possibilities of this kind that influenced the author at a later date to develop the approaches and models described in Chapters 8, 9 and 10.

The study of cereals has shown that buying policies using only the prices offered as the basis of revising the price forecasts, could have given good purchasing results for buying periods ranging from 2 months to 4½ months. The decision price break purchasing policy derived from the dynamic programming formulation of the tactical commodity purchasing problem can be successfully applied to practical situations. Standard simple statistical forecasting systems are an adequate basis to generate the various probability distributions for the price offers at the differing buying opportunities, at least for purchasing maize and wheat.

12.9 THE MULTILOT PURCHASE STRATEGY FOR CEREALS

As described in the LFF case study of Chapter 3.3, the buyers tend to purchase their requirements in two or three or four small lots rather than in one lot. The idea behind this, is that with the uncertainty in future price movements this procedure will minimise the possibility of a bad buy and hence reduce their long-term average purchasing costs. The theoretical dynamic programming model of Chapter 11 showed that in an unrestricted market the best policy was to make only one purchase at what the model anticipated was the lowest price offer. Since the forecasting systems, even though updated after each price offer, may be grossly in error it may well be that in practical situations the multilot purchase strategy of the buyers is better. This procedure may be particularly useful on occasions such as when the new crop becomes available and introduces step changes in the price movements part way through the offer period, for example as apparently occurred for July and August maize.

For this reason it was decided to evaluate the effect of buying each month's requirements in several small lots. This is a direct application of the restricted supply/purchase quantity model of section 11.11 in the previous chapter. The best results for buying in one lot had been obtained with the adaptive exponential smoothing model. This is a convenient model to use and update, and with its error limits it is particularly useful to indicate the occasions when the buyers should manually override the automatic forecasts. This model was therefore used in this situation for purchasing several small lots.

The detailed results showing the price paid for each month for 2, 4 and 8 small lots of maize, and 2, 4 and 8 lots of wheat are shown in Table 12.8. The results show that the overall average price for purchasing maize in two lots and four lots are virtually identical, being about 1.7 shillings per ton above the best result for a single lot model, and in fact better than the second best for the single lot model. It is also worth noting that for August maize where the introduction of the new crop affected the single lot policy badly (see Table 12.6) both the 2 and 4 lot models bought at a lower price

TABLE 12.8 Performance of the multilot purchasing strategy

	MAIZE			WHEAT		
Delivery Month	2 Lots	4 Lots	8 Lots	2 Lots	4 Lots	8 Lots
June	485.4	485.7	487.1	510.5	507.2	508.8
July	478.0	478.8	480.6	490.4	494.2	498.2
August	471.7	476.0	478.8	484.4	481.0	485.6
September	477.9	477.9	472.0	439.8	444.2	450.2
October	452.9	452.0	456.2	441.0	438.4	436.8
November	443.0	442.7	442.4	447.5	446.6	442.2
December	442.1	442.7	444.1	449.2	443.8	439.8
January	520.8	520.2	522.2	498.7	486.2	492.6
February	522.1	523.3	522.6	483.7	486.2	487.8
March	520.9	518.4	521.6	471.2	480.6	484.6
Overall Average	481.6	481.8	482.8	471.6	470.9	472.7

than the single lot model. The average saving for those 2 months was almost 15 shillings per ton for the 2 lot and 10 shillings per ton for the 4 lot model. Thus if either the 2 or 4 lot model were used for those months, where it might have been suspected that the new crop would cause large changes in the price movements, and the single lot model elsewhere further savings than those shown in Table 12.6 could have been achieved.

The wheat results are quite interesting. The apparently paradoxical result is obtained that buying in 4 lots is on average better than buying 2 lots. This average result is mainly due to the relative purchases made for January delivery. This is because both the two lot and the single lot model bought at the very end of the offers, when prices rose to a temporary peak, whilst the four lot model bought at least two lots earlier at a much lower price. January wheat prices rose to a temporary peak at the end of December. A better policy is to adopt a composite approach of normally buying requirements in one large lot but splitting them into two or three lots whenever the buyer has reason to believe that external factors such as the introduction of a new crop, devaluation, currency uncertainties, etc., will produce an immediate step change in the future price series at sometime in the offer period.

Appendix

The Decision Price Break Recurrence Relations for the Chief Buyer's Forecasts

The simple price forecasting model used to convert the buyer's forecasts into a series of probability distributions, see Fig. 12.1, is as follows:

1. Let Q^* be the initial start price. Then Q^*+a and Q^*-a are the upper and lower initial price limits.
2. Q^*+Nm is the mean forecast price at the final deadline, the last opportunity at which to make a purchase. m is the daily price trend, assuming one buying opportunity per day. N is the number of days in the buying period.
3. Q^*+Nm+A and Q^*+Nm-A are the upper and lower limits on the forecast of the final price.
4. At the jth buying opportunity

   the mean price $$M_j = Q^* + jm$$

   the upper price limit $$U_j = Q^* + a + jm + \frac{j(A-a)}{N}$$

   the lower price limit $$L_j = Q^* - a + jm - \frac{j(A-a)}{N}$$
5. The range of the price movements at the jth buying opportunity is thus

$$R_j = 2a + \frac{2j(A-a)}{N}.$$

At the start of the buying period for each of the consumption requirements, the relevant central buyer's forecasts in the published surveys were analysed to determine the appropriate value for Nm. As mentioned this was either $+10$, -10 or 0. The value for N was then worked out and hence the value of m calculated. The parameters a and A were kept at the same constant value for all the purchase orders evaluated in the initial feasibility study.

For the case of purchasing a single lot prior to some known deadline, the recurrence relations for the decision price breaks with nonstationary probability distributions for future prices become

$$P_{k+1} = \int_0^{P_k} p\phi_k(q)\, dq + P_k \int_{P_k}^{\infty} \phi_k(q)\, dq \text{ for } k = 2,3, \ldots, N$$

$$P_2 = \int_0^{\infty} q\phi_1(q)\, dq$$

where $\phi_k(q)$ is the price distribution for the kth period prior to the deadline at the end of the buying period. The storage cost is assumed to be zero. The kth period before the deadline is the $(N-k+1)$th buying opportunity forward from the initial offer. Hence $\phi_k(q)$ has a mean value M_{N-k+1} and lies between upper and lower limits U_{N-k+1} and L_{N-k+1}.

Rectangular Distribution for Price Fluctuations

If $\phi_k(q)$ is a rectangular distribution between the finite upper and lower limits given then

$$\phi_k(q) = 1/R_{N-k+1} \qquad \text{if } U_{N-k+1} \geqslant q \geqslant L_{N-k+1}$$

$$= 0 \qquad \text{otherwise}$$

If $\quad P_k > U_{N-k+1} \qquad$ then $\int_{P_k}^{\infty} \phi_k(q)\, dq = 0$

so that

$$P_{k+1} = \int_{L_{N-k+1}}^{U_{N-k+1}} q\phi_k(q)\, dq = M_{N-k+1}$$

$$= Q^* + (N-k+1)m$$

Hence $P_{k+1} < P_k$

If $U_{N-k+1} \geqslant P_k \geqslant L_{N-k+1}$

$$\text{then } P_{k+1} = \int_{L_{N-k+1}}^{P_k} q\phi_k(q)\, dq + P_k \int_{P_k}^{U_{N-k+1}} \phi k(q)\, dq$$

$$= P_k - \frac{1}{2R_{N-k+1}} (P_k - L_{N-k+1})^2$$

Hence $P_{k+1} \leqslant P_k$

If $P_k < L_{N-k+1}$ then $\int_0^{P_k} q\, \phi_k(q)\, dq = 0$ and $\int_{P_k}^{\infty} \phi_k(q)\, dq = 1.0$

Hence $P_{k+1} = P_k$

It has been shown above that for all values of k P_{k+1} is less than or equal to P_k. However, P_2 is the forecast final price offer since

$$P_2 = \int_0^{\infty} q\, \phi_1(q)\, dq = M_N = Q^* + Nm.$$

Hence all of the decision price breaks are less than or equal to the forecast of the price offer at the final buying opportunity.

Furthermore, if $U_{N-k+1} \geqslant P_k \geqslant L_{N-k+1}$

$$P_{k+1} - L_{N-k+1} = P_k - L_{N-k+1} - \frac{1}{2R_{N-k+1}} (P_k - L_{N-k+1})^2$$

i.e.
$$P_{k+1} - L_{N-k+1} = \frac{(P_k - L_{N-k+1})}{2R_{N-k+1}} (2R_{N-k+1} - P_k + L_{N-k+1}).$$

The first term in brackets on the right-hand side is positive by definition of the range of values for P_k. In addition

$$(2R_{N-k+1} - P_k + L_{N-k+1}) = (U_{N-k+1} - P_k) + (U_{N-k+1} - L_{N-k+1}).$$

On similar reasoning both of the terms in brackets on the right-hand side of the above expression are positive. Thus

$$P_{k+1} - L_{N-k+1} \geqslant 0.$$

Thus if $P_k \geqslant L_{N-k+1}$ it follows that $P_{k+1} \geqslant L_{N-k+1}$, provided $P_k \leqslant U_{N-k+1}$ and if $P_k < L_{N-k+1}$, $P_{k+1} = P_k$, see the previous page.

If the price trend m is negative then the lower limit L_j decreases as j increases, or alternatively L_{N-k+1} increases as k increases. The minimum forecast price level is L_N, the lower limit forecast for the final price offer. Furthermore since $U_{N-k+1} > M_N$, the mean final forecast price level, and the decision price breaks are all less than or equal to M_N it follows that $P_k \leqslant U_{N-k+1}$. We can therefore apply the result just derived above. As shown P_2 is equal to M_N and hence $> L_N$. It follows immediately applying the above result that

$$P_3 > L_N$$

Applying the same result to P_3 it follows that P_4 and all subsequent decision price breaks are greater than L_N.

In the case of a negative trend all the decision price breaks have values lying between the final forecast mean price and the final forecast lower price limit.

Normal Distribution for Price Fluctuations

If the price distribution $\phi_k(q)$ is assumed to be a normal distribution then it is most reasonable to set the buyer's upper and lower limits at two standard deviations above and below the mean value. Thus for the jth buying opportunity forward in time:

$$\mu_j = \tfrac{1}{2}(U_j + L_j) = M_j$$
$$\sigma_j = \tfrac{1}{4}(U_j - L_j) = \tfrac{1}{4}R_j.$$

The upper and lower limits of the buyer's forecasts are now disregarded. Since the normal is a continuous distribution a small possibility of prices occurring outside the upper and lower limits has been introduced into the situation. Since in practice μ_j/σ_j is usually reasonably large the approximation

$$\int_0^x \phi_j(q)\,dq = \int_{-\infty}^x \phi_j(q)\,dq$$

can be used, even though it introduces negative prices. The likelihood of negative prices for the large values of μ_j/σ_j is infinitesimal.

$$\text{Let} \quad \psi(x) = \frac{\exp(-\tfrac{1}{2}x^2)}{\sqrt{2\pi}} \quad \text{and } \Phi(y) = \int_{-\infty}^{y} \psi(x)\,dx$$

$$\text{and let} \quad \eta = \frac{P_k - \mu_{N-k+1}}{\sigma_{N-k+1}}$$

Then after some algebraic manipulation the recurrence relation for the decision price breaks becomes

$$P_{k+1} = P_k - \sigma_{N-k+1}\,\psi(\eta) - (P_k - \mu_{N-k+1})\,\Phi(\eta)$$

Note that

$$\mu_{N-k+1} = Q^* + (N-k+1)m$$

$$\sigma_{N-k+1} = \frac{a}{2} + \frac{(A-a)(N-k+1)}{2N}$$

It is seen that the recurrence relations for the decision price breaks are different for the normal and rectangular distributions. So although they have the same starting value P_2 equal to M_N the decision price breaks soon diverge in value.

13

Practical Implementation of the Decision Price Break Policy. Case Study 2—Daily Copper Buying

13.1 THE PURCHASING PROBLEM

This case study describes the application of the decision price break model to the copper buying situation of Bowland Metal Products, (BMP), discussed earlier in Chapter 3. The name BMP is of course fictitious. However, this chapter describes a study of a real copper buying situation for a company. All the details given, including the results, are taken from the real situation. Only the company name has been changed. The study was carried out over 1968. It is described exactly as it was carried out, on a semihistorical basis, for the reasons given at the beginning of Chapter 12.

In many ways BMPs problem is the simplest commodity purchasing situation. The copper consumption requirements to be met are a regular 20 tons per day. Each day a price offer is made by the supplier at 2.00 p.m. BMP then has until 5.00 p.m. to decide whether to make a purchase at that price. BMPs purchasing director can buy as little or as much as he wishes. The supplier will then deliver the purchase order made by BMP in several small lots on consecutive days or in a single delivery, according to BMPs preference. Once the company has decided its strategic policy as to the maximum purchase and stock cover it will carry at any time, the situation is reduced to the tactical daily purchasing problem with no real difficulties in determining the scheduling of purchase order deliveries.

A major difference between this study and the cereals study was that the purchasing director of BMP had never made price forecasts in any form. Furthermore he stated that he did not feel able to do so in the future. It was therefore necessary from the very start of the study to derive and compare alternative price forecasting systems.

13.2 HISTORICAL SURVEY OF COPPER MARKETING AND PRICING POLICIES

During the early sixties there were two markets for copper, the major copper mining companies and the London Metal Exchange. The London Metal Exchange (LME) is the major world commodity market for copper, followed by the COMEX market in New York. However the major copper producers sold their refined and semirefined new copper directly to industrial consumers at a fixed price. This was adjusted on only rare occasions. Most U.K. consumers obtained about 75% of their requirements direct from the producers at the fixed price. The remainder was bought either directly through the LME free market or from suppliers on the basis of the LME prices. This copper was provided by smaller mining companies, usually in countries outside the major producers such as Chile, Zaire, Zambia and North America, or was reclaimed copper scrap. Reclaimed copper scrap is a very significant source of copper in the Industrialised countries. Much of it is provided by scrap dealers who have always used the LME prices as the basis of the prices at which they offer copper scrap each day to potential buyers.

Unlike the fixed producer price, prices on the LME could vary as supply and demand varied. Prices fluctuated by large amounts, both up and down. Due to increased industrial use, over 1965 and 1966 the LME prices reached very high levels, almost twice the producer prices. The governments of the producing countries where the mines were situated, were paid royalty payments by the mining companies as a percentage of the price at which the companies sold their copper. These governments became concerned that they were losing potential revenue because of the companies low fixed producer price relative to the LME prices. Economists argued that if all copper were sold through the LME, price fluctuations would be lessened and the LME prices would stabilise at a lower level. The mining companies expressed concern that higher copper prices would lead to substitution of copper in some of its end uses by aluminium or plastics. Nevertheless, under the various pressures exerted, the mining companies agreed to end their fixed producer price system. The producers would continue to sell their copper direct to the industrial consumers and would not change to selling through the LME. They would, however, use the LME prices as the pricing basis. As mentioned in the opening chapter, the London Metal Exchange deals in two different prices, for each form of copper traded. One is a spot or cash price for copper bought and sold for immediate delivery. The other is a 3 months future price. This is for copper to be delivered in 3 months' time, but charged at the fixed price today. The 3 months future price is a rolling future, since it is always for delivery 3 months after the day on which the future was bought and sold.

The LME has a 5 minute trading session for copper each morning. In an open public session, brokers acting on behalf of buyers and sellers shout to each other offering to buy or sell specified amounts of copper at particular prices. When another broker accepts the offer a deal is made. The prices reached at the end of the 5 minute trading session are recorded by the exchange and widely publicised immediately by telex, etc., as that morning's 'official' settlement prices.

Following the ending of their fixed price system, the copper producing companies agreed to price their copper on a daily basis using the LME morning settlement prices. They chose to price their copper as a weighted average of the morning settlement cash price (30%) and the morning settlement 3 months future price (70%) on the LME. The higher weight given to the future price was to moderate large fluctuations in the cash price caused by temporary shortages of immediately available copper or speculation, or so it was believed. Furthermore it reflected the lead time for delivery of copper from the mines in Africa and South America to Europe of near to 3 months. This change to the selling policy was made in April 1966. *A priori*, daily price movements on the LME after this change are likely to be different to those before the change, since they cover all copper rather than the marginal supplies.

BMPs supplier sold the copper wire that BMP used on the basis of the weighted average of the LME settlement prices plus a known manufacturing margin. His prices thus changed every day by the same amount as the LME prices. The problem of buying processed copper is thus effectively the same as trying to buy copper daily at the LME morning settlement prices. The LME also has a second trading session during the afternoon and brokers also deal direct with each other by telephone towards the end of the afternoon. These are 'unofficial' prices but they are still widely circulated to copper producers and users. The BMP supplier, as is the case for most suppliers of copper and copper products, did not change his price offer on a day if the afternoon price was different to the morning settlement prices. However as will be seen later, the existence of these afternoon LME prices could be used to improve the purchasing policies for BMP.

13.3 STATISTICAL ANALYSIS OF PAST COPPER PRICES

Hence unlike the cereals purchasing problem of the first half of Chapter 12, a continuous daily price stream exists for copper. It is therefore possible to carry out a statistical analysis of a long series of the past daily prices to determine which forecasting systems are likely to be appropriate. However because of changes to the copper marketing system this analysis had to be confined to a recent 2 year time series of copper prices.

To ensure that the price forecasting models derived would be appropriate to the new selling situation, the statistical analysis was confined to the daily prices over the period 28 April 1966 to 16 April 1968. The prices are shown in Fig. 13.1. The Box Jenkins method of time series analysis of the autocorrelation coefficients was used. The autocorrelation coefficients measure how closely values 1, 2, 3, ... days apart are related to each other. Naturally since the price changes over several days are usually small relative to the absolute levels of prices, the prices 1, 2, 3, ... days apart are usually closely connected in value so that the autocorrelation coefficients are high. Attention is usually concentrated on the first difference of prices, i.e. the price change from yesterday to today, and the second difference of prices, the difference between the price change from yesterday to today and the price change from the day before to yesterday. The results of the analysis showed that the autocorrelation coefficients for the first and second differences of prices were all small and insignificantly different from zero, except for the lag 1 coefficients. The lag 1 coefficient measures how closely the values only one day apart are related to each other. These results suggest that the underlying process generating prices and hence the model for forecasting prices was most likely to be an exponentially weighted average of the past prices only. There was some slight evidence that the exponentially weighted average model should also contain a trend component as well as a level component. These terms were defined in sections 12.4 and 12.5 of the cereals case study. In fact the analysis showed

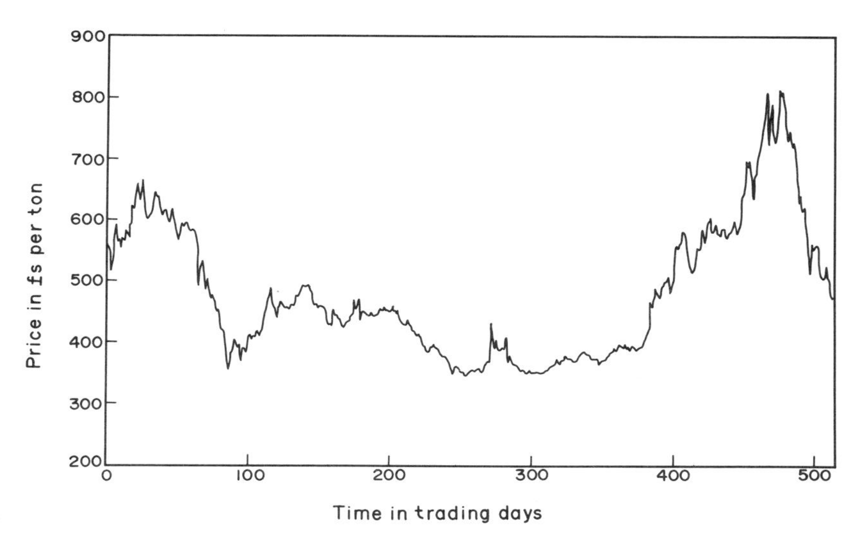

FIG. 13.1 Daily cash copper prices on the London Metal Exchange 28 April 1966 to 16 April 1968

that the most appropriate forecasting models for future prices would be the same linear exponential smoothing models as evaluated in the cereals case study. The analysis gave estimates for the smoothing constants on these models, α and β, as $\beta = 0.1$ and α in the range 0.75–0.95. However, when the price series was split into four parts, which by eye appeared to be different types of price movement, different best values of α and β were estimated for the four parts.

Thus unfortunately the statistical analysis did not give a clear cut result to identify one particular type of forecasting model as distinctly better than the others. The results furthermore suggested that the Box Jenkins model was likely to be rather insensitive to changes in the time series. It was therefore decided, as in the cereals buying study, to carry out an evaluation of the decision price break policy for the Type 1 tactical purchasing situation, i.e. for buying to meet a regular daily consumption, using several alternative statistical price forecasting systems.

Subsequent studies have shown that the best forecasting system for use in the purchasing policy can only be determined from a buying evaluation. The different effects on purchasing costs of over and underforecasting errors mean that the best forecasting system is not necessarily that which gives the lowest squared forecasting errors. Classical time series analysis of a long stream of prices should be used to give a general indication of the type of forecasting methods that should be evaluated. It cannot indicate which method will give the best inputs to the decision price break policy.

13.4 RESULTS OF THE INITIAL STUDIES

Initially a buying evaluation was performed using a simple linear exponential smoothing model for alternative maximum purchase and stock covers of 5, 10, 15, 20, 40, 50 and 100 days' consumption. This showed that as the maximum cover was increased the purchasing costs rapidly decreased, reaching their minimum value at 20 days, remained much the same up to a 40 day cover and then began to increase slowly. After 40 days the costs of the errors in this forecasting system due to the extreme fluctuations in the market, far outweigh the potential savings.

At this point BMP stated that they would only consider adopting a purchase and stock cover of 15 days' consumption as a maximum at any time. Even though larger values would have greater savings, they thought the 'risks' associated with larger maximum stock cover levels were too great. So most of the evaluations were run with a 15 day cover. However, a 40 day cover was evaluated for some models to see what savings the company was foregoing by their policy.

Note that in this copper buying situation the company could have generally bought a regular 20 tons of copper per day. They could thus have paid the average daily price. Any proposed policy should do better than this. Simple linear exponential smoothing models gave small savings of about £2–3 per ton, 0.4–0.6% of the average daily price. This was achieved when both the forecasting system and the decision price breaks were updated daily. If the decision price breaks were only updated weekly then losses on average of £2–3 per ton, relative to the back-to-back policy, were incurred. Apparently copper prices are much more volatile than either maize or wheat prices, where it was found that the decision price breaks only needed to be updated fortnightly.

Brown's quadratic exponential smoothing forecast system gave an average purchasing cost of only a few pence below the average price. With larger stock covers the quadratic model was distinctly inferior. This result was anticipated since the Box Jenkins model identification procedure showed that a quadratic model was over parameterised. A pure random normal walk model, where the forecast for all future prices is today's price, gave a loss of over £5 per ton on average. On analysing the results it was found that the main faults of the forecasting systems used so far were that they were either too sensitive to isolated price changes or not sensitive enough to real changes in trend and price level. With the linear smoothing systems, the Brown model, with a low value of α as used generally for sales forecasting, avoided overreacting to isolated price changes but missed out on adapting quickly to real changes in price level and trend, whilst the Box Jenkins model, with a high value of α, reacted quickly both to isolated price changes and to real changes. A system which attempts to react satisfactorily to both types of situations is the adaptive linear exponential smoothing model of Trigg and Leach, used with success in the cereals buying problem. With this system the parameters α and β are related to the smoothed cumulative errors. Thus if the smoothed cumulative errors become large α and β become large and the system reacts quickly to real changes in the price movements. If the errors are small, α and β are small since the system is forecasting adequately. Using this forecasting system gave a 50% improvement in savings over the best of the simple linear exponential smoothing models.

An important result with this adaptive smoothing forecasting model was that a 40 day cover was substantially better than a 15 day cover. This is very different to the earlier results with the other forecasting systems. The purchasing costs were still decreasing at a 40 day cover. It appeared that the purchasing costs began to increase slowly again as the maximum cover was increased beyond 60 days. However because of the company's attitude to risk no evaluations were performed for a maximum cover longer than 40 days. Their attitude was still that a 15 day cover was as much as they were prepared to accept.

13.5 FURTHER IMPROVEMENTS TO THE COMBINED PRICE FORECASTING/PURCHASING MODEL

13.5.1 *The Maximum Allowable Daily Trend*

It was found that with this adaptive smoothing forecasting system, the decision price break purchasing policy at times overreacted and behaved very much like a nervous buyer. It panicked as soon as a high upward trend occurred. Often over a period of 3 or 4 days a high upward trend of say £15 per ton per day might be maintained. The forecasting system is thus predicting that prices will rise by £225 per ton over the length of the 15 day maximum cover. Inevitably a very large purchase is made. Yet over the 15 days of the maximum cover the average upward trend might be only £5 per day, and for longer periods even smaller. The 3 or 4 days of increasing prices might only be a short temporary situation due to some rumour, found soon to be false. Prices could then fall to below the earlier prices. Any large purchase made when the high trend began would then be very costly. Over the first 3 of 4 days of the development of the high trend, the variance (the up and down fluctuations) of the daily prices and the price changes is also increasing. As shown in Chapter 11 the decision price breaks depend on the variance of future prices as well as the mean level and trend. A high variance with only a moderate daily price trend would probably indicate that no purchase should be made, at least until the situation is clarified. One way to achieve this is to restrict the forecast daily trend to be used in calculating the decision price breaks to be less than some fixed maximum value. The forecasts of the adaptive smoothing model are updated in the usual manner. The values of the price levels and daily trend, etc., are then used to generate the probability distributions for the prices on successive future days required in the formulae to calculate the decision price breaks. However, if the forecast daily trend is larger than the chosen maximum value, the forecast daily trend is reduced to this maximum level for estimating these probability distributions for future prices. The actual value of the forecast daily trend is, however, used in the updating of the forecasting systems. Different levels for these maximum values were evaluated. The results for the adaptive smoothing forecasting system for a 40 day maximum cover are shown in Fig. 13.2. It was found that the best value was approximately equal to the largest trend sustained over any period equal to the maximum cover, that occurred over the 2 years of price data available. Different maximum levels will usually occur for different maximum cover values. This procedure gave further savings in purchasing cost over those achieved with the adaptive smoothing forecasting model unmodified. In fact this procedure improved the savings in purchasing costs by around 50% for all the various forecasting systems evaluated earlier. However, since the adaptive exponential smoothing

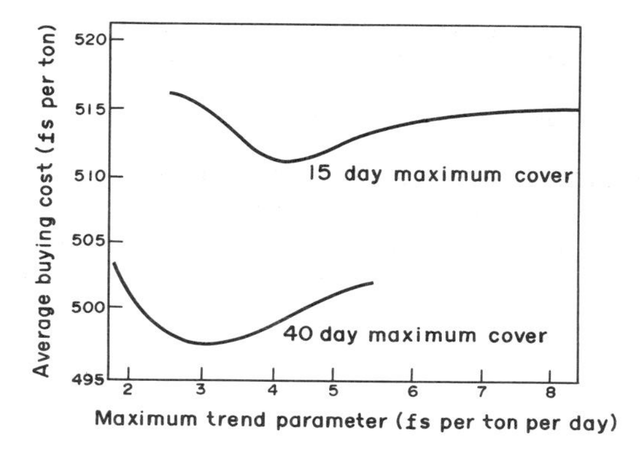

FIG. 13.2 Effect of varying maximum allowable daily trend on average purchasing costs

model was apparently so superior to all other forecasting systems all subsequent work was based on this model.

13.5.2 *The Smoothing Parameter, γ, of the Adaptive Smoothing System*

An important parameter in this adaptive forecasting model is γ, which is used to smooth the forecasting errors and the mean absolute deviation (see section 12.5). It is also important in updating the estimate of the variance of the one step ahead forecasts. Varying the value of γ varies the amount by which large errors are smoothed out. If γ is small, isolated errors are effectively damped out and hence oscillations in the forecasts are reduced, but at the same time real changes in the trend are also damped. Real changes must be present for some time before they are detected and the forecasts changed. If γ is large real trends are detected more quickly but isolated errors are not damped sufficiently. The parameter γ thus controls the rate at which the forecasts are adapted to the recent price changes. The right value for γ must be a compromise between these two extremes so as to give the lowest purchasing costs over a long period of time.

The effect on the average purchasing cost of varying γ over the range 0.05 to 0.25 for a typical 40 day cover model is shown in Fig. 13.3. A 15 day cover model gives the same type of results. The optimal value occurs at 0.2. But because there is a sharp increase in purchasing cost if γ increases to 0.25 it is probably better to give γ a value somewhere between 0.15 and 0.20.

The parameters α and β determine the amount of the one step ahead forecast error that is absorbed into updating the average price level and the trend of price respectively. In the original model of Trigg and Leach both α

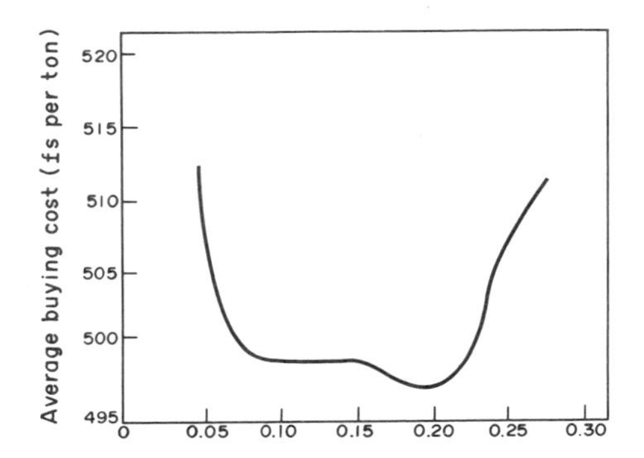

FIG. 13.3 Effect of varying the smoothing parameter, γ, on the average purchasing cost for a 40 day cover

and β are continually updated. For this copper price series, the value of α was generally in the range 0.55 to 0.9 and β in the range 0.2 to 0.3. However, if α was kept at a constant value of 0.7 the increase in purchasing cost was quite small. This point played an important role at the implementation stage.

Control limits have been devised for the ratio of the smoothed mean error to the absolute deviation as a function of γ. Provided the ratio is within $\pm$ 0.60 for $\gamma = 0.1$ and $\pm$ 0.80 for $\gamma = 0.2$ the forecast system is under control. If there is an isolated large one step ahead error caused by an unusually large increase or decrease in the daily prices, then the smoothed error increases relatively more than the smoothed absolute deviation, so that the ratio may go outside the control limits. This indicates that the past price was unusually high or low. If it is abnormally low this suggests a good buying opportunity. On the other hand it may indicate that the market is about to enter a period of falling prices. If the forecasting system is automatically updated, with no manual intervention, there is a danger of the purchasing system overreacting to these isolated price changes. In the practical operation of the system a good purchasing manager might be able to discriminate which of these isolated prices are merely temporary and which indicate a sudden fall or rise in the market. The temporary prices can be assessed as good or bad buy points by the purchasing manager and then left out of the updating of the forecasts. This procedure has not been included in any of the evaluations since inevitably any such assessment would use knowledge of the price development after this point. However, it will be important to include this procedure in practice. Whether a purchasing manager is generally good at assessing these isolated prices as good or bad buying points can only be determined by evaluating his performance in a real time situation. It is impossible to make any meaningful judgements from an experiment on the last price series.

It is seen from Figs. 13.2 and 13.3 that, because of the way the purchasing costs change with increasing values of the maximum trend parameter and γ, it may be more satisfactory to choose a value for each different from the optimum. Similar deviations from the optimum point above and below do not give the same increase in average purchasing costs.

Each of them has been discussed separately. It is of course the optimal set for each maximum cover that is important. The optimal set may be different for different planning horizons. Thus for each maximum cover value an evaluation over the two dimensional space of γ and the maximum daily trend parameter is required.

13.6 PERFORMANCE OF THE MODIFIED MODEL—APRIL 1966 to APRIL 1968

The prices used in the evaluations are the company buying prices, the weighted average of the cash and 3 months future prices on the LME. The pound sterling was devalued by 14% against all other currencies in November 1967. Prices immediately rose by over £50 on the day following the announcement of the devaluation. A bad purchasing policy over most of the time, which by chance gave a large purchase just prior to devaluation, would appear to give an overall reasonable performance. To avoid such problems all prices before devaluation were increased by 14% to give a consistent price series over the 2-year period, when used in the evaluations.

The average daily price over the period was £517.2 per ton. The average purchasing cost for the decision price break model with the adaptive forecasting system would have been £511.5 per ton for a 15 day maximum cover and £498.8 per ton for a 40 day maximum cover. The actual purchases made for the 15-day cover, in terms of the number of days' consumption, are shown in Fig. 13.4. A more detailed analysis of the results of the policy with a 40 day cover is given in Table 13.1. This shows for approximately 2 monthly periods the average price paid for material consumed in that period. It is assumed that material is consumed on a first bought first used basis. Hence part of the material used in a period may have been bought in the previous period. The price paid for the material used in a period is compared with the average market price over that period. It is seen that, although the overall average saving is £18.4 per ton, there is a wide variation in the purchasing performance between the individual periods. The maximum saving during a 2 monthly period was £88 per ton (18%) but in two periods there were large losses of £30 and £23 per ton (6%). These losses occurred at times when the 'bottom dropped

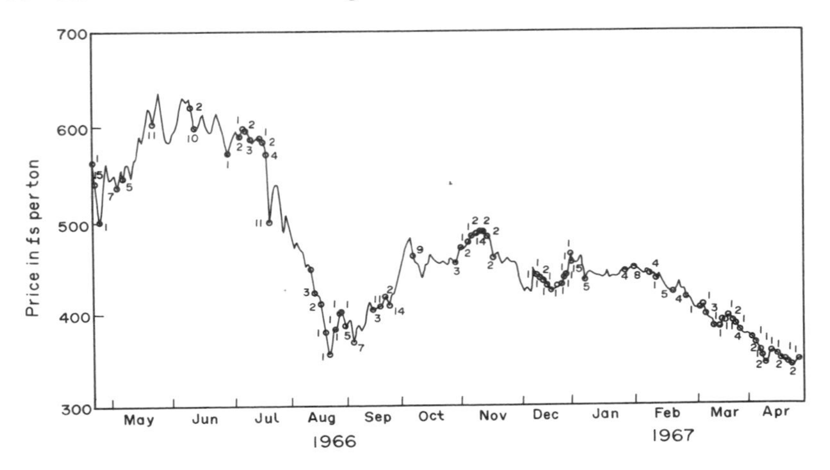

FIG. 13.4 Purchases made (in day's requirements) with a 15 day maximum cover using adaptive smoothing price forecasting model

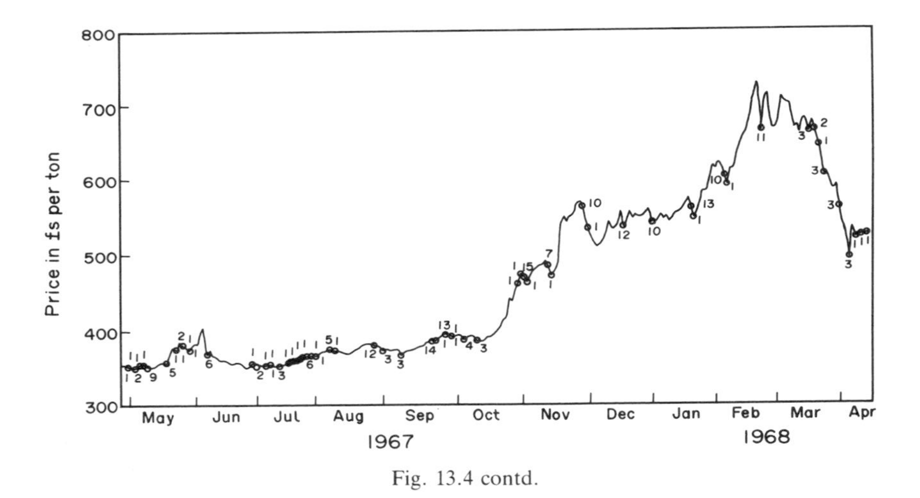

Fig. 13.4 contd.

out of the market' unexpectedly, shortly after a large purchase was made because the forecasting model was predicting a continual large increase in price.

Any purchasing policy that attempts to reduce costs will inevitably give such variations in performance over time as shown in Table 13.1. It is not possible to win all of the time, particularly in volatile markets such as copper. Some losses over short time periods are inevitable. Over the long run, these are more than counterbalanced by the more frequent savings.

TABLE 13.1 Purchasing performance of a 40 day maximum cover model in 2 monthly periods

Time period	Days	Average market price over time period in £s per ton	Average buying price of units consumed in th time period in £s per ton	Average saving in £s per ton
250466–310566	26	644.9	612.5	32.4
310566–250766	39	678.2	634.6	43.6
250766–60966	30	501.5	531.4	−29.9
60966 –61066	22	481.9	446.2	35.7
61066 –101166	25	529.5	465.4	64.1
101166–30167	34	508.8	532.4	−23.6
30167 –280267	39	500.9	505.4	− 4.5
280267–80567	47	423.9	426.7	−2.8
80567 –270767	57	413.0	408.0	5.0
270767–270967	43	426.9	412.1	14.8
270967–301167	46	502.3	452.4	49.9
301167–240168	36	544.3	541.9	2.4
240168–230268	22	630.6	542.2	88.4
230268–160468	34	622.7	594.1	28.6
Overall performance				
250466–160468	500	517.2	498.8	18.4

N.B. The prices shown in the table are the adjusted values as used in the evaluations.

The lowest cost that could have been achieved, in hindsight, was £490 per ton for the 15 day maximum cover and £462 for the 40 day cover. The purchasing efficiencies were thus 21% and 33% for the 15 and 40 day covers respectively. These are not as good as those achieved for maize and wheat. The copper market is much more volatile than wheat and maize. Sharp reversals in trend occur more frequently in copper prices. The assumptions used in deriving the nonstationary decision price break model, see Chapter 11, section 11.11, are thus more often broken for copper prices. Nevertheless the example shows that substantial savings can be achieved with a systematic purchasing policy that uses only past prices as input.

The company's actual purchases were available for most of the period. Their actual average purchase cost was £9 per ton above the average daily price. It was noticed that the buyer seemed to operate often with a constraint of a maximum inventory of 20 days' consumption rather than the 15 days that BMP had specified as their strategic policy. Indeed on one occasion a purchase of 27 days' supply was made. A simulation of the decision price break policy with a 20 day cover was carried out for the same time period. This gave a saving of £7.5 per ton below the average daily price offer. Thus if this policy had been used over the preceding 2 years the company would have saved over 3% on their purchasing costs, and 5% if a change to a 40 day cover had been made. This would have increased company profits by at least one third.

On comparing the decision price break policies with the company's actual performance, it was apparent that they gave large savings in periods of long upward trends and long downward trends. It is particularly on the long downward trends that the company buyer made large losses. In no month did the buyer do better than the decision price break policy.

13.7 OTHER FACTORS AFFECTING THE MODEL

13.7.1 *Use of the LME Afternoon Closing Prices*

As mentioned at the beginning of this chapter in section 13.2, the London Metal Exchange has both a morning and an afternoon market. Bowland Metal Products are also in a special situation, which does not apply to all raw copper consumers, in that their suppliers allow them until about 5 o'clock in the afternoon to make a decision on how much to buy at the morning settlement price. This means that the company is able to obtain the afternoon closing price before they take their decision on purchasing at the day's morning settlement (i.e. closing) price. Thus they have further information which has not been taken into account in the models described previously.

One way of using this information is to include the afternoon prices as well as the morning prices in updating the price forecasting system. An alternative approach is to see whether the afternoon price can help to give better predictions of tomorrow's morning price. It was found that on 549 out of the 712 days up to August 1968, the direction of the morning to afternoon price change was the same as the change from the morning price to the following morning price. Thus if the LME afternoon price is down on the morning settlement price the supplier's next price offer, tomorrow, is likely to be lower than today's since both are based only on the LME morning settlement prices and tomorrow's settlement price is likely to be lower than today's settlement price. Thus any purchase recommended for today's price offer should be delayed, if the LME afternoon closing price is less than the morning settlement price since a lower price can be expected from the supplier tomorrow if the buyer waits. A purchase should only be made if stocks of copper have fallen to the minimum safety levels. Further evaluations showed that this delay mechanism could give further reductions in purchasing cost as will be seen later.

13.7.2 *The Distribution of the Price Forecasting Errors*

In the adaptive exponential smoothing forecasting system, the standard deviation of the one step ahead forecasting error, σ, is

$$\sigma = 1.25 \times \text{Smoothed Mean Absolute Deviation}$$

The standard deviation of the forecasting errors L steps ahead is then σ_L, where

$$\sigma_L = \sigma \{1 + \alpha^2(L-1)\}^{\frac{1}{2}}$$

The future prices L steps ahead, are then assumed to be normally distributed about the mean forecast with a standard deviation σ_L.

The twenty step ahead forecast errors were computed over the period April 1966 to December 1968. The assumption that future prices were normally distributed about the mean forecast with a standard deviation, σ_L, was found to be invalid. The distribution of the errors had extremely long tails, The range of the distribution of these errors was approximately +£120 to −£120, with the majority of the errors in a central region within the range ± £30. This central region was approximately normal though there were far more observations in the tails than a normal assumption would imply. When the twenty ahead step errors were measured as the number of standard deviations, σ_L, about the mean forecast the vast majority lay within the range ± 1, far more than the two-thirds implied by the normal assumption. This indicates that σ_L is too large an estimate of the standard deviation. This error distribution also suggests that the forecasts are either very good or very bad. When the forecasts of trend are good, prices appear to be concentrated in a narrower range about the trend than the forecasting model as outlined above assumes. If σ_L overestimates the price variation about the 'accurate' trend, then the purchases will be put off until later. In the case of an increasing trend a much higher price than necessary will have been paid. If the price trend is downwards no purchase will be made until a substantial upward recovery has taken place.

This suggests that a revised formula for σ_L might be more appropriate. One obvious revision is to estimate the standard deviation of the one step ahead error, σ, by the formula

$$\sigma = \lambda \times \text{Smoothed Mean Absolute Deviation}$$

where $\lambda = 0.5$ or 0.75 rather than 1.25 as in the original model. When the forecasts are very bad the smoothed mean absolute deviation will be large.

In such cases $\lambda = 0.5$ rather than 1.25 will not affect σ and σ_L sufficiently to change the purchasing decisions.

13.8 FINAL FORM OF THE PURCHASING MODEL

The various factors discussed above were incorporated into the models. These modifications to the model, particularly when the afternoon prices were included, implied that a re-estimation of the optimal set of all the model parameters was required.

The long upward price movement over the second half of 1967 (see Fig. 13.1) was due to a 9 month strike of U.S. copper miners. Cash copper prices climbed to over £800 per ton, a record level for postwar copper, although some of this very high price was partially due to the devaluation of the pound sterling. The premium of the cash price over the 3 months future price, known as the backwardation, widened to over £100 per ton. Basically over the whole period of the U.S. copper miners' strike the market was continually expecting the strike to be settled in weeks rather than months. It was believed that prices would then fall rapidly. The producing countries again became concerned that they were losing potential extra royalty payments. They persuaded the mining companies to change their selling policy once again. Thereafter copper was to be sold on the basis of the daily morning cash settlement price alone. The change took place in January 1968. BMPs supplier also adjusted his pricing policy to this new basis so that from 12 January BMP bought their copper from him on this new basis.

The evaluations to give the final model were performed on purchasing copper at the London Metal Exchange morning cash price, for consumption at a regular rate over the period April 1966 to December 1968. Three basic purchasing models were evaluated for the different ways the afternoon prices can be incorporated into the purchasing models. The purchasing results for each of these three models are given in Table 13.2. For each model type two sets of parameters are illustrated, the optimal set and the original set used for the modified model discussed in section 13.5.

The Actual models are as follows:

Model (1a) This is the basic model used earlier for the evaluations of section 13.6 on the company buying price. The forecasting system uses only the morning cash price. The maximum trend parameter is taken as £3 per day whilst the parameter, λ, as described above, takes the value 1.25 as in the original model.

Model (1b) This is the same as Model (1a) but with a maximum trend parameter of £2 per day and a value for λ of 0.5. An additional constraint is

TABLE 13.2 Average unit purchasing costs for different policies with different maximum cover values

Model Type	Maximum Purchase and Stock Cover in Days			
	10	20	30	40
(1a)	476.7	473.7	470.2	468.3
(1b)	475.1	469.3	465.8	463.2
(2a)	473.0	469.1	464.7	459.5
(2b)	470.8	464.9	460.2	456.7
(3a)	476.9	474.3	474.2	475.5
(3b)	476.8	473.3	472.9	474.9

imposed that, if the forecast daily trend is greater than twice the maximum trend parameter, a purchase should be made immediately to raise the stock to the maximum purchase and stock cover level.

Model (2a) This model makes use of the morning and afternoon prices as a delay mechanism. If the afternoon closing price is less than or equal to the morning settlement price then no purchase should be made. If the afternoon price is greater than the morning price then Model (1a) is used to determine the appropriate purchasing decision.

Model (2b) This is the same as Model (2a) but with a maximum trend parameter of £5 per day and a value for λ of 0.5.

Model (3a) In this model both the morning and afternoon prices are used in the forecasting system. Buying decisions are made only at the morning price, however. The maximum trend parameter is taken as £3 per half day, whilst λ has a value of 1.25.

Model (3b) This is the same model as (3a) but with a maximum trend parameter of £1 per half day and with $\lambda = 1.0$.

The mean price offer for copper consumed over the period April 1966 to December 1968 was £479 per ton. One immediate result from the values in the table is the poor performance for all values of the maximum cover of models (3a) and (3b). These models use both the morning and afternoon prices directly in the forecasting system. Apparently including the afternoon prices directly in the updating of the forecasting systems causes undue variations in the forecasting parameters. Although the direction of the change from the morning to the afternoon prices are usually the same as the change from the morning to the following morning price, the size of the changes from the morning to afternoon are usually greater in magnitude than the change from the morning price to the following morning price. Including the afternoon price in the forecasting system will thus generally well over or well underforecast tomorrow's morning price.

For both models 1 and 2 the purchasing costs are still decreasing at a 40 day cover. Eventually an 'optimal' cover value will be reached, after which the purchasing costs begin to increase again. For a 40 day cover, model (1b) shows a saving below the average price offer of about 3% and model (2b) a saving of about 5%. Thus substantial reductions in purchasing costs could have been achieved if these decision price break models had been used over the period indicated, particularly when it is recalled that the buyer had been paying more than the average daily price offer over the earlier part of the period.

Models (1b) and (2b) are for two different purchasing situations. In order to make use of model (2b) a purchaser must be able to obtain the afternoon price before having to make a decision on what to buy at the prior morning price. The buyer of Bowland Metal Products Ltd. was in this situation. This will not be so for many copper purchasers. Such purchasers will have to use model (1b) or attempt to negotiate with their suppliers to increase the time limitation on their decision making at each price offer.

13.9 PRACTICAL IMPLEMENTATION

By December 1968 the company was satisfied that the decision price break purchasing system would be of value to them. They were keen to implement the system as an online test in direct comparison with the buyer as soon as possible. The policy was written as a set of computer programs, which took only a few seconds to give the decision required at each buying opportunity. Bowland Metal Products Ltd. did not have a computer. They could of course hire a terminal from a computer bureau. The company was loath to do this. It was also felt by the study team that there were many advantages in having a manually-operated system that could be run by the company buyer himself. This would enhance the likelihood of the buyer adopting the system and overcome his suspicions of being 'dictated to' by an impersonal machine.

The daily calculation procedure required by the purchasing system falls into two parts, updating the forecasting system and calculating the decision price breaks and working out the stock level to achieve after a purchase at today's price offer. The updating of the forecasting system is relatively simple and takes less than 5 minutes with the assistance of any small calculating machine or a slide rule. The main computational effort is in estimating the price breaks giving the purchasing policy for the current forecasting values, level, trend and error variance. The buying decision is then taken by comparing today's price offer with the price breaks to determine how many days' requirements should be covered at today's price offer. It was soon clear that manual calculation of the decision price breaks was not a feasible proposition. Experiments showed that, even with experience, over 2 hours would be required for calculating twelve decision price breaks or more. With such lengthy and complex calculations mistakes are likely to be made. The significance of the two hours is that this is the effective time allowance that the supplier gives the buyer to telephone back with an order at the price offered before the 5 o'clock deadline, allowing for the time to obtain the price offer, consider the recommendations of the policy and contact the supplier to place a purchase order.

One way to overcome this problem is to derive a set of tables giving the values of the price breaks of the theoretical purchasing policy for all conceivable parameter values. This will overcome the decision time problem and provide an accurate enough policy to obtain the potential savings in purchasing costs.

For a linear statistical forecasting system the decision price breaks, the solutions to equations 11.11 of Chapter 11, will be the initial price level plus some complex function of the trend value, the standard deviation of the one step ahead forecasting errors and the smoothing parameter α. Thus the tables can be derived in terms of an arbitrary initial price level and easily modified to take account of each day's estimate of the initial price level. However the tables will still need to be defined in terms of the values of three parameters. Clearly the size of the tables would be large. The complex interpolation required to find intermediate values would not be conducive to easy usage.

The first two parameters, the trend and standard deviation of the forecasting errors, both vary widely. The parameter α generally varied in the range 0.55 to 0.9. Obviously one simplification is to make one of the parameters constant so that the set of tables required is only two dimensional. With the above results an obvious choice is to make α constant. It was found that compared to an adaptive α, giving α a fixed constant value of 0.7 showed an increase of only £1 per ton in the average purchasing costs. Thus the loss in using a constant $\alpha = 0.7$ is quite small and is acceptable because of the enormous simplification in the form of the purchasing policy for practical implementation.

The decision price breaks are now specified in terms of only two parameters, the forecast trend and variation. It was found that a step length of 0.5 in the trend and 1.0 in the standard deviation was adequate. The price breaks associated with intermediate values of these forecasting parameters could easily be derived from the tables using linear interpolation. The errors in this procedure were less than £0.1. Since the price offers are always quoted to the nearest £0.5 this size of error is irrelevant.

Thus a simple practical system for implementing the theoretical purchasing policies has been derived which enables almost all the potential savings in purchasing costs to be realised. This practical system takes at most 10 minutes' clerical work a day and can be easily operated by the buyer himself. The procedure also included the calculation of the ratio of the smoothed mean errors to the mean absolute deviation. The times when the forecasts are out of control can thus be indicated. These are the occasions when the buyer himself should intervene to override the automatic updating of the forecasts and the times when he should treat the buying recommendations with care.

The development of relatively inexpensive mini and microcomputers since this copper study was carried out has transformed the implementation situation. The daily calculation of the updated forecasting parameters and the decision price breaks can be performed easily on such machines. In view of the size of the potential savings and the convenience of deriving each day's buying recommendations, the expenditure of several hundred pounds on buying a microcomputer on which to update the forecast system and calculate the price breaks is well worthwhile.

13.10 THE SUBSEQUENT PERFORMANCE OF THE MODEL IN PRACTICE

Although BMPs purchasing director accepted that there was some value in the decision price break policies, he was not entirely convinced. The company therefore decided to carry out a simulation of the policy over the next 6 months, side by side with the buyer each day, as the price offers arrived. This simulation was performed by the company themselves, not the study team. The policy simulated was for the 20 day maximum cover.

An internal company report on the results of this exercise is given in its entirety in an appendix to this chapter. This report was produced by the Management Services department of the company who carried out the simulation exercise for the purchasing director. The simulation was performed in close contact with the copper buyer and the purchasing director, who saw the model results on most days. The exercise showed that the decision price break policy produced an average saving of £7.4 per ton compared to daily buying of the copper consumed each day over the period, see the appendix. This amounted to £15,000. The model ended the period with a high stock of copper at a time of rapidly increasing prices. This copper would have been used over the next 8–10 days. Taking account of this a cost saving of £20,000 would have been made relative to a daily buying hand-to-mouth policy over the 6 month period.

Following the successful 6 months exercise the recommendations of the combined adaptive forecasting system and decision price break policy were now used as an essential input of information to the copper buyer. He used this together with other market information and his own experience to make his daily purchasing decisions. The policy gave him guidance rather than instructions. He did not always accept the recommendation of the policy. The daily simulation was continued within the company. In addition the performance of a 40 day cover model was simulated to examine whether increasing the constraints on the maximum stock cover would be desirable. The results over the next 12 months to the 29 May 1970 are shown in Table 13.3. This is again taken from an internal company report. This shows that the good results suggested by the study have continued. The 20 day cover model gives a better performance than was actually achieved by the buyer, so that the buyer is over-ruling many of the

TABLE 13.3 Copper buying performance for the model and the buyer
2 June 1969–29 May 1970

Average Prices Paid per tonne*

	Copper Consumed (4395 tonnes)	Copper Purchased	
		tonnes	at Price
Lancaster 40 day Model	£657.6	4485	£656.7
Lancaster 20 day Model	£664.8	4485	£663.7
Buyer	£669.4	4436	£669.1
Daily Buying	£675.5	4395	£675.5

Comparative Purchasing Performances

			Savings in £s	
			Total	Per tonne
40 Day Cover Model	versus	Day-to-Day Buying	78,670	17.9
" " " "	"	Company Buyer	51,861	11.8
" " " "	"	20 Day Cover	31,644	7.2
20 Day Cover Model	versus	Day-to-Day Buying	47,026	10.7
" " " "	"	Company Buyer	20,217	4.6
Company Buyer	versus	Day-to-Day Buying	26,809	6.1

*During this time period the London Metal Exchange changed from a long ton to a metric ton (tonne) basis. All purchases and usages over the period were converted to a common basis of tonnes.

1 long ton = 1.016 tonnes.

model's recommendations. The buyer on average has saved about 1% on the average daily price offer during this 12 month period. Since over the previous 3 years he was generally making losses compared to the average daily price it is reasonable to conclude that the daily recommendations of the decision price break model have helped him to improve his purchasing performance.

As part of his research Taylor simulated the effect of model 1a of Table 13.2 over the 6 year period from 2 January 1969 to 12 December 1975. As discussed in Chapter 2 this covers the period of traumatic changes in commodity markets of 1973/4 and a rise in copper prices to over £1300 per ton. The average daily price of copper over the period were shown in Fig. 2.5. The 20 day and 40 day cover models gave average savings of £4 per ton (0.7%) and £9.5 per ton (1.6%) respectively. This is smaller than the savings over the earlier period of the study, 1966–68. However it is a very good performance and shows the robustness of the models. The purchasing efficiencies are also quite small at 13% and 21% for the 20 and 40 day cover respectively. In view of the extreme volatility of copper prices at times over the late sixties and seventies low efficiencies are not surprising. It is pleasing to record that a model devised to fit the circumstances of the mid sixties would have continued to give good purchasing policies for the early and mid seventies also.

Appendix
Internal Company Report on Online Buying Simulation

Notes on Copper Buying Performance
(Made by a member of the company's management team)

1. Tables 1 to 3 show the performance of the Lancaster Model (the decision price break policy), the Buying Office and a daily buying strategy from the introduction of the model on 4.12.68 up to and including 5.6.69. Table 1 shows it in terms of average price per ton and Table 2 gives total values. The simulations of the model and daily buying, and the estimate of

Copper Buying Statistics at 5 June 1969

TABLE 1. Averages per ton

	Copper Consumed 2027 tons	Copper Purchased tons	at
Lancaster Model	£547.3	2190	£551.6
Buying Office	£547.7	2005	£541.6
Daily Buying	£554.8	2027	£554.8

TABLE 2. Total Values

	Copper Consumed 2027 tons	Copper Purchased tons	for
Lancaster Model	£1,109,413	2190	£1,207,539
Buying Office	£1,110,131	2005	£1,097,925
Daily Buying	£1,124,491	2027	£1,124,491

TABLE 3. Saving over Daily Buying

	£	£ per ton
Lancaster Model	£15,078	£7.4
Buying Office	£14,360	£7.1

savings derived from them are subject to errors due to the fact that they cannot exactly mirror the pattern of actual copper consumption. A reasonable estimate of this error would be ± £250, or ± £0.125 on the average prices per ton. On top of this all averages are rounded off to the nearest £0.1.

On assessing performance one must look at the position from both the point of view of copper consumed and of copper purchased. Any difference between the two is a stock which has a profit or loss potential depending on whether future prices rise or fall.*

2. Table 3 shows the Lancaster Model has saved marginally more than the Buying Office, although this difference of £718 is not very significant in view of the limits of error stated above, and both have made considerable gains over daily buying.
3. It would be unfair to the Lancaster Model if one were not to point out that the performance of the Buying Office has been enhanced by two factors:
(a) With such a short period under review any large fortuitous purchase, or one based on good local knowledge will have a large influence on the performance figures. Such a purchase was made on 30.12.68 with the prospect of an 'eleventh hour', but certain settlement in the Chilean Miners' Strike. The effect of removing the influence of such a purchase is seen in Tables 4 and 5 which give performances since 23.1.69, when the effect of the purchase of 30.12.68 would have ended.
(b) Even though it has not been regularly heeded, the Buying Office has had the benefit of the advice of the model. Unfortunately it is not possible to quantify the effect of this on their performance.
4. The evidence indicates that a blend of traditional buying practice and the mathematical forecasting technique will bring about the greatest savings. This strategy would rely most of the time on the recommendations

TABLE 4. Averages per ton

	Copper Consumed	Copper Purchased	
	1557 tons	tons	at
Lancaster Model	£556.8	1779	£562.4
Buyer	£561.1	1565	£561.3
Daily Buying	£566.7	1557	£566.7

*The extra stock of the Lancaster model in fact gave a large profit in this case. It would have cost an extra £102,000 if it had been purchased daily over the next few days. The average daily price would be increased by almost £6 above the value shown in Table 1.

TABLE 5. Savings over Daily Buying

	£	£ per ton
Model	£15,746	£9.9
Buyer	£ 8,998	£5.6

of the model, followed much more faithfully and with much more commitment than at present. The Buyer would only intervene if he had *certain* information on the future of the market, not merely hunches or feelings. It is true to say that in most ordinary circumstances (and that includes most of the experience of the past 6 months) the model is as good a predictor of the future of the market as any human being and a wiser buyer, due to its ability to systematically appraise each buying opportunity in an objective manner.

14

Organising a Study of Raw Material Purchasing

14.1 PLANNING A PURCHASING STUDY

The question naturally arises both for the purchasing manager and the operational research practitioner as to how a study of the company's raw material purchasing operations should be organised and what areas should first be tackled. The major condition for success to be achieved is that any study must be a joint co-operative venture by the buyers and the OR practitioners together. The area is full of pitfalls and temptations to oversimplify situations. Thus success without the buyers playing an active role in the study is highly unlikely. This means that the potential value of changes to existing policies has to be demonstrated and confidence established that the OR people have the knowledge and competence, with the co-operation of the buyers, to bring any studies to a successful conclusion. This is best achieved via a series of feasibility studies which will achieve both of the above objectives.

It is clearly highly desirable for the operational researcher new to the area to learn something of commodity markets and how they operate before beginning serious discussions with the purchasing manager and the buyers about the particular problems of interest. There is a new 'language' for him to learn. It is obviously somewhat of a drawback if the members of the OR study team have to keep asking the buyers to stop and explain new terms at frequent intervals. Some of the commodity exchanges and several of the larger brokers, particularly those that are American owned, have produced pamphlets and booklets introducing commodity exchanges and commodity trading methods. There is a useful book edited by Granger (1974), *Trading in Commodities*, which provides a good introduction to many of the terms commonly used as well as the basic operation of commodity markets. Rees (1972) has written a detailed well researched book giving the history and development of the U.K. commodity exchanges, which is well worth dipping into for the particular raw material of interest. The *Commodity Yearbook* published annually by the Commodity Research Bureau Inc., of New York has over the years contained several good articles on various aspects of commodity market operations. A visit to one of the major brokers of the material of interest and to the exchange is also another way to help gain a good introduction to the subject.

The publications by brokers and the Granger book give much emphasis to the trading of commodities, both buying and selling of futures contracts to make a profit. The brokers make their money from a commission on the business they transact and naturally they try to foster trading on the exchange to improve the turnover and hence their commissions. Many U.K. companies do not involve themselves in trading on the exchanges, rather suspiciously regarding it as speculation with all the pejorative connotations of that word. The first time reader should pick up the market terminology and market background situation and take care to avoid a conversion at this early stage to the notion that successful improvements to the purchasing policies must include a substantial amount of trading on the exchange.

Given this introductory background the next stage is for the study team to thoroughly familiarise itself with the current purchasing operations and procedures. As well as numerous discussions it is useful for a member of the team to sit in the buying office for a couple of weeks or so observing what is going on. A joint report should then be produced describing the planning and operational procedures currently being used. This report should attempt to structure the current system in the form of the various stages of the purchasing decision specified in Chapter 4 of this book. The manner in which the purchasing function has set itself up to cope with each of the problems identified at each stage should be discussed. The report should consider in some detail what alternative courses of action could be taken, compared to the current system, for each aspect of the problem, with particular reference to the deadlines required for deliveries prior to usage, the deadlines for purchasing particular delivery orders, etc. The report should also include a discussion on the areas where the buyers feel some more formalisation or quantitative assessment would be helpful to them. A further important aspect to cover is to examine what control procedures are in routine use, whether what actually happens in practice is routinely compared with the initial plans and whether updating of decisions is on any formal basis or is merely *ad hoc*. This first report is essentially a background paper describing the purchasing context and situation and laying the foundations for investigating which areas seem most worthy of detailed investigation.

Given this understanding of the current way the purchasing function is organised it is now possible to carry out feasibility studies of the various problem areas so as to assess more quantitatively the potential for improvements in purchasing performance. Unless there are very strong views expressed at the previous stage in producing the initial report the feasibility studies should encompass all the aspects of the problem identified in Chapter 4. To some extent the feasibility studies can be regarded as performing an audit of the current purchasing performance. The following paragraphs indicate the general lines upon which the feasibility studies should proceed.

If the company is buying one single material which, however, has alternative grades available, then a study should attempt to estimate the range of the price variations between the grades that has occurred in the past, the relative availabilities of the different grades and then compare the market price differences with the likely production cost and yield differences if different grades are used by the company. From this an estimate of the relative number of occasions when varying the grades purchased would be beneficial and a rough estimate of the annual expected cost savings produced. If it is a blending problem situation, of buying competing partially substitutable materials, then developing some type of formal linear programming model (see Chapter 6) is a major priority. In most U.K. situations this will already have been done. To assess how well the system being used is working the average prices paid for each unit of each material in each usage period over the past year should be calculated. The various linear programs for each usage period should then be rerun with these calculated average prices to determine what amounts of each material should have been used. The resulting total costs of buying all the material requirements for each usage period can then be compared with the actual total costs that were incurred. If there are significant differences then this implies that problems exist in predicting the future prices likely to be paid and that there is a need to introduce better and more regular feedback between the price changes over time and the determination of amount of each material to be bought for each usage period.

A feasibility study of the planning and scheduling of future orders should concentrate on the level of stocks carried over time and the economies of scale that may exist for differing delivery quantities. The deliveries of material made over the last 2 years should be compared with the way the material was actually used in production. From this the level of raw material stocks over time can be calculated. This can be converted into a stockholding cost either on the basis of a simple capital interest rate charge or by using whatever is the general inventory holding charge rate used elsewhere in the company. At this initial stage there is little need to undertake a detailed study into determining the inventory holding charge exactly. The stock profile over time can then be examined to see if excessive stockholding appears to be taking place. A rough estimate of the likely savings in this area can be made by reducing the excessive stocks to the level of the average stocks over the 2 year period. The nature of production demands should be examined also. For example are they scheduled fairly precisely well in advance or do they demonstrate large random variations. This latter information is necessary to give some indication of the complexity of the problem to be solved and the effort likely to be required. The changes in purchasing cost due to the transportation component for differing sizes of delivery quantity should also be examined. If these differences are significant then it is an area worthy of detailed research but again likely to require a lengthy and complex study. Economies in transportation for differing delivery quantities is more likely to be relevant to semigovernmental organisations or multinational large companies and probably not worth a great deal of attention for most medium-sized manufacturing companies. As Guimaraes has shown in his research into the problems of a governmental cereal importing agency, proper scheduling of deliveries, proper choice of the right vessel size and the size of the port storage silos can make a difference in transportation and storage costs of millions of dollars per year.

The actual performance achieved should be compared to the initial planned delivery schedules. This will show, as in the case study described in Chapter 7, whether improvements should be made to the initial planning process or in the updating and modifying of the schedules over time as the circumstances change, e.g. the production usages, vary from the values forecast earlier. A major part of the benefit in this aspect of the purchasing problem is likely to come from the setting up of formal procedures for updating and modifying the future plans in appropriate ways. It is more difficult to do this in the most economic way if plans are produced by rule of thumb methods rather than by formal models.

A feasibility study for the remaining stages of the purchasing problems as specified in Chapter 4, would best take the form of an assessment of past buying performance as described in the case study of Chapter 5. The 'savings' achieved relative to a notional hand-to-mouth buying policy, the average market price, and the 'Purchasing Efficiencies' for the last 3 to 5 years should be calculated. Care must be taken in calculating the purchasing efficiency values to only include prices offered within the freedom of action allowed to the buyers by the company policy at the time. As part of the study the buying period for each specified delivery order should be derived, i.e. the length of time and/or number of buying opportunities at which each purchase order could have been made. If the average 'savings' over the study period are negative then this implies that a study of the day-to-day tactical buying problem will be a prime candidate for an initial stage of the main research study. If the company allows the buyers considerable freedom of action in the sense of a long buying period for an order then such a study will include the medium-term price forecasting problem implicit in the concept of the active buying period introduced in Chapter 4. If the average purchasing efficiency achieved is less than 20% then again valuable improvements to the tactical buying policy can probably be achieved.

Two exercises should be carried out as part of the feasibility study in addition to the above. As discussed in the case study of Chapter 5, a notional target purchasing performance should be determined by discussion with the purchasing manager and the buyers. This is a value to aim to achieve and not a figure to be used as a yardstick by which to measure the buyer's performance. Any measures of this sort must be left until the very end of the implementation phase of the main studies. The reductions in purchasing costs that would have occurred if this notional target performance had been achieved consistently over the past 3 to 5 years should be calculated. This then indicates the level of potential savings from a study of this area. This can be compared to the likely costs of a study to determine the expected return. If the buyers in the past have had a buying period for each delivery order of more than 2 months then a study to assist in improving performance will have to examine both the medium-term price forecasting situation as well as tactical buying systems, i.e. the aspects covered in Chapters 8 to 10 as well as in Chapters 11 to 13. With a long buying period in operation it is likely that it is the determination of the active buying period, the shorter period of time when a purchase should be made for a delivery order, that is the major headache for the buyers. This implies that the development of econometric/behavioural models for medium-term price forecasting should be given higher priority than tactical buying systems.

A second exercise in this feasibility study is to examine the effect of increasing the buying period, if the current purchasing efficiency is positive. The reduction in purchasing costs that would be achieved assuming the same purchasing efficiency is maintained for the longer buying periods should be calculated. If there are potentially substantial savings from such a change in the length of the buying period then the major priority would again be the development of an econometric/behavioural model for medium-term price forecasting.

Once the various feasibility studies have been completed an assessment of the potential benefits from investigating the alternative aspects of the problem is available. The relative size of the benefits obviously plays a major role in deciding which aspects to research first. The opinions of the purchasing manager and his buyers as to which they see of prime importance and also at this point in time which they feel the operational research practitioner can most help them with are of equal importance. Success in this area will only come from a joint effort between the purchasing people and the OR practitioner. It is unwise of the OR practitioner initially to push strongly for a particular area if the buyers are highly doubtful at this time of the usefulness of his approach to problem solving. If he is new to the area then the OR practitioner will probably himself feel more confident in tackling something nearer to the type of problems he has encountered in other management functions. He is thus likely to feel most at home in tackling either the problem of determining a schedule for the size and delivery of future purchase orders or in the medium-term price forecasting area. He would probably be wise to nominate either of these areas as the ones in which he is most confident of success. These are also areas where the purchasing people are likely to feel that the operational research practitioners can contribute most since they are concerned with the preplanning rather than detailed buying decision areas. They do not impinge so directly on to what is traditionally regarded as the skill buyers need to develop and exercise, the short-term tactical buying. In tackling the scheduling of delivery orders the OR practitioner must remember that he must eschew the concepts of classical inventory control models that he usually tries to apply to purchasing and production planning for manufactured items. This approach is inappropriate. In most circumstances he will have to apply the Materials Requirement Planning approach. He should therefore ensure he is familiar with this approach before starting any study.

Any work on the tactical day-to-day buying policy should be mainly based on the general models discussed in Chapter 11. However each raw material undoubtedly has particular idiosyncracies in its price movements arising from the particular characteristics of its own market structure or the policy of suppliers of that material. The buyers will have developed various maxims and rules of thumb from their long experience with that market. It is a good idea to encourage the buyers to discuss any such rules or any specific patterns of price movements they watch out for which can be used to help make good purchases. These should then be formally specified by the research team so that they can be statistically tested over the past years to assess their reliability or to determine the circumstances in which they apply. It may be possible then to develop an *ad-hoc* buying policy to use on its own or as an extension to one of the general models discussed in Chapter 11.

14.2 IMPLEMENTATION OF STUDY RESULTS

Any studies undertaken are likely to take longer than is frequently the case with other management problem areas. Usually the OR practitioner will have to maintain a continuing contact with the problem after successful policies have been devised and initially implemented. The early months of the implementation stage will really be part of the study period since it may often be found necessary to make modifications to the proposed policies which were difficult to foresee in the main study period.

Simulation of proposed policies and forecasting systems over past prices and operations is possible and necessary for validation of the suggested improvements. However, certain simplifications are necessary in order to set up such a simulation. For example a suggested new tactical day-to-day buying policy would be run usually on the basis of using a single market price offer each day. The new policy proposed may take the form of presetting a price level for each day as the decision criteria to determine whether to buy or not. The simulation would generally compare the lowest price offer recorded by the buyer or the market closing price with this preset level and then make any purchases for that day at the price recorded by the buyer or the market closing price. There are usually several price offers during the day. Hence in practical operation the purchase of such a policy would naturally be made at the first price offered that is less than the pre-set level. This will probably be different to the closing price or what the buyer records as the lowest price offer for the day. This might have a worsening effect on the average results achieved compared to the simulation results. Alternatively if the policy is based on comparing what the buyer feels is the lowest price offer on a day with some pre-set level, during the few minutes the buyer takes to decide he feels it is the lowest price offer for that day, that offer may be taken up by someone else and no longer available to our buyer. Yet in the simulation it was assumed that purchases could always be made at the price the buyer recorded as the lowest real price offer. Again if it is an informal market where buyers can attempt to negotiate reductions on the broker's initial price offer, since he does not try to negotiate a reduction every day the simulation can only be made on the basis of crude assumptions as to what possible reductions can be made at any time on the broker's initial offer. Many practical problems of this type can arise. Although their effect on the overall average purchasing costs may be small they may require necessary modifications to the proposed new policy. Because of such practical problems it is strongly recommended that any new policy proposed should be simulated on a real time basis, side by side with the buyer's current operations as a paper exercise for a period of time, certainly at least 6 months before it is implemented. Only if the results over this trial period demonstrate an improvement in performance should the proposed policy be implemented. Adopting this procedure ensures that no difficulties will crop up due to unwise (with hindsight) simplifications in the simulation.

Exactly the same procedure should be adopted in using econometric/behavioural models for medium-term price forecasting. Here, however, the trial period should cover at least one whole crop year. The same situation applies, in that numerous practical problems arise in generating regular forecasts from an econometric/behavioural model in a real time basis, that do not manifest themselves at the research stage of building a model to explain past price behaviour. The initial trial year of making forecasts will be a year of further learning which may lead to certain modifications or extensions to the model or to the processes by which the numerical values for the expectation for stocks, production, consumption and exports, etc., are generated.

An example of this learning process occurred with the maize price econometric/behavioural model described in Chapter 10. Over the period of time initially analysed in the study the U.S. government's agricultural support programme had not apparently affected Chicago maize prices. Hence no formal models of the effect of the support loan rate, etc., on prices had been required. The first forecasts made in August 1977 (see Chapter 10) predicted that prices would be lower at times than the loan rate. It was therefore necessary to quickly derive formal models for the effect of the loan rate on market prices and combine this with the original model. With the unexpected increase in the loan rate in September 1977 the loan rate model became the major determinant of maize prices over the first few months of the crop year. The use of the original models for routine price forecasting led to an extension to the model. However, if the forecasts had not been made by the research team and the need for a formal modelling of the loan rate effect perceived, the organisation concerned might have found the original model's forecasts unsatisfactory and therefore abandoned them as unhelpful. This example demonstrates well the need for the operational research practitioners to maintain an involvement with the routine use of the models and systems he has devised and for a longish trial period in real time to ensure that the original research analysis covered all the practical problems and aspects of the situation that can arise.

14.3 ECONOMETRIC AND/OR TIME SERIES PRICE FORECASTS

One aspect of the conventional approach to price forecasting taken by market analysts that has been implicitly covered in this book should also be mentioned, since it is relevant to the way a study of raw material purchasing should be carried out. This is the division into two schools of thought, the technical analysts and the fundamentalists. The viewpoint of the technical analyst is that *all* the information about the state of the market situation is included in today's price. All analysis must therefore concentrate only on examining past price movements. This school of thought then leads on to the Chartist approach discussed at the beginning of Chapter 11. Price predictions must, and indeed can only, be made from the time series of past market prices. The fundamentalists on the other hand say that one must examine and analyse all the supply and demand factors and relate these to prices. Future price predictions will come only from estimating the future state of supply and demand. As the discussion in the book has shown the choice between these alternative schools of thought is a false dichotomy. It has arisen because of a lack of understanding of the different decisions that must be taken in a successful raw material purchasing policy. The two approaches basically refer to two different time scales, the short-term for the technical analyst versus the medium-term for the fundamentalist.

Much of the conflict between the two schools has come about because of contradictory research results as to the nature of price movements, see Chapter 2. Analysis of past daily prices has until recently confirmed the random walk model for daily commodity price movements. This model states that systematic and persistent trends in prices do not exist for any length of time period whatsoever. Yet on the other hand econometric/behavioural models have been constructed which demonstrate the existence of systematic trends in prices over the medium-term. Conventional time series models have been unable to reconcile the conflict between these two results. This fundamental difficulty has now been resolved by the new model for daily commodity price movements put forward by Taylor and Kingsman, see the appendix to Chapter 2. Their work has shown that the random walk model is an inadequate approximation. A model for daily price movements explicitly including a systematic trend term has been devised and shown to fit the observed pattern of daily prices observed in practice.

The classic operational research approach would now be to attempt to combine these two different types of forecast at any time together into a single forecast. The forecasts made from an econometric/behavioural model as discussed in Chapters 8 to 10 become the *a priori* forecast. This forecast should be modified and updated daily with the actual prices using bayesian methods. Whenever new information becomes available, usually monthly, to modify the econometric/behavioural medium-term price forecasts either the whole processs should be started again or these new forecasts combined with the current forecasts in a bayesian manner. In the author's opinion there is little to be gained by this combination process since the two types of forecast are required for two different types of decision. The two types of forecast should be brought together via the active buying period concept. The forecasts given by the econometric/behavioural models should be used to determine the active buying periods of about 4 weeks usually, for each scheduled delivery order. A tactical buying policy, one of the types discussed in Chapter 11, should then be used to decide at which price or on what day a purchase should be made within the active buying period. These generally require only the use of short-term statistical forecasting methods such as adaptive exponential smoothing or the Taylor price trend model.

14.4 AN OVERVIEW OF THE BOOK

The aim of this book has been to cast some light on a major significant problem area of management currently much neglected. The buying situations described in Chapter 3 and the methodology of the commodity purchasing decision proposed and analysed in Chapter 4 have shown the wide ranging aspects of the situation that need to be considered. Often the commodity purchasing decision has been perceived to be the rather narrow question of whether today's price offer from a particular supplier is likely to be lower or higher than the price offer he will make tomorrow. The reasons for this view of the problem as mentioned in Chapter 1, are partly historical, partly due to the market structure, the mystique surrounding financial markets, the stochastic nature or unpredictability of the many events that affect prices, the lack of formal analysis and research, etc. A major lesson for purchasing managers and companies from the analysis presented in this book is that the problem is in fact very much broader than this conventional view. The traditional skill that individual buyers have been expected to develop and exercise, of judging individual price offers on a short-time scale basis, needs to be used within a purposely determined and well-laid out medium-term purchasing plan.

It is in carrying out this planning process or setting up systems to enable this planning to be performed on a routine basis that the operational research practitioner and management scientist has a great deal to contribute. It is precisely this aspect of the problem that operational research has been concerned with successfully in other management function areas such as production, marketing, distribution, etc. Chapters 6 to 10 of the book in particular are concerned with the development of models and policies for this purchasing planning. The various case studies described show that success can be achieved in this area. The methods of operational research can help by providing more formal methods for carrying out on a routine basis the planning that currently is done by simple rule of thumb or by giving guidance on the likely general level for future prices.

Furthermore the book has shown that various methods and models have been derived that can assist the buyer with his traditional activity, the tactical day-to-day buying problem. It should be stressed that these methods are only giving guidance and perhaps a better starting point for the buyers to exercise their skill, judgement and exercise. They cannot replace the buyer's skill and experience. They only claim to assist him in using those skills effectively, by processing in an appropriate and systematic way much of the quantitative information that comes to him. There are still numerous imponderables and stochastic events to consider which need his acquired skill, experience and judgement. Even if the methods only help avoid some of the occasional 'mistakes' or formally specify some of the risks they will help the purchasing function to improve the purchasing function themselves.

Another major problem faced by buyers is the sheer quantity of the data that is routinely available to them. Brokers and other organisations provide all sorts of information services on a daily basis often via an online telex system. Generally all the news and the items of information are presented as they come with little priority ordering or weighting of their relative importance. A major problem is that of separating the wheat from the chaff, of determining which are the important items that will really affect current and future prices and which are merely random background noise. The problem is not just that of picking out the important bits of information but also that of processing the information into a form such that its impact and consequences can be assessed and assimilated. Several changes may be occurring in the market at the same time which on the surface are somewhat contradicting each other. Ways of combining these together to give a clearer indication of their resulting overall impact need to be determined. Different pieces of information are more relevant to the medium-term planning than to the tactical daily buying and vice versa. Determining what information is important for what purposes and packaging it in various ways, has the advantage of removing from the buyer the need to sift through all the miscellaneous redundant information and should give him more time for trying to assess the consequences of particular changes on his operations.

Chapters 8 to 10 and Chapters 11 to 13, in their different ways, are in a sense very much concerned with this particular problem. Chapters 8 to 10 describe how to construct behavioural/econometric models that explain past price behaviour and can be used to make medium-term forecasts. They are essentially examining how to consider and combine together all the various pieces of information on production, consumption and stocks and relate them to prices so as to predict and anticipate the general levels for future prices and any major changes from the current levels. The models are saying what are the most important pieces of information and what factors the buyer should be keeping a close eye on. Chapters 11 to 13 are concerned with the shorter term movements of prices. The models discussed in these chapters are essentially taking the price movements over the recent past and combining together in a single measure, the decision price break, any systematic current trend plus the fluctuations around the trend to estimate what are the lowest prices that can be expected over differing lengths of future time, assuming that the pattern of recent price behaviour is continued over the future time period. The complex information provided by recent price movements is presented to the buyer in a simple form. If he agrees that the recent pattern of price movements will continue he should generally accept the buying recommendations. If based on his own experience and knowledge of suppliers and other participants in the market he feels a change to the pattern will occur, possibly due to some event due to happen in a few days' time, he should naturally override them. Clearly any successful models of the type described in Chapters 8 to 10 may also assist in this process of deciding whether a change in the pattern of price movements is likely in the near future.

To the operational research practitioner the book should give the confidence that the methods and approaches he has used with success elsewhere can also be applied to the raw material purchasing problem with benefit to the management concerned. The immediate reaction of some buyers to the OR practitioner is that prices are subject to such a wide variety of influences, with so many uncertain and unexpected events and effects happening, that prices are completely random and entirely unpredictable. The models proposed and the case studies described show that this is far too negative a reaction. The effect of many of the influences and events can be modelled or anticipated. Reasonable attempts at predicting future prices in both the medium- and short-term can be made. The predictions will of course be wrong on occasions. The question is one of trying to improve the decision making, not to develop super accurate price forecasting systems. Very accurately forecasting a price on such and such a specific date in the future may be well nigh impossible. However this book has shown that price forecasting systems can be developed that are sufficiently accurate to make good purchasing decisions most of the time.

A sceptical attitude among buyers is right initially, since this will ensure that the research concentrates on deriving policies and improvements that are suitable for practical use. However the purchasing manager's natural scepticism should not lead him to rejecting blindly the possibility that a joint research effort can lead to successful results. It is hoped that the methods and case studies described in this book demonstrate to both the purchasing manager and the OR practitioners that such a joint effort is a worthwhile venture.

References

Bachelier, L. (1900) *Theory of Speculation*. Pages 17–78 of Cootner (1964).

Beale, M., Coen, P. J. and Flowerdew, A. J. F. (1965) Separable Programming applied to an Ore Purchasing Problem. *Applied Statistics*, **XIV**, Nos. 2 and 3 (Series C).

Bellman, R. (1957) *Dynamic Programming*. Princeton University Press, Princeton.

Blattenberg, R. C. and Gonedes, N. J. (1974) A Comparison of the stable and student distributions as statistical models for stock prices. *Journal of Business*, **47**, 244–280.

Box, G. E. P. and Jenkins, G. M. (1970) *Time Series Analysis, forecasting and control*. Holden Day, San Francisco.

Brown, R. G. (1962) *Smoothing, Forecasting and Prediction of Discrete Time Series*. Prentice-Hall.

Brown, R. G. (1967) *Decision Rules for Inventory Management*. Holt, Rinehart & Winston, New York, N.Y.

Carlson, P. G. and Corrigan, J. J. (1962) Procedures for the Economic Purchase of Raw Materials. Guides to More Effective Purchasing. American Management Association Report No. 68.

Cargill, T. F. and Rausser, G. C. (1975) Temporal Price Behaviour in Commodity Futures Markets. *J. Finance*, **30**, 1043–1053.

Chappell, E. (1974) Linear Programming cuts costs in Production of Animal Feeds. *Opl. Res. Q*, **25**, No. 1, 19–26.

Clark, P. K. (1973). A subordinated stochastic process model with finite variance for speculative prices. *Econometrica*, **41**, 135–155.

Cohen, J. B. and Zinbarg, E. D. (1967) *Investment Analysis and Portfolio Management*. Irwin.

Cootner, P. (1964) *The Random Character of Stock Market Prices*. The M.I.T. Press.

Desai, M. (1966) An econometric model of the World Tin economy 1948–61. *Econometrica*, **34**, 105–134.

Fabian, T., Fisher, J. F., Sasieni, M. W. and Yardeni, A. (1959) Purchasing Raw Material on a Fluctuating Market. *Opns Res.*, **7**, No. 6, 107–122.

Fama, E. F. (1965) The Behaviour of Stock Market prices. *J. Business*, **38**, 34–105.

Goss, B. A. and Yamey, B. S. (1976) Economics of Futures Trading. Macmillan Press.

Granger, C. W. J. (1974) Trading in Commodities, Woodhead Faulkener, Cambridge, England.

Granger, C. W. J. (1975) A survey of empirical studies in capital markets. In 'International Capital Markets' (E. T. Elton and M. J. Grube, Eds) pp. 3–36. North Holland/American Elsevier.

Guimaraes, R. M. C. (1981) National Planning of Commodity Import Operations: The Case of Portugal's Feed Grain Imports. Ph.D. Thesis University of Lancaster.

Hadley, G. and Whitin, T. M. (1963). *Analysis of Inventory Systems*. Prentice-Hall.

Jarrett, F. G. (1965) Short term forecasting of Australian Wool prices. *Australian Economic Papers*, June–December. pp. 93–102.

Kingsman, B. G. (1969) Commodity Purchasing. *Opl. Res. Q*. **20**, No. 2, 59–79.

Kingsman, B. G. (1969) Commodity Purchasing in uncertain fluctuating price markets. Ph.D. Thesis, Lancaster University.

Kingsman, B. G. and Jex, C. J. (1972) Some experiences in developing commodity price forecasting models. Lancaster University Forecasting Conference.

Kingsman, B. G. (1974) Commodity Purchasing—A case study: purchasing strategy and tactical buying for linseed oil. Opl. Res. Soc. Conference. Brighton.

Kingsman, B. G. (1974) Forecasting and research for supply markets—commodity buying systems. *Long Range Planning*, **7**, 24–38.

Labys, W. C. and Granger, C. W. J. (1970) *Speculation, Hedging and Commodity Price Forecasts.* Massachusetts: Heath Lexington.
Lawrence, J. P. and Flowerdew, A. D. J. (1963) Economic Models for Production Planning. *Opl. Res. Q,* **14,** No. 1, 11–29.
Leuthold, R. M. (1972) Random Walks and Price Trends: the live cattle futures market. *J. Finance,* **27,** 879–889.
Mandelbrot, M. (1963) The variation of certain speculative prices. *Journal of Business,* **36,** 394–419.
Mandelbrot, M. (1967) The variation of some other speculative prices. *Journal of Business,* **40,** 393–413.
Morris, W. T. (1959) Some Analysis of Purchasing Policy. *Mgmt. Sci.,* **5,** 443–452.
Naik, A. S. (1970) *Effects of Futures Trading on Prices.* Bombay: Somaiya Publications.
Oliver, B. (1966) British Non Ferrous Metals Research Association—Research Report No. A-1583.
Orlicky, J. (1975) *Material Requirements Planning: the new way of life in production and inventory management.* McGraw-Hill Inc.
Praetz, P. D. (1975) Testing the Efficient Markets Theory on the Sydney Wool Futures Exchange. *Aust. Econ. Papers,* **14,** 240–249.
Praetz, P. D. (1976) On the methodology of testing for independence in futures prices. *J. Finance,* **31,** 977–979.
Rees, G. L. (1972) Britain's commodity markets.
Rule, P. (1970) *Purchasing and Melting an Economic Mix.* International Metals Conference, Amsterdam.
Shishko, I. (1965) *How to forecast Cocoa prices. Guide to Commodity Price Forecasting.* Commodity Research Bureau Inc., 98–107.
Taylor, S. J. and Kingsman, B. G. (1977) Comment—An autoregressive forecast of the World Sugar Future Options Market. *J. Fin. Quant. Analysis,* 883–890.
Taylor, S. J. and Kingsman, B. G. (1978) Non-stationarity in sugar prices. *J. Opl. Res. Soc.,* **29,** 971–980.
Taylor, S. J. (1978) *Time series properties and models of commodity prices.* Ph.D. Thesis, Lancaster University.
Taylor, S. J. and Kingsman, B. G. (1979) An analysis of the variance and distribution of commodity price-changes. *Austral. J. Management,* **4,** 135–149.
Taylor, S. J. (1979) Empirical evidence for trends in capital markets. *Economic Letters,* 271–274.
Taylor, S. J. (1980) Conjectured Models for trends in Financial Prices, Tests and Forecasts. JRSS Series A, **143,** No. 3, 338–362.
Tewles, R. J., Harlow, C. N. and Stone, H. L. (1974) *The Commodity Futures Game.* McGraw-Hill Inc., New York.
Trigg, D. W. (1964) Monitoring a Forecasting System. *Opl. Res. Q,* **15,** 271–274.
Trigg, D. W. and Leach, A. G. (1967) Exponential Smoothing with an Adaptive Response Rate. *Opl. Res. Q,* **18,** 53–59.
Wagner, M. M. and Whitin, T. M. (1958) Dynamic version of the Economic Lot Size Model. *Mgt. Sci.* **5,** No. 1, 89–96.
Weymar, F. H. (1968) *The Dynamics of the World Cocoa Market.* M.I.T. Press.
Williams, H. P. and Redwood, A. C. (1974) A Structured Linear Programming Model in the Food Industry. *Opl. Res. Q,* **25,** No. 4, 517–527.
Working, H. (1958) A theory of anticipatory prices. *American Economic Review,* 48 Papers and Proceedings 188–199.
Working, H. (1963) Futures Markets under Renewed Attack. *Food Research Institute Studies,* **4.**

Sources of Trade Information Published Regularly

GENERAL

There are several regular sources of information on all commodities and raw materials. Particularly useful are the daily telex services offered by Reuters, commodity telex or by individual brokers such as Merill Lynch and Sherson, etc. A most useful and important source of regular daily information, particularly on prices, is the *Public Ledger*, a daily newspaper for commodities. A similar source, less comprehensive, is the commodities page of the *Financial Times*. Most of the large brokers for most materials offer some kind of written weekly or monthly short report.

SPECIFIC MATERIALS

The following is a list of data sources on a regular basis offered for particular materials. They are ones that the author himself has found useful at varying times. They are not in any sense meant to be an exhaustive list.

Cereals: The major source of useful data here is the United States Department of Agriculture. The two most relevant publications are their Foreign Agricultural Circular—Grains and their Agricultural Supply and Demand Estimates.

Cocoa: Gill and Duffus Ltd.

Non Ferrous Metals: World Bureau of Metal Statistics; International Tin Council, Rudolf Wolf & Co.; Amalgamated Metal Corporation Ltd.

Sugar: Czarnikow (London) are sugar brokers who issue regular weekly reports. O. O. Light are a German sugar broker who issue monthly reports and an excellent annual review, probably the most comprehensive set of statistics on production, stocks, consumption, imports and exports published.

Oilseeds: Frank Fehr and Co. Ltd (London). Annual review of Oilseeds, Oils, Oilcakes and other commodities. *Oil World*—a weekly paper published by Ista Mielke Ltd of Hamburg. Again the United States Department of Agriculture is an important source via their Foreign Agricultural Circular—Oilseeds.

Name index

Subject index